Research Methods for the Biosciences

research methods for the biosciences

SECOND EDITION

D. Holmes

P. Moody

D. Dine

OXFORD

UNIVERSITY PRESS

OXFORD

UNIVERSITY PRESS

Great Clarendon Street, Oxford OX2 6DP

Oxford University Press is a department of the University of Oxford.
It furthers the University's objective of excellence in research, scholarship,
and education by publishing worldwide in

Oxford New York

Auckland Cape Town Dar es Salaam Hong Kong Karachi
Kuala Lumpur Madrid Melbourne Mexico City Nairobi
New Delhi Shanghai Taipei Toronto

With offices in

Argentina Austria Brazil Chile Czech Republic France Greece
Guatemala Hungary Italy Japan Poland Portugal Singapore
South Korea Switzerland Thailand Turkey Ukraine Vietnam

Oxford is a registered trade mark of Oxford University Press
in the UK and in certain other countries

Published in the United States
by Oxford University Press Inc., New York

British Library Cataloguing in Publication Data
Data available

Library of Congress Cataloging in Publication Data
Data available

Typeset by Newgen Imaging Systems (P) Ltd., Chennai, India
Printed in Italy
on acid-free paper by
LEGO SpA—Lavis TN

ISBN 978–0–19–954576–6

10 9 8 7 6 5 4 3

Contents

Detailed Contents

Boxes outlining the methods for the statistical tests included in this book

Non-parametric ANOVAs (including the Kruskal–Wallis test) and multiple comparisons

Preface

Getting the most out of this book

We write this section with some uncertainty since we rarely read a preface in a book and wonder if anyone is going to read what we've written here. Nonetheless, there are several important things we need to explain to you to enable you to get the best out of this book and its Online Resource Centre, so we'd encourage you to take a few moments to read on.

Who should read this book?

This book is primarily written for undergraduates who wish to develop their understanding of designing, carrying out, and reporting research. We anticipate that graduates, although familiar with most of this material, will find this book and its Online Resource Centre to be a useful reference too. We have therefore used examples almost entirely drawn from real undergraduate and some graduate research projects. We are indebted to all our students who have allowed us to use their ideas and data in this book.

Content

The content is designed to take you through all the steps you need to follow when choosing, planning, evaluating and reporting undergraduate research. The content is therefore laid out in a number of sections:

How to choose a suitable topic for undergraduate research (Appendix a)

Section 1: Planning an experiment (Chapters 1 – 4)

Section 2: Handling your data (Chapters 5 – 10)

Section 3: Reporting your results (Chapter 11).

In addition, in the appendices, we have included an explanation of the mathematical processes used in the chapters (Appendix e), a decision web to help you plan your research (Appendix b), and a summary of the information you need to choose the correct statistical test to test hypotheses (Appendix c).

There is a close and essential link between planning research and understanding how statistics fit into this process. We have demonstrated these links

in Section 1 by cross references to content in Section 2. In Section 2 we have included an overview at the end of each chapter on statistics explaining how this relates to experimental design.

Getting started

You may be coming to this book with very little in the way of training in maths. If you do not know how to calculate this sum $(4\ 3)^2/2$, or if you do not recognize the symbols $<$ or $>$, then we suggest you first look at Appendix e and the additional examples on the Online Resource Centre.

You may wish to analyse data you have gathered from an experiment carried out in your course. For this we suggest you start with Appendix c. which will then direct you to the correct sections in Chapters 5 to 10.

You may wish to prepare a critique of published research. The chapter that considers this is Chapter 2, with cross-referencing to earlier and some later chapters.

If you wish to design a research project and you are familiar with terms such as variable, parametric, aim, hypothesis, etc., then go to Appendix a and Chapter 2. If you are not familiar with these terms still refer to Appendix a, and then continue from Chapter 1.

Learning features

Sign posts

There are a number of ways we have tried to help you quickly find the information you need. We have included chapter ('In a nutshell') and section ('Key points') summaries. At the top of each chapter in Section 2 we have included both an introduction and a guide to choosing the correct statistical test for that chapter. All these chapter specific guides are integrated in Appendix c.

> **7.1 Chi-squared goodness-of-fit test**
>
> Key points In some experiments, you may have a reason for expecting your data to be explained by an equation or particular ratio. You may analyse data to examine whether your data fit this *a priori* expectation. If your data are counts or frequencies, one of the most common statistical tests used to test this type of hypothesis is the chi-squared goodness-of-fit test.
>
> There are three types of investigations where you may use a goodness-of-fit test:
>
> 1) You can reasonably argue that all samples should have the same

Key terms

We know that most people will dip in to this book and so we have included a glossary of most terms that you need to be familiar with towards the end. Key terms are shown in coloured type the first time they appear; definitions of these key terms then appear in the glossary at the end of the book.

Boxes

In our experience we have found that students best understand the statistics element of this book if they first use a calculator to work out the calculation.

BOX 7.6 How to calculate a 2 × 2 chi-squared test for association	
GENERAL DETAILS	**EXAMPLE 7.5**
	This calculation is given in Resource Centre. All value five decimal places.
1. Hypotheses to be tested H$_0$: There is no association between the two variables. H$_1$: There is an association between the two variables.	1. Hypotheses to be tested H$_0$: There is no association species (*Cepaea nemoralis* habitat (hedgerow and wo H$_1$: There is an association species (*C. nemoralis* and (hedgerow and woodland)
2. Have all the criteria for using this test been met?	2. Have all the criteria for us Yes (see answer to Q4).
3. How to work out expected values In chi-squared tests for association, you have no a priori expectation against which to compare your observed data. Instead, the expected values are	3. How to work out expecte Look first at the expected 1 in Table 7.12. To calcul take the column total for

Therefore we have arranged the statistical information in boxes with general details and a worked example for you to follow.

In the calculations we've included in this book we have rounded all values, usually to five decimal places. However, we carried out all the calculations using all decimal places (as you should). This means that some of our sums do not appear to quite add up. Any minor differences in the calculations should be the result of this rounding of values.

In the boxes some the mathematical steps have not been included. We have therefore included the full calculations in the Online Resource Centre. We have also included an explanation of how to use SPSS, Excel, and Minitab to carry out the same calculations in the Online Resource Centre.

be an integral part of the investigation and must be carried out at the same time as the rest of the investigation.

Q2 In the experiment described in Example 2.2 on the antibacterial properties of triclosan and tea-tree oil, are any controls needed? If so what?

A control may not always be necessary. For example, in an investigation into the effect of temperature on the behaviour of *Oniscus asellus*

Questions

To help you check your understanding of the topics covered by this book we have included a number of questions. The answers are provided at the end of each chapter. More questions are included in the Online Resource Centre.

Online Resource Centre

Research Methods for the Biosciences is more than just this printed book. The *Research Methods for the Biosciences* Online Resource Centre features extensive online materials to help you really get to grips with the skills you need to carry out research work. This can be found at:
www.oxfordtextbooks.co.uk/orc/holmes2e/

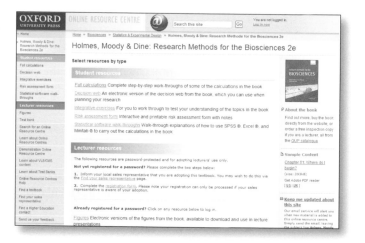

The student area of the Online Resource Centre includes:

- full details of all calculations in this book: every step in each calculation is shown so you can see exactly how we reach the answers shown in the book
- walk-through explanations of how to use SPSS, Excel, and Minitab to carry out these calculations
- interactive tasks for you to work through to test your understanding of the topics in this book, and hone your research methods skills.
- an electronic version of the decision web which you can use when planning your research
- an electronic risk assessment which may be used to help you complete this process

You'll see the Online Resource Centre icon throughout the book. This icon tells you that the part of the text you are reading has online materials to accompany it.

online
resource
centre

The lecturer area of the Online Resource Centre provides additional materials to make the book easier to teach from. These materials include:

- figures from the book, available to download for use in lecture slides
- test bank of questions

Simply go to www.oxfordtextbooks.co.uk/orc/holmes2e/ and register as a user of the book to gain free access to these materials.

And finally ...

Just in case OUP ask us to produce a third edition, we would like to hear about any errors (we hope there are none) and any suggestions you have for improvements. You can contact us by using the 'Send us your feedback' option in the Online Resource Centre.

Planning an experiment

Where do I begin? 1

In a nutshell

This chapter gets you started with your research and explains some of the terms you will need to be familiar with when designing or reviewing experimental work. First you need to define an aim and at least one objective; then you will need to decide what you are measuring and whether you are collecting these measurements (observations) from the statistical population or a sample. Finally, you need to decide if your data can be analysed by testing hypotheses.

You may be using this book because you are starting to plan a piece of independent research, you have been carrying out an experiment designed by someone else, or you are reading about published research. This book is written for those of you who need to have an understanding of experimental design and, as part of this, need to develop your statistical skills. We begin in this chapter by introducing you to some of the basic principles and terms that you will come across in your research. If you are about to embark on a piece of research and need help choosing a research topic, we have included some suggestions in Appendix a. If you are working through this chapter, it should take about 1 hour to complete all the exercises. The answers for these exercises are at the end of the chapter. All our examples are based on real undergraduate research projects. If these examples are not in your subject area, you will find more in the Online Resource Centre.

Research The process by which you try to find out more about a particular topic

 online resource centre

1.1 Aims and objectives

Key points Research is guided and explained in terms of aims (a general overarching statement about the research) and objectives (specific explanatory statements about elements within the research). These terms are often confused.

When designing research, you start by choosing an aim and usually then subdivide the work you are doing under the heading 'objectives'. It is a common failing to confuse these two terms. This can lead to a lack of direction in the research and research report. All investigations require at least one aim and at least one objective. You therefore need to understand the distinction between these terms and why this difference is important.

1.1.1 The aim

Most research is developed through a series of investigations. All the investigations are carried out within a specific area of interest. This area of interest is the aim.

Aim A generalized statement about the topic you are investigating

---

EXAMPLE 1.1 The effect of the introduction of new individuals on behaviour in an established herd of wildebeest

An undergraduate had links with a local safari park and was interested in the changes in behaviour of animals when new individuals were introduced into an established herd.

In Example 1.1, the student's aim was therefore to 'investigate the effect of the introduction of new individuals on behaviour in an established herd of wildebeest'.

EXAMPLE 1.2 An undergraduate wanted to study the capacity of *Calluna vulgaris* to take up heavy metals from soils contaminated by wash from electricity pylons. How might the aim be phrased?

The aim in some investigations can be very broad and may require several experiments to cover all aspects of this topic. Each experiment or group of experiments is described by an objective.

1.1.2 The objective

An objective Describes a specific investigation where data will be collected. Objectives include specific experimental details. There will be one or more objectives for each aim

To thoroughly investigate the topic described by an aim can require one or more experiments, each of which will examine one aspect of the overall topic. The experiments will have a particular design and record particular data (1.2). It is therefore important to have objectives that provide additional information about each step of the overall investigation. This enables the researcher to keep track of their work, and anyone reading their reports can see how the separate experiments relate to each other and to the aim.

EXAMPLE 1.3 The effect of cell culture volume on parameters of growth of tobacco cells in liquid culture

Researchers were interested to see whether they could effectively culture tobacco cells in microfuge tubes rather than as larger cultures. The effectiveness of culturing in a smaller volume container was tested for a number of different temperatures and different subculturing routines. To assess the effects of these parameters on culturing tobacco, the cell viability and cell number were recorded.

The aim for Example 1.3 was therefore to examine the effect of culture volume on tobacco cells. The objectives were:

1) To examine the effect of temperature after 24 hours on the viability of tobacco cells in 5ml and 0.5ml (microfuge) cultures.

2) To examine the effect of temperature after 24 hours on the number of tobacco cells in 5ml and 0.5ml (microfuge) cultures.

3) To examine the effect of the frequency of subculturing on the viability of tobacco cells in 5ml and 0.5ml (microfuge) cultures.

4) To examine the effect of the frequency of subculturing on the number of tobacco cells in 5ml and 0.5ml (microfuge) cultures.

i. Clumpers or splitters

Objectives can be 'clumped' or 'split'. In Example 1.3, it would be possible to rephrase the four objectives outlined so we only have two:

1) To examine the effect of temperature after 24 hours on tobacco cell viability and cell number in 5ml and 0.5ml (microfuge) cultures.

2) To examine the effect of the frequency of subculturing on tobacco cell viability and cell number in 5ml and 0.5ml (microfuge) cultures.

When reading, these two objectives are much neater and less repetitive than the previous four and can therefore be a more efficient way of reporting your investigation. However you may wish to use your objectives as a basis for reporting your results. In our first version when our objectives are split into four statements, this works well as each objective has only one data set (either cell viability or cell number). In the second 'clumped' version, there are two data sets for each objective (cell viability and cell number). Using only two objectives as a basis on which to present your results is therefore less suitable. In this book, we will therefore be 'splitters' and have more rather than fewer objectives. You may like to discuss this aspect of any report you prepare with a member of staff and either be a 'splitter' or a 'clumper' according to their advice.

ii. Only one objective

In undergraduate work, it can be difficult to appreciate the need for objectives, as many investigations of a topic only involve a single experiment.

This is particularly true of undergraduate practicals. You may not therefore have many opportunities to see the real value of having objectives. Where there is only one experiment carried out to examine an aim, the aim looks almost identical to any objective. Hence the confusion that can arise in understanding what is an aim and what is an objective.

iii. Personal and experimental objectives

Objectives can be personal or experimental. For example, in the experiment on the tobacco cells in cultures (Example 1.3) there are four objectives. These all describe experiments. However, you could add two more objectives to this list:

5) To analyse the data.
6) To write a report.

These are personal rather than experimental objectives. In most undergraduate reports and any publications where objectives are included, you will only see experimental objectives. However, when you are asked to include objectives, you may need to check that only experimental objectives are wanted.

1.2 Data, items, and observations

Key points In experiments, you generate data, which are observations derived from items.

Data A number of observations or measurements on the subject you are investigating

Observation A single measurement taken from one item

Item One representative of the subject you wish to measure

 online resource centre

When carrying out an investigation, you will generate data (singular—datum). Data consist of a number of measurements or observations on the topic you are investigating.

The individual or sample from which the measurements are taken is usually called the item. We use these terms throughout the book; however, in our Online Resource Centre, we also show you how to use the statistical software SPSS to carry out calculations. In this software, the term 'case' is used in place of the word 'item'.

EXAMPLE 1.4 The effect on heart rate of watching live and recorded football matches in self-declared football fans and non-football fans

One undergraduate wanted to assess the effect of watching live and recorded football matches on the heart rate of self-confessed football fans and those who were not football fans. Having identified a number of volunteers, their heart rate was recorded first during a recorded football match and then during a live match.

In Example 1.4, each person taking part is the item and each recorded heart rate is an observation. All the results together are the data.

> **Q2** Examine Example 1.3. What are the items and observations?

1.3 Populations

Key points The term population has a number of meanings. In statistics, a population is all the items that are the subject of your research.

There are several different definitions for the word **population**. In ecology, the term usually means the individuals that can be observed within an identifiably discrete group; for example, the number of individuals in a wood, river, or city. In genetics, a population usually means the grouping in which all the individuals of a particular species mate at random. When you are using statistics to analyse your data, the term population is used in yet another way.

In Example 1.1, the statistical population would be all the wildebeest in the population at the safari park. You might argue that safari park conditions are sufficiently similar that the statistical population might be extended to all safari park populations of wildebeest. However, it would be difficult to argue a case for extending the statistical population to all wildebeest including those in the wild.

The statistical population reflects your aim. Sometimes a statistical population will be the same as an ecological population. For example, if your aim was to investigate the characteristics of individuals of one species within a wood, then the ecological population and the statistical population are the same. However, if your aim was to examine the characteristics of a particular species worldwide, then the statistical population will be all individuals of that species on the planet. The ecological population will still be the group of individuals of that species in each wood.

Identifying the statistical population is of great importance to your work. Any hypothesis testing revolves around the statistical population (Chapter 2 and onwards). Similarly, when you come to discuss your results you will automatically apply your findings to the statistical population. For example, having studied the wildebeest behaviour (Example 1.1), the student presented some suggestions on management of such populations. She needed to make it clear in her report whether her recommendations were just for the safari park she studied, for safari parks in general, or for the management of all wildebeest. Her previous consideration of the 'statistical population' meant she was sure where her results could be justly applied.

Population All the individual items that are the subject of your research

In Example 1.4 where the effect of watching live and recorded football matches on the heart rate of self-confessed football and non-football fans was being examined, the student decided to limit his study to undergraduates in an English university with an age range of 18–30 years and no known health problems. What is an appropriate statistical population?

1.4 Sample

Key points It is common practice to sample from your statistical population. It is usual to try to obtain a representative sample. This can be achieved in a number of ways (random, systematic, and random stratified).

It is often not possible to study the whole statistical population because of practical and economic constraints. The testing procedure may also be destructive and you may therefore wish to limit your investigation. For example, if you wanted to carry out taste testing on cheese and onion crisps, it would not help the factory managers if you tasted all the crisps. Usually a subset of the population is examined. This is the sample.

For example, in the investigation into the effect on heart rate of football and non-football fans when watching live and recorded matches (Example 1.4), the student might study only five Manchester United football supporters and one of their non-football-fan friends. But is this a representative sample?

1.4.1 Representative samples

If it is not possible to measure the whole statistical population, then a representative sample is required. A representative sample should be obtained by a clearly defined and consistently applied method. The observations recorded by this method should closely match the result that would be obtained if every item in the population was measured.

In Example 1.4, if only five Manchester United fans and one non-football fan were studied, this sample is not representative of the number and range of football fans and non-fans. To resolve this, the student might decide to test the non-fan five times as often to obtain a similar number of observations as the observations of the five football fans. The measurements on the non-fan will be more similar to each other than would be expected to be obtained from five different people. This is an example of a lack of independence. The five measures on the non-fan are not independent of each other, whereas the single measures on each of the five fans are independent of each other.

Sample A number of observations recorded from a subset of items from the statistical population

Independent observations Where each measurement in the data set has a similar likelihood of showing variation to the same extent. Measurements that are not independent invariably arise when the same item is measured more than once

A mix of independent and non-independent measures will also not lead to a representative sample. (Another example of non-independent measures where this element has been allowed for in the experimental design is considered in 9.3).

What should be done in the case of Example 1.4?

1.4.2 How do you obtain a representative sample?

There are a number of sampling methods that may be used. Most of these are designed for particular circumstances and to produce a representative sample. We describe the common methods here.

i. Random sampling

Here each item must have an equal chance of being sampled each time. This method avoids conscious and unconscious bias. On average, random sampling does produce a representative sample and is therefore in common usage.

Random sampling Each item must have an equal chance of being sampled each time

How do you randomly sample? In ecology, a common error is to believe that throwing quadrats over your shoulder is going to provide you with a random sample. This is untrue: you will only sample within a small area around yourself and will tend to throw forwards or backwards because this is how arms work. A better approach is to use a random number table to identify an item within your population.

For example, there are a number of practical approaches the undergraduate might use to sample football and non-football fans (Example 1.4). One would be to survey as many people as possible to identify those willing to take part, who are within the age range 18–30 and where there are no known health problems. If these are then grouped into self-confessed football fans and non-football fans, it would be possible to allocate each person a number and use a random number table to select individuals from this pool of volunteers. The number of volunteers (i.e. sample size) is something we will come back to (2.2.7 and 3.5).

ii. Systematic/periodic sampling

Systematic/periodic sampling is a regularized sampling method, where an item is chosen for observation at regular intervals. This method can be suitable, for example, for examining the general distribution of plants across a footpath using a line transect where data are collected at regular intervals. This method is also commonly used when studying the distribution of flora and fauna along a shore from cliff to sea.

Systematic/periodic sampling A regularized sampling method where observations are recorded from items chosen at regular intervals

iii. Stratified random sampling

This is a combination of the two methods outlined above, where random sampling occurs at regular intervals. Thus, an area may be divided into

sections, and random sampling occurs within each section, the numbers sampled reflecting the relative size of the section. An example of this would be where soil samples are collected from a contaminated site and where it is important that all parts of the site are represented within the sample. The site can be divided into a number of discrete areas and random samples are taken within each area.

iv. Homogeneous stands

In some work, you may need to be selective and to identify areas for your investigation that are homogeneous, i.e. similar. In the UK, this approach is most common when collecting data about a plant community and assigning a National Vegetation Classification (e.g. Rodwell, 1991). Full details about this approach, including suitable quadrat sizes for the canopy, ground, and field layer, are given within this series of books.

Q4   Anthills are dispersed throughout a grassland. You wish to study the percentage coverage of specific plant species on the anthills. What sampling strategy would you use?

1.5 Population parameters and sample statistics

Key points Mathematical methods for summarizing data and testing hypotheses differ depending on whether you have measured every item in the population or only a sample. To make it clear whether you are using population parameters or sample statistics, different notation is used.

Later in this book we introduce you to mathematical methods that you may apply to your data: both the methods used to summarize data (Chapter 5) and those used to test hypotheses (Chapters 7–10). Where the answers from these calculations relate to a population, they are usually called parameters. Where they relate to a sample, they are called statistics. Most research, of necessity, collects observations from samples, and therefore all our chapters focus on designing investigations and evaluating data from samples. Even so, there are frequent occasions when you will come across terms that relate specifically to a statistical population or to a sample.

1.5.1 Mathematical notation for populations and samples

In statistics, we use mathematical notation. For those not familiar with this notation, we have included further explanation in Appendix e. Different symbols are used if we are thinking about populations or samples. The most common symbols are shown in Table 1.1. Different symbols are used because there are important differences in the meanings and methods involved when calculating parameters of populations or statistics of samples. More information, including an explanation of the terms mean, variance, and standard deviation, is given in Chapters 5 and 6.

1.5.2 Calculations

Some mathematical tests exist in two forms, one suitable when all population values are known and one when a sample is being studied.

Table 1.1. Symbols used to indicate population parameters and sample statistics

Term	Population	Sample
Mean	μ	$\bar{x}$
Variance	σ^2	s^2
Standard deviation	σ	s

One value you may often use is called a variance. We look at these equations in Chapter 5 onwards. The variance you calculate may be a population parameter and in this instance the formula for the variance from a normal distribution (5.2.1) can be written as:

$$\sigma^2 = \frac{\Sigma (x - \mu)^2}{N}$$

If the variance you are working out is for a sample, then the equation you use is different and is written using notation relating to the sample:

$$s^2 = \frac{\Sigma (x - \bar{x})^2}{n - 1}$$

(The meanings of the symbols Σ, $\bar{x}$, x, N and n are explained in Appendix e.)

Calculators and computer software invariably have the facility to calculate both population parameters and sample statistics, and it is therefore important to ensure that you are using the correct version. Calculators and computer software often indicate the version being used either by the symbols shown in Table 1.1 or by the symbols n (population) or $n - 1$ (sample). This is examined again in the Online Resources Centre.

 online resource centre

1.6 Treatments

Key points There are two valid definitions of the term 'treatment'. The first relates directly to the variables under examination and the second is broader and relates to the experimental design.

Treatment (definition 1) When carrying out an experiment, your items are exposed to a particular environment that is manipulated by you, the investigator

Treatment (definition 2) A statistical term in which any samples or groups of observations that are being compared are called treatments

Treatment is a word that has two valid meanings and both are used in this book. The examples we use here appear in Chapters 11 and 7, respectively.

EXAMPLE 11.1 The response of tobacco explants to auxin

In an undergraduate investigation into the effect of different concentrations of auxin on the growth of tobacco in tissue culture, the relative increase in diameter (mm) of leaf explants was recorded after 2 weeks.

In this example from Chapter 11, the tobacco explants were *treated* by exposure to different concentrations of auxin.

EXAMPLE 7.3 Shell colour in *Cepaea nemoralis* in coastal and hedgerow habitats

An investigation was carried out into the frequency of banding and colour patterns in snail shells in two different habitats (a coastal region and a hedgerow).

In this example from Chapter 7, although the investigator has not imposed these habitats on the snails, when the data are analysed each sample is called a treatment. Thus, the coastal region is one 'treatment' and the hedgerow is a second 'treatment'.

The word treatment is often found coupled with other terms; hence, you will come across treatment variable, non-treatment variable, and treatment effects. In all these contexts, it is the broader definition of 'treatment' (definition 2) that is used.

Q5 In Example 9.2, representative samples of *Littorina obtusata* and *Littorina mariae* were collected from Porthcawl in 2002 and their shell height (mm) recorded. The investigators wished to test the proposal that there is no difference between shell height of the two groups of periwinkles (*L. obtusata* and *L. mariae*). What is/are the treatment(s)?

1.7 Variation and variables

Key points Most data collected in an experiment varies—not all the values are the same. In general, the factor(s) that is being studied is the variable. When selecting statistical tests, the key variables are those factors whose *effect* you wish to analyse. All other factors that may affect your experiment but are not being studied are known as confounding or non-treatment variables.

1.7.1 Variation

In any investigation, you will record observations. Usually these observations do not all have the same value—they vary. A set of observations is said to show variation.

Variation When the observations within your data set do not all have the same value

1.7.2 Variables

As an extension of this, we also have the term variable. There are two definitions for this term. Both are correct. In the first definition, all the factors that you are studying are called variables.

Variable (definition 1, general use) The factors that are a part of your experiment

Variable (definition 2, when choosing statistical tests) The factors whose effect you wish to examine

As we explained in Example 1.3, the investigators wished to see whether tobacco cells could be cultured in a microfuge (0.5ml) tube rather than in larger culture volumes (5ml). There were a number of objectives outlined for this investigation indicating different experiments (1.1.2). The first of these objectives (1) was to examine the effect of temperature after 24 hours on the viability of tobacco cells in 5ml and 0.5ml (microfuge) cultures. In this experiment, the tobacco cells were grown for one day at four different temperatures. After 24 hours, an aliquot was taken and the percentage cell viability recorded. These three factors (culture volume, temperature, and percentage cell viability) are of interest to the investigator and as such are variables.

The second definition of the term variable is used when you are selecting statistical tests. Here, a variable is any factor in your experiment whose effect you wish to evaluate.

In Example 1.3, the student wished to examine the effects of culture volume and temperature. Therefore, for the purposes of choosing and using statistics, there are two variables. There is no 'effect' of percentage cell viability. To describe percentage cell viability as having an effect does not make sense and therefore, although this is a variable in the experiment, it is not counted in the variables whose effects are to be analysed statistically. Sometimes the word 'dependent' variable is used to describe variables such as percentage cell viability and we return to this in Chapter 8.

Being able to identify all the variables in your experiment and those whose effects you wish to analyse takes practice. We return to this explicitly in Chapter 2 and in general in Chapters 7–10. In addition there are more exercises in the Online Resource Centre.

**online
resource
centre**

Q6 In Example 1.4, 50 volunteers took part: 25 were self-confessed football fans and 25 were not football fans. All volunteers were fitted with a heart rate monitor and as a group were shown a recorded football match. Then all the volunteers attended a live football match. Maximum heart rate was recorded for each volunteer. What are the variables for the experiment? Which variables are relevant when choosing a statistical test?

1.7.3 Partitioning variation and confounding (non-treatment) variables

Variation between observations can be due to many factors. For example, the variation in human height is due to genetics, diet, and disease. If you wanted to investigate the effect of diet on human height, how can you decide how much of the variation is due to diet and how much

variation is due to other factors? As an investigator you want to be able to 'partition' the variation in your data:

Variation in your data = variation due to the factor of interest
+ variation due to other factors

In our example this would be:

Variation in human height = variation in height due to diet
+ variation in height due to all other factors

Using the second definition of the term treatment (1.6), this relationship can be written in more statistical terms as:

Variation in your data = treatment variation
+ variation due to confounding variables

In our example, the treatment variable is 'diet' and all the other factors that contribute to human height would be **confounding or non-treatment variables** that result in non-treatment variation.

Confounding or non-treatment variation All factors outside the scope of your experiment that may influence the results

In the rest of this book, we develop these ideas and show how experimental design and data analysis can be used as powerful tools to allow you to examine the effect of the treatment within your research. In Chapters 2–4, we begin by considering good practice in experimental design, and in Chapter 6 we focus on hypothesis testing and data analysis in relation to treatment and non-treatment variation. In Chapters 5 and 7–10, we return to this topic when we review experimental designs such as the Latin square.

 Q7   In Example 1.4, the maximum heart rate of a total of 50 volunteers (aged 18–30) was examined while they watched a recorded football match and a live match; 25 volunteers were self-confessed football fans and 25 were not. What might the confounding variables be?

1.8 **Hypotheses**

Key points When mathematically assessing your data, you may either see which trends are present (information-theoretic models) or test your data against a statistical (null) hypothesis.

As part of the process of designing your experiment, you need to give some thought as to how you will analyse your data. This is really important

as it can help you decide how many samples or observations you need and ensures your experiments are likely to produce data that can be analysed (Chapter 2). There are two main approaches to analysing data. Either you choose one null hypothesis to test your data against or you analyse your data to see which of a number of alternative models best fits it (information-theoretic approach).

1.8.1 Null hypotheses

Experiments set out to test a specific question. This question may be written in a format called a **hypothesis** (plural hypotheses). Chapter 6 introduces you to hypothesis testing and explains in more detail how to write hypotheses and why this format is used.

For Example 1.3, objective 1, the hypothesis would be:

H_0: There is no significant difference in percentage cell viability of tobacco cells in 5ml and 0.5ml (microfuge) cultures in response to temperature (°C).

H_1: There is a significant difference in percentage cell viability of tobacco cells in 5ml and 0.5ml (microfuge) cultures in response to temperature (°C).

From the example we include here, you can see that hypotheses come in pairs. These two hypotheses are called the null hypothesis (H_0) and the alternate hypothesis (H_1). The word 'significant' has a specific meaning in statistics that we will discuss later (Chapter 6). For the moment, you can think of it as being used in the 'everyday' sense of 'big enough to be important'.

In many experiments, you may chose to identify a null hypothesis and then test your data to see if this proposed model appears to probably explain the data. Most often, this null hypothesis is a general statement that there is no difference between sample A and sample B that cannot be accounted for by chance. For example, an undergraduate wished to examine the differences in seed quality between seed sold by a number of different companies. After carrying out a number of seed purity, viability, and vigour experiments, the student tested her data against a general null hypothesis that there was no difference in the quality of seed from the different companies.

You may wish to devise a null hypothesis other than 'there is no difference'. We introduce you to some 'specific' hypotheses in Chapters 7–10. However, most statistical tests included in this book are designed to test general null hypotheses. In Chapters 2 and 3, we discuss the design of investigations. There are more examples of developing hypotheses in these chapters. Exercises for you to test your understanding are included in the Online Resource Centre.

Hypothesis The formal phrasing of each experimental objective; includes details relating to the experiment and the way in which the data will be tested statistically

online resource centre

1.8.2 **Information-theoretic models**

In our experience, this alternative approach to analysing data is most often encountered in studies where little is known about the topic you are investigating. For example, you may be surveying a particular location such as woodland to see which species are present and the relative abundance of each species. There are mathematical techniques that will allow the data to be grouped into observations that are similar. For example, a forensic science undergraduate carried out a survey of diatom species at two rivers and a number of sites along these rivers during the year. She used cluster analysis to see whether there were distinct groupings within the data. She was then able to see whether these grouping related to location, season, or other factors. This is a more open approach to evaluating your data. You do not compare your data to the one model you devised before you analysed your data. Instead, you *allow* the data to indicate a suitable model. This is known as an information-theoretic model.

There is often a concern that all experiments must be tested against a null hypothesis. This is not the case for many types of investigation. There is also an argument that, by only testing against one null hypothesis, the data analysis is constrained and less information can be discovered by this approach than by taking an information-theoretic model approach. This topic is discussed further by Lukacs *et al.* (2007) and Stephens *et al.* (2005, 2007). At present, the use of an information-theoretic approach is rare for undergraduate research other than when surveys of species are carried out. We have not therefore covered this approach, but refer you to Fielding (2007) and Quinn & Keough (2002). Even if you do take this approach, you may also wish to summarize your data (Chapter 5) and use tables and/or figures to communicate your findings (Chapter 11).

Information-theoretic models
When data are collected and analysed to see whether there are significant groupings. Only then may hypotheses about the factors causing these groupings or the nature of these groupings be tested, i.e. models are constructed from the observations

Summary of Chapter 1

- To provide a focus for your research, you will have an aim and objectives. These are not the same thing, and when designing your investigation you will need to distinguish between them (1.1 and developed in Chapter 2).

- Within your investigation, you will record observations for particular items. You may collect observations for every item in a population or you may sample (1.3, 1.4, and 1.5, and developed in Chapter 2).

- When sampling, your intention is usually to obtain a representative sample that reflects what is happening in the population. Statistical tests, including those programmed in calculators and computer software, use different symbols when referring to populations and samples (1.3, 1.4, and 1.5, and developed in Chapter 2).

- The term 'treatment' is used when designing and evaluating experiments. There are two definitions

and both are used in this book (1.6 and developed in Chapter 2).

- In most investigations, the observations do not all have the same value—they vary. This variation is central to examining the effect of treatments (1.7 and developed in Chapter 2). The factor(s) that you are studying is called the variable(s). The factors that may affect your data that you are not studying are called confounding or non-treatment variables (1.7 and developed in Chapter 2).

- Some investigations start out with a question (hypothesis). When you use statistics to test the hypothesis, then the hypothesis is phrased in a specific way (1.8 and developed in Chapter 6).

- The Online Resource Centre includes interactive exercises that test your understanding of this chapter with other topics, particularly those considered in Chapters 2 and 6.

 online resource centre

Answers to chapter questions

A1 Aim: The capacity of *Calluna vulgaris* to take up heavy metals from soils contaminated by wash from electricity pylons.

A2 The items are the tobacco cells in their particular cultures. The observations are the number of viable cells and the total number of cells in samples taken from each culture.

A3 There are two statistical populations here. Self-confessed football fans, aged 18–30 with no known health problems and self-confessed non-football fans aged 18–30 with no known health problems.

A4 There may be several practical and effective answers. Here is ours. Take numbered plant pot labels with you and place one in each anthill. Number a corresponding set of tickets and place in an opaque container. In planning your investigation you will already have determined how many anthills need to be sampled. Draw out

this number of tickets and these will give you your random sample of anthills.

A5 For this example you would use definition 2 to explain your use of the term treatment. The treatment here is 'species of *Littorina*'.

A6 The experimental variables are live versus recorded matches (variable 1), football fans versus non-football fans (variable 2), and maximum heart rate (variable 3). When choosing a statistical test, the student wishes to examine the effect of live versus recorded matches (variable 1) on football versus non-football fans (variable 2).

A7 There are a number of potentially confounding variables. You may identify others. Confounding variables might include: gender; an assumption that volunteers correctly identified themselves as fans or not fans; social interaction between volunteers; personal preference or dislike for particular teams; and resting heart rate.

Planning your experiment 2

In a nutshell

In this chapter, we take you through the process of planning your experiments. First we consider how to critically review published experimental designs. The first step in this process is to identify the aim and objectives and then to consider the strengths and weaknesses of the work. In doing this, you will develop your ability to be a critical reader and extend your understanding of what makes a competent experimental design. In the second part of this chapter, we take you through the process of designing your own experiments using the decision web that is available in this book and online. If you are unsure about a topic to investigate, refer to Appendix a. As we take you through the process of planning an experiment, you are introduced to a number of important concepts and terms such as the use of replication and controls.

The best way to learn about the design of investigations is to look at what other people have done and to have a go yourself. In this chapter, we take you through both these approaches. The aim is to show you that you are already very able to understand the strengths and weaknesses of an experimental design and identify the ways in which the design might be improved. We also provide information about the process of designing experiments and the things you need to consider.

We have developed a decision web that you can use to help you plan an experiment (Appendix b). This is provided in an interactive version in the Online Resource Centre.

online resource centre

Designing an experiment is an iterative exercise. It requires an understanding of certain design principles, but also an understanding of data and how they are to be analysed. This means that, although in this chapter we cover all the steps you need to consider when designing an investigation, you may not fully appreciate the importance of some of our comments until you have become more familiar with the content of the other chapters in this book. One of the most common problems in

designing research that is a direct consequence of this need for a rounded approach is that many people ignore the importance of considering how the data will be analysed before they begin their investigation: they only come to this when the data are collected and they then find out that the data cannot be analysed. To avoid this, you must accept that this chapter, although providing an overview of experimental design, does not stand alone and must be used with subsequent chapters. If you work through this chapter, it should take about 2 hours to complete all the exercises. The answers for these exercises are at the end of the chapter. All our examples are based on real undergraduate research projects. If these examples are not in your subject area, you will find more in the Online Resource Centre.

online resource centre

2.1 Evaluating published research

Key points Research is guided by an aim and objectives. When reviewing published experiments, you should consider both the strengths and the weaknesses of the work. The weaknesses might be faults that could be revised without changing the aim and limitations that cannot be changed without changing the aim.

Example 2.1 is based on a short research paper that appeared some time ago in a prestigious journal. Read it through and then answer the questions at the end.

If possible, discuss this example of an experimental design with your friends and combine your ideas. Think about the following three questions:

1) What are the aim and objective(s)?
2) What are the strengths of this experimental design?
3) What are the weaknesses of this experimental design?

2.1.1 What are the aim and objective(s)?

The aim describes the broad area being examined by the researcher, and the objectives break this down to reflect the various investigations that were carried out (1.1). We identify the aim and objectives for the study described in Example 2.1 as follows.

Aim: The influence of habitual intake on morning-after breath alcohol concentrations in men.
Objectives:

1) To quantify the habitual intake of alcohol in 58 men.
2) To assess morning-after breath alcohol levels in 58 men.

EXAMPLE 2.1 Social drinking and morning-after breath alcohol levels

The investigator wished to study the level of morning-after breath alcohol concentrations after an evening of social drinking to see how these levels were affected by the levels of habitual intake. An important element of this investigation was that real social drinking behaviour was examined rather than effects observed in a laboratory trial.

Fifty-eight men aged 20–50 years took part in the study. Habitual alcohol intake was estimated by a 4-week prospective drinking diary or accounts of the frequency of drinking and quantity per session in a typical month. The participants were asked to count the number of drinks they consumed at an evening social event where they anticipated drinking heavily. Other people present were asked to verify the amount drunk and the time that drinking stopped. All the men ate shortly before or during drinking. The morning-after alcohol concentrations were then recorded in each person's home 7–8 hours after drinking had stopped (Wright, 1997).

Compare our aim and objectives with your own. How, and in what way, did they differ?

2.1.2 Strengths of the experimental design

It is always easy to be critical, and sometimes overcritical, of research. It is therefore best if you start by considering what is good about the experimental design before considering the experiment's limitations. What strengths have you listed for this experiment? We identified the following strengths:

- The experiment was carried out in the environment in which this activity usually takes place.

- Attempts were made to keep the effect of the experiment on people's behaviour to a minimum.

- Fifty-eight is a reasonable sample size (1.4, 2.2.7iv, 2.2.7v).

2.1.3 **Weaknesses of the experimental design**

There are three types of weakness that you may identify in any report of
an investigation.

i. Faults

When you are considering published research, the first type of weakness
is a fault that could be avoided through a change in the design of the
investigation without changing the aim. In any publication or report,
these faults should be acknowledged and should be taken into account
when evaluating the results (Chapter 11). If you design and carry out a
piece of research, then careful preparation and planning should ensure
that there are few, if any, faults in your work.

ii. Limitations

The second type of weakness, known as a limitation, can arise in an
investigation, but cannot be overcome without changing the aim. These
limitations should also be taken into account when the results are evaluated (Chapter 11). If you are evaluating published research, then your
task is to identify these limitations, as you clearly cannot change the aim.
If you were designing an investigation and were aware before you started
of significant limitations in your design, then you could consider changing your aim to enable you to carry out an investigation without these
weaknesses. You should also show that you are aware of any limitations
in your design, when reporting on your research.

iii. Communication

The third weakness is one of communication. Omissions and lack of clar-
ity when reporting research can make it difficult to fully comprehend the
design of an investigation. The format of the paper required by a journal
can sometimes exacerbate this.

   If you look at Example 2.1 you will quickly appreciate that there is
plenty of scope for discussion here about what is a fault in the experimental design, what is a limitation, and what is due to omissions in the
journal article. We think the following fit into these three categories.
What did you list?

a. Fault in the design   The age of the men varied considerably and it is not
known how this may affect the results. There is a need to demonstrate that
this sample is truly representative of the average male drinking population,
and if it were, then this range in age is a strength more than a weakness.

Alternatively, using statistics to confirm that 'age' as a factor did not influence the results would resolve this concern.

It seems unlikely that the drinker and friends will remember accurately how much was drunk and over what time period. This may be a factor that affects the heavy drinkers more. The author does acknowledge this weakness in the article. Could a bartender not be recruited as an independent recorder?

Two methods for establishing habitual intake were used. Will the drinkers be honest about how much they habitually drink? There is a need to test these methods of recording in advance to see how reliable these forms of reporting are and which is the better method, and then to be consistent in the use of just one method.

b. Limitation   The times of eating varied and may have affected alcohol metabolism. If eating was controlled so that certain items were eaten at certain times during the drinking session, this may resolve the concern over the impact of eating on alcohol metabolism, but would almost certainly interfere with the normal behaviour of the participants.

The rate at which the alcohol was consumed was not recorded. Quickly drinking three pints and then nursing the fourth may result in a different rate of alcohol metabolism than a steady drinking pattern. It is difficult to see how this can be either controlled or recorded without affecting the participants' behaviour.

c. Communication   The published report did not indicate the length of the drinking period or what was eaten by the participants during their drinking session. It was not clear how the two methods for establishing habitual intake were used. Were they alternative methods or used in conjunction with each other?

What weaknesses have you identified? How do they compare with those we have included? Do you disagree with any parts of our evaluation? Why?

By thinking about this investigation, we have introduced you to some of the important elements you need to consider when designing an investigation. To start with, these include knowing your aim and having clear objectives. There were two treatment variables: 'the amount drunk habitually' and 'the level of breath alcohol when measured on the "morning after"'. You then identified a number of other factors that were either limitations or faults in the design of the investigation. All these factors that are not treatment variables are called confounding or non-treatment variables and can result in sampling error (2.2.4 and 6.1.3).

2.2 **Have a go!**

Key points We take you through the steps that are common when planning an experiment that will test a null hypothesis. To aid this process, we provide a decision web. There are nine key steps in the planning process: identifying an aim and objective(s), determining the statistical population, identifying the treatment and confounding variables, replication, the use of controls, the role of statistics when planning experiments, compliance with UK law, making assumptions, and causes of bias.

online resource centre

In our Online Resource Centre, we have a range of different topics for you to have a go at planning an experiment. They cover a broad range of subject areas, so there should be one that is closely related to your area of study. Alternatively, think about the following topic: 'The effect of wind strength on maximum seed dispersal in the common creeping thistle (*Cirsium arvense*).' There are many acceptable experiments that could be devised. We have provided you with one version to illustrate each of the design steps. You may find it helpful to use the decision web provided with this book in Appendix b or the interactive version in the Online Resource Centre.

When designing an investigation you will find that you need to be prepared to go through the process several times, first drafting your design and then redrafting as all the elements fall into place. So many factors need to be considered that, as you adjust one, you will need to go back and re-think all the others again. We have organized this process under the following ten subheadings, 2.2.1–2.2.10 (Fig. 2.01).

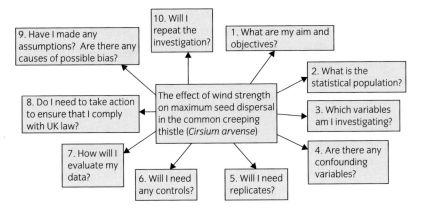

Fig. 2.01. Overview of the steps to be taken when planning the experiment: the effect of wind strength on maximum seed dispersal in the common creeping thistle (*Cirsium arvense*).

2.2.1 What are my aim and objectives?

You can realistically only write an aim after quite a lot of preliminary planning and discussions. You may be given a topic to research or you may need to choose one of your own. If you are not sure where to begin, we make some suggestions in Appendix a. Having found an area that interests you, you then need to read about the subject.

i. Background information

You cannot gather too much background information. It provides the following critical elements for the researcher:

- A biological context.
- The justification.
- Practical information.
- A guide to other designs and the way in which similar data have been analysed.

Background information is so critical, yet often the time given to this stage in planning is seriously underestimated. You need to carry out a sufficient review of literature relating to your topic to ensure that you have found all key references (usually those most frequently cited). One approach is therefore to use a citation index or search facility to identify the most significant publications in your area of interest.

a. The context   Having an understanding of what is known about the bio-logical system you are investigating is critical. It helps you to see what has

already been discovered and what is unknown or uncertain. You are then able to identify a real, meaningful research question. Reading to establish a context for research is often wide-ranging such as understanding what is known about the ecology of a species, the effectiveness of different methods used to replicate DNA *in vitro*, the different media used to grow a number of different bacteria, or the anatomy of the leg.

One problem that can arise from background reading, however, is that you may believe you know what the outcome of your research should be. We call this a 'prediction'. In our experience, if you have a prediction in mind at the outset, you are likely to carry out poor research as a result of conscious or unconscious bias. Therefore, although you need to know the background information to plan an experiment, you still need to approach your work with an open mind.

b. The justification   When you come to report on your investigation, you will need to explain the scientific reasons for carrying out a research project. Just because you can record or measure something does not make it appropriate research. For example, you may be able to record the effect of adding whisky to soil on the growth rate of daisies. But why investigate this? Do you have a sensible reason? Background reading at the early stages of planning an experiment is therefore critical to ensure that you do have a sensible justification for carrying out the research.

c. The practicalities   Many topics may be affected by seasonal factors, such as climate, timing of growth in plants, or behaviour in animals. If you were intending to examine the effect of temperature on the behaviour of *Oniscus asellus* (woodlice), what temperatures do they normally experience in their habitat? What temperature is likely to kill them?

Background information can come from a variety of sources, most often from literature reviews and asking other researchers. Do not rely on abstracts, such as those easily available on the Internet. This will not provide you with sufficient detail to be useful when planning an experiment. You may also carry out a preliminary investigation or pilot study to provide relevant information. This can help you to identify the appropriate scope of a treatment or the timescale that the experiment may need to run. For example, in an investigation on the survival of bacteria on different types of chopping boards, you will need to find out the numbers of bacterial colonies you expect in your samples so that you can use appropriate levels of dilution. If you are interested in the leaching of nutrients into an adjacent stream from fertilizer applied to a meadow, you would need to know the cation exchange capacity of the soil. (Further comments on practicalities are included in Appendix a: How to choose a research project.)

What background information would you need to develop an investigation into the effects of wind speed on maximum seed dispersal in *Cirsium arvense*? We suggest the following:

The context and justification

An understanding of the reproductive biology of this species and its ecology would enable you to put your work into a broad context. For this investigation, a key reference is: Klinkhamer & de Jong (1993). Biological flora of the British Isles: *Cirsium vulgare*. *Journal of Ecology 81*: 177–91. This reviews the ecology of the species. Other studies on wind dispersal of seed in other species would demonstrate the areas of investigation where wind dispersal is relevant and would provide a justification; for example, niche breadth (e.g. Thompson *et al.*, 1999); the effect of grazing (e.g. Bullock *et al.*, 1994) and conservation (e.g. Westbury *et al.*, 2008).

Practical implications

This investigation requires 'live' material; consequently there are many practicalities to consider. These include: when the seeds are produced; the normal range of wind speeds experienced by the plants; the equipment we would need to examine wind speed; whether *Cirsium arvense* can be grown in pots; whether it is known from other studies what the natural dispersal of thistle seed is; and the factors that are known to affect seed dispersal in general and *Cirsium arvense* in particular. In the review by Klinkhamer & de Jong (1993), it is clear that seed production per plant varies considerably but is reported usually as hundreds to thousands of seeds per plant. These seeds are large enough to be handled individually (each seed weighs between 1.3mg and 1.8mg) and can travel between 3.7m and 32m when dispersed naturally. This dispersal is affected by the height of the inflorescence. The first ripe seeds are known to be found by the end of July.

Previous designs and data analysis

How have other researchers carried out similar studies in this or related wind-dispersed species? How were the data analysed? What was the sampling strategy and sample size? For example, in a study by Fresnillo & Ehlers (2008), dispersibility was studied by recording the time taken for more than 900 *Cirsium arvense* seeds to fall 1m. These data were analysed using a nested ANOVA to allow the comparison of seed from mainland and island populations. In a second study (Skarpaas *et al.*, 2004),the authors released about 50 seeds taken from ten plants of *Crepis praemorsa* manually in two locations and recorded wind speed, seed characteristics, and distance travelled. In addition, these researchers also collected 101 plants with mature seeds, placed them in a mown field, and recorded the distance the seeds travelled. These results were transformed and analysed using a linear regression.
(For details on nested ANOVA, see 5.9, 8.4, and Chapter 9.

d. Previous designs   Background reading can be really helpful in showing you not only know how similar previous studies have been designed but also how the data were then analysed. Previous studies can also indicate suitable sample sizes. If you use other studies in this way, the authors' work must be cited.

You also need to be a critical reader and see if there are ways in which the design and/or analysis can be improved for your own work.

ii. Aim and objectives

Having completed your literature review and collected other background information, you should now be able to write a draft aim and objective(s). Aims and objectives are discussed in 1.1.

For the investigation into the effects of wind speed on maximum seed dispersal in *Cirsium arvense*, the aim and objective could be:

Aim: To investigate the effect of wind speed on maximum seed dispersal in *Cirsium arvense*.

Objective: To examine the maximum distance travelled by the seeds of *Cirsium arvense* when placed in a wind tunnel and exposed to wind speeds reflecting the range and average wind speed in the natural population (Fig. 2.1).

2.2.2 What is the statistical population?

The term population has different meanings depending on whether you are using the term, for example, in ecology, genetics, or statistics (1.3). When starting to plan your investigation, you may be using any one of these definitions. You need to be clear, therefore, as to what your

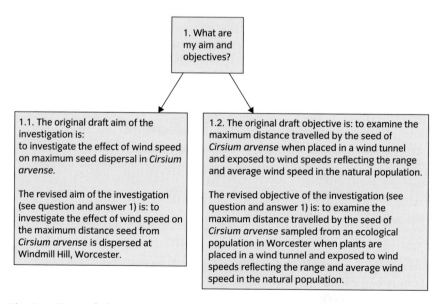

Fig. 2.1. Aims and objectives for experiment: the effect of wind strength on maximum seed dispersal in the common creeping thistle (*Cirsium arvense*).

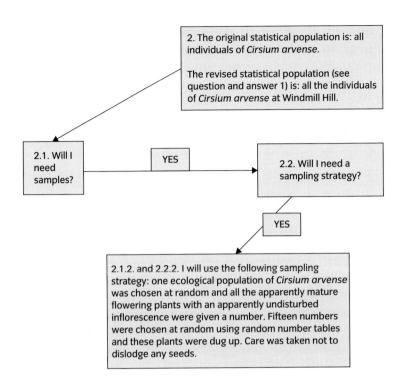

Fig 2.2. Statistical populations, samples, and sampling strategy for the experiment: the effect of wind strength on maximum seed dispersal in the common creeping thistle (*Cirsium arvense*).

statistical population is. In addition, it is rarely possible or even of any value to measure every individual in your population. You will usually sample. Samples need to be representative and you will therefore need a suitable sampling strategy (1.4).

Population

Our aim indicates our interest in maximum seed dispersal in the species *Cirsium arvense*. All individuals of this species then form our statistical population (Fig. 2.2).

Sample

In an investigation on seed dispersal, you cannot realistically examine every plant in the statistical population; therefore, you will need a representative seed sample. For our sampling strategy, we have decided to use whole plants, as this will then mean that the seeds are held at the same height as they would be in the field and within an inflorescence, as they normally would be (Fig. 2.2 and 2.2.7).

Therefore, in our design one ecological population of *Cirsium arvense* was chosen at random and all the apparently mature flowering plants with an apparently undisturbed inflorescence were given a number. Fifteen numbers were chosen at random using random number tables and these plants were dug up. Care was taken not to dislodge any seeds. (The rationale for choosing this sample size is reviewed in 2.2.7.)

Having identified the statistical population and considered a draft sampling strategy, you will need to decide how many items (1.2) you will need in your sample. Your choice depends on the type of data you will collect (5.1) and how you will analyse this (Chapters 6–10). This is not an easy task. We therefore spend more time on this topic in 2.2.7.

 Q1   Do you think that our design so far will provide us with a representative sample?

A1   It is difficult to say with any confidence that this will be the case, as a sample of plants from a single ecological population may not reflect this species across its range. We must consider revising our experiment, and there are two ways forward at this stage. We can either extend the sampling into a number of ecological populations over a broad range of the species distribution, or we can revise the aim and objective so that they reflect the limited nature of this study. We will take the second course. Therefore, the revised aim and objective are:

Aim: To investigate the effect of wind speed on the maximum distance seed from *Cirsium arvense* is dispersed at Windmill Hill, Worcester (Fig. 2.1).

Objective: To examine the maximum distance travelled by the seeds of *Cirsium arvense* sampled from an ecological population in Worcester when placed in a wind tunnel and exposed to wind speeds reflecting the range and average wind speed in the natural population (Fig. 2.1).

The statistical population is now all the individuals of *Cirsium arvense* at Windmill Hill (Fig. 2.2).

2.2.3 Which variables am I investigating?

This may seem to be an unnecessary step in our planning process. However, there are so many terms that relate to the factors we investigate, such as 'variable', 'treatment', and 'sample', that it can become confusing as to which is which. The identification of the variables that we are interested in also allows us to determine with more confidence the factors that we are not investigating in our experiment (2.2.4). You may examine the effect of a variable by recording another variable (1.7). We can see this in our example.

In our *Cirsium arvense* investigation, the factors we are investigating are the distance travelled by the seeds and the wind speed (Fig. 2.3). The wind speed is the variable whose effect we are examining. The distance travelled by the seed is the variable whose response to this treatment we then measure.

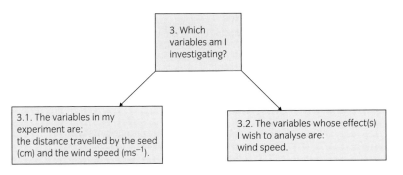

Fig. 2.3. The treatment variables for the experiment: the effect of wind strength on maximum seed dispersal in the common creeping thistle (*Cirsium arvense*).

2.2.4 Are there any confounding variables?

When you record a set of measurements in an experiment, you usually see variation in these measurements. This variation may be due to the factors you are investigating, i.e. the treatments (1.6 and 1.7) or other causes, i.e. confounding (non-treatment) variation (1.7.3). Variation due to the action of confounding variables is commonly known as sampling error (6.1.3). If the variation in your data caused by confounding variables is considerable, it can mask the effect of the factors you are investigating. Therefore, it is important to be clear about what you are investigating (2.2.3), identify possible confounding variables (2.2.4i), minimize the effect of these confounding variables (2.2.4ii), and if possible mathematically separate out (partition) the variation in your data due to the factors you are investigating and the variation due to the confounding variables (2.2.4iii).

i. Identifying confounding variables

Identifying the confounding variables has to be your next step as you cannot minimize the effect of something if you are not aware that it might affect your experimental system. Background or preliminary investigations, common sense, and reading help in identifying causes of this non-treatment variation.

EXAMPLE 2.2 The antibacterial properties of triclosan and tea-tree oil

An undergraduate took swabs from the hands of volunteers working in the medical profession. These were the 'before' measures. She then washed the volunteers' hands. One hand was washed with triclosan and one with tea-tree oil. Having allowed these washed hands to air dry, the student took a second swab from each hand. These gave the 'after' results. At the laboratory, the swabs were used to inoculate plates. As the bacterial count could be high, serial dilutions were also made from the swabs and plates were then inoculated. After 2 days' incubation, the numbers of bacterial colonies on the plates were counted.

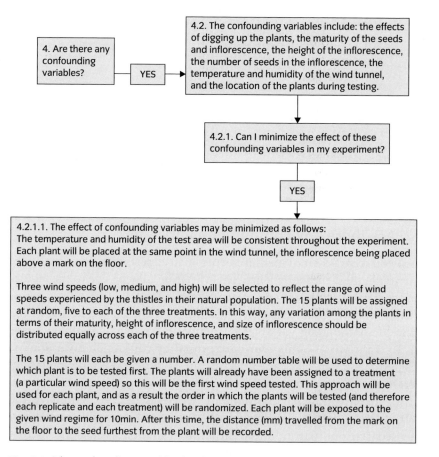

Fig. 2.4. The confounding variables for the experiment: the effect of wind strength on maximum seed dispersal in the common creeping thistle (*Cirsium arvense*).

In Example 2.2, there are a number of potential confounding variables. These include the method used to wash the hands, the method used to swab the hands, the consistency of the poured agar plates, the method used to inoculate the plates, the temperature of the incubator, the time of incubation, and the method used to count the resulting bacterial colonies. In the investigation into seed dispersal in *Cirsium arvense*, there are also many possible causes of non-treatment variation. These include: the effects of digging up the plants, the maturity of the seeds and inflorescence, the height of the inflorescence, the number of seeds in the inflorescence, the temperature and humidity of the wind tunnel, and the location of the plants during testing (Fig. 2.4).

ii. Minimizing the effect of non-treatment variation

In your investigation, you will be collecting observations from a number of items (1.2). These data will almost certainly show variation (1.7). Often your items are grouped in some way, such as different samples. If

you wish to gain an understanding of the degree to which the variation in your data is due to the factors you are investigating, you need to ensure that all other non-treatment effects are equalized in some way across all the items in all the groups. This can be achieved in two ways: 'equalizing the effect' and 'randomization'. Where possible, both approaches should be incorporated into your design.

a. Equalizing the effect   One approach towards minimizing the effect of confounding variables is to ensure that the causes of this non-treatment variation are made as constant as possible across your experimental system. In doing so, the real effects should be visible above the now-consistent background noise.

In Example 2.2, the student ensured that she was consistent in her use of all methods throughout the experiment. In addition, she made up a single batch of agar and poured all the plates at same time. All the plates were placed in the same locations, e.g. the same flow cabinet or the same incubator. You may have come across this idea before as part of the concept of designing a 'fair test'.

To minimize the non-treatment effects in the experiment on seed dispersal, the temperature and humidity of the test area were consistent throughout the experiment. Each plant was placed at the same point in the wind tunnel, the inflorescence being placed above a mark on the floor (Fig. 2.4).

---

b. Randomization   A common practice that, on average, will act to reduce the impact of confounding variables is that of randomization. The nature of the randomization will depend on your investigation and the confounding variables. Most often, randomization occurs in relation to treatment, location, or time of day.

In Example 2.2, to ensure that the effect of confounding variables was kept to a minimum, the inoculated plates were arranged at random within the incubator. In this way, all the 'triclosan' plates were not in one part of the incubator whilst all the 'tea-tree' plates were in another part and thus subject to the effects of any micro-variation in temperature within the incubator. Thus, all replicates will have an equal chance of exposure to the confounding variable—another example of designing a 'fair test'.

In the experiment we have been planning on *Cirsium arvense*, three wind speeds (low, medium, and high) were selected to reflect the range of wind speeds experienced by the thistles in their natural population. The 15 plants were assigned at random, five to each of the three speed treatments. In this way, any variation between the plants in terms of their maturity, height of inflorescence, and size of inflorescence should be distributed equally across each of the three treatments (Fig. 2.4).

iii. Mathematically separate out (partition) the variation in your data

Another approach to dealing with non-treatment variation is to design your investigation in such a way that you can mathematically obtain an estimate of the variation in your data that is due to the effect of confounding variables. The simplest method of achieving this is by using replicates (2.2.5). There are specific designs that allow for simple mathematical partitioning of the variation due to confounding variables. This aspect will be covered in Chapter 9.

2.2.5 Will I need replicates?

i. When should you use replicates?

Replication *Where you have more than one of something. It can be more than one item being exposed to a treatment or more than one group of items exposed to one treatment or more than one item in a sample*

Replication should be used whenever possible for two reasons: firstly, to increase the reliability of your estimate of the population parameters; and secondly, to allow mathematical estimates of variation due to confounding factors.

a. To increase the reliability of your estimate of the population parameters   In Example 2.2, the student could carry out her investigation on one person, where one hand is washed in triclosan and one in tea-tree oil. If the aim is to investigate the effectiveness of these two hand washes on people working in a medical profession, then this is clearly a very small sample and any values from a single individual are not likely to closely match the population parameters. It would be better to increase the sample size and to do it in the same way for both treatments. In fact, 25 people took part in this investigation. Each hand washed in triclosan was a replicate within the treatment 'triclosan'. Each hand washed with tea-tree oil was a replicate within the treatment 'tea-tree oil'.

b. To allow mathematical estimates of non-treatment variation   Using replicates may allow you to use certain statistical tests, such as an analysis of variance or ANOVA (Chapter 9). These tests estimate the variation in your data due to the effects of the confounding variables and the variation due to the factors you are investigating. They make allowances for the non-treatment variation when testing hypotheses.

An understanding of the level of variation within your population may be critical. For example, in the study on the effectiveness of tea-tree oil, the results showed that in some cases tea-tree oil reduced bacterial counts, in some cases the counts were increased, and in some cases counts were too low or too high to be measured effectively by the method used. In a practical setting such as this, the variation demonstrated between people (replicates) is crucial to gauging the real effectiveness of a bactericide.

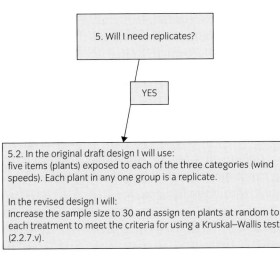

Fig. 2.5. Replication in the experiment: the effect of wind strength on maximum seed dispersal in the common creeping thistle (*Cirsium arvense*).

In the thistle experiment, five plants were exposed to each of the three wind speeds. Each plant in any one group was a replicate (Fig. 2.5). The 15 plants were each given a number. A random number table was used to determine which plant was to be tested first. The plant had already been assigned to a treatment (a particular wind speed) so this was the first wind speed tested. This approach was used for each plant. As a result, the order in which the plants were tested (and therefore each replicate and each treatment) was randomized. Each plant was exposed to the given wind regime for 10min. After this time the distance (mm) travelled from the mark on the floor to the seed furthest from the plant was recorded.

ii. Features of replicates

Replicates must be an integral part of the experimental design and must all be established and examined within the single experiment and not subsequently. Any work carried out later is a new experiment (2.2.10). Replicates must also be 'independent' of one another (1.4.1). This means that you should not use the same animals, plots, etc. as replicates. For example, if you were investigating the response of plants to a watering regime, you might measure the growth of several leaves on the same plant exposed to one treatment and several leaves on another plant exposed to a different watering regime. This is an example of pseudo-replication. Clearly the leaves on any one plant are not independent of each other: they share the same genotype and microclimate. A more appropriate method for replication would be to expose many plants assigned at random to each treatment and either randomly or using a stratified method (1.4.2) sample one leaf from each plant.

Another example of pseudo-replication frequently arises when items are grouped in the same Petri dish, pot, etc. For example, in an investigation into the effect of fertilizer on the growth of marigolds, there were ten plant pots each containing four marigolds. Five of these pots were treated with a new fertilizer and the other five pots were treated with the standard plant food. After six weeks, the dry mass of the shoots and roots from each plant was measured. A common mistake is to treat each plant as a replicate and therefore believe that there are 20 replicates for each treatment, but the four plants in a pot are not independent of each other. (If you do design an experiment like this, you should refer to 9.13. as the four plants are 'nested' within the five pots.)

iii. How many replicates?

One of the hardest things to establish is how many replicates are needed in any particular investigation. An indication of the number of replicates comes from the requirements of the particular type of statistical test to be used and the extent of the variation due to confounding variables. It may be necessary to carry out a preliminary investigation to identify the degree of non-treatment variation before choosing the number of replicates for the main experiment. We look at this thorny issue again in relation to sample sizes in 2.2.7.

2.2.6 Will I need any controls?

In laboratory experiments it is usually relatively easy to construct a control. In field experiments, if a control is necessary then you will need to select an area that is as similar as possible to the one used in the rest of your experiment. In complex experiments where you are investigating the effects of more than one variable, you may require more than one control.

A **control** is most often included in an experiment to provide data to confirm that an aspect of your experiment is working as it should. Alternatively, a control may act as an experimental baseline against which any effects of the treatment(s) may be compared. A control should be an integral part of the investigation and must be carried out at the same time as the rest of the investigation.

Control An experimental baseline against which any effects of the treatment(s) may be compared. It must be an integral part of an experiment

Q2 In the experiment described in Example 2.2 on the antibacterial properties of triclosan and tea-tree oil, are any controls needed? If so what?

A control may not always be necessary. For example, in an investigation into the effect of temperature on the behaviour of *Oniscus asellus*

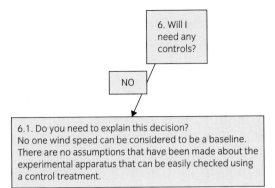

Fig 2.6. Controls in the experiment: the effect of wind strength on maximum seed dispersal in the common creeping thistle (*Cirsium arvense*).

(woodlice), you might select temperatures of –5, 0, 5, 10, 15, and 20°C as your treatment range. There is no one temperature that could be called a control.

> The investigation into the effects of wind speed on seed dispersal in *Cirsium arvense* is similar to the *Oniscus asellus* example in that there are no controls (Fig. 2.6).

2.2.7 How will I analyse my data?

This critical stage of experimental design is the one most frequently left out of the planning stages, and in our experience is the most frequent cause of disappointing research. You will see that we feel very strongly about this as we keep drawing your attention to it. This step is very much a forwards and backwards process. Your draft design is used to determine which is the most appropriate statistical test to use, but this choice of test may then influence factors such as how many observations you should have in each sample or how many replicates. There are four elements covered under this topic: choosing your statistical tests; reconsidering your experimental design; finalizing your aim and objectives; and writing your hypotheses—these take you through both the 'forwards' and the 'backwards' steps. Completing these steps requires an understanding of some of the concepts we do not cover until later in this book. Therefore, you will need to jump forward in places to read these sections where indicated. Most terms are also defined in the glossary. A full summary to help you choose between any of the statistical tests we have included in this book can be found in Appendix c.

In Chapter 1, we explained that there are two approaches to data analysis: information-theoretic models and hypothesis testing. In the first, you collect your data and then determine whether there is any evidence

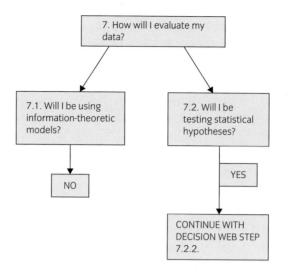

Fig 2.7. Planning the analysis of data for the experiment: the effect of wind strength on maximum seed dispersal in the common creeping thistle (*Cirsium arvense*).

for particular groupings within the data. If there is, you may then proceed to test hypotheses. If you are setting out to test a null hypothesis, this question determines your experimental design. Thus, the first step you need to take is to decide whether you are starting out with a hypothesis or not.

> To examine the effect of wind strength on maximum seed dispersal in *Cirsium arvense*, we will test a hypothesis. We have posed the question: does the wind speed affect the maximum distance travelled by the seed? (Fig. 2.7)

We are not able to cover information-theoretic models in this book. Our example and the following chapter therefore deal with hypothesis testing. However, we have included information-theoretic models on the decision web (Fig. 2.7.1) so you may continue to plan your experiment if this is the approach you are taking, although for some details you may need to refer to other text books.

i. Which type of hypothesis am I testing?

The next step is to decide which type of hypothesis you are testing and it is about now that you will feel inclined to give up on this chapter because here we present an overview of the next six chapters before you have read them. To succeed at this stage, you need to read on without worrying too much about detail. Aim to get a sense of the steps involved. We would not expect you to be able to repeat this yourself at this stage until you become more familiar with what follows in the rest of this book.

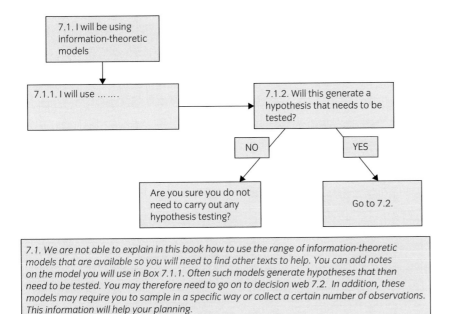

Fig 2.7.1. Using an information-theoretic model when planning your experiment.

There are three types of hypothesis you may encounter:

1) Do the data match an expected ratio?

2) Is there an association between two or more variables?

3) Do samples come from the same or different populations?

In our experiment, we have no reason at the outset to expect the data to match a particular mathematical outcome (Chapter 7). We are examining the effect of only one variable or treatment (wind speed). However, we do want to know whether all the samples behave in the same way, i.e. is the maximum distance travelled by a seed the same for all wind speeds? The hypothesis for our experiment therefore falls into the third group (Fig. 2.7.2).

When we test hypotheses, we use a particular statistical test to determine for example whether there 'is' or 'is not' a significant difference between our samples. The next step you need to complete is therefore to decide which is the correct statistical test to use. There are many different tests to choose between. We cover the most common ones in Chapters 6–10. The three steps you follow to enable you to choose a statistical test involve deciding (i) what type of data you have, (ii) how many variables you wish to test, and (iii) how many observations you have. Each statistical test has a number of criteria such as the data need to be measured on an interval scale.

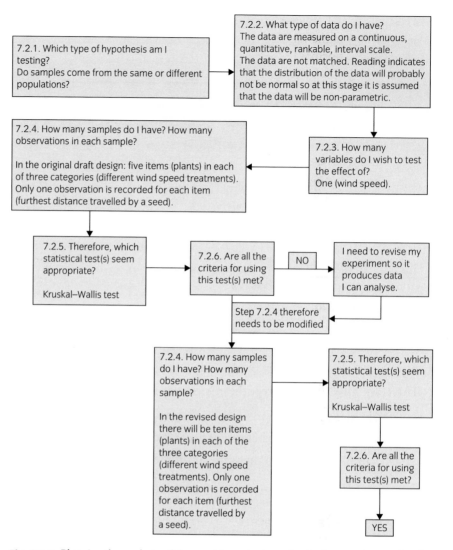

Fig. 2.7.2. Planning the analysis of data and hypothesis testing in the experiment: the effect of wind strength on maximum seed dispersal in the common creeping thistle (*Cirsium arvense*).

You compare the information about your experiment with the criteria relating to each statistical test to find which is the most appropriate test to use for your experiment. We take you through this step by step in 2.2.7ii–v.

ii. What type of data do I have?

Types of data are described in 5.1. Read about these terms and then think about the units you will be recording. Are they mm, pH, percentages, etc.?

You need to decide what scale you are using (i.e. interval, nominal, or ordinal); whether the data are matched (9.3), and whether the distribution is possibly normal and the data may therefore be considered to be parametric (5.8).

In this experiment, we will be recording the distance travelled by the furthest seed (in mm). Length measurements like this are measured on a continuous, interval scale: they are numbers and are therefore quantitative, and these numbers can be arranged in a specific order so they are rankable (5.1). There is only one observation for each item (i.e. the distance travelled by any one seed has only been recorded once), so the data are not matched (9.3). It is difficult until you have carried out your experiment to be certain whether the data will be parametric (5.8). Background reading about seed dispersal indicates that the data are unlikely to be parametric, so this is what we will assume in the planning stage, but we must then confirm this after the experiment has been carried out (Fig. 2.7.2).

---

iii. How many variables do I wish to test the effect of?

We know from 1.7 and 2.2.3 that you may have a number of variables that are part of your experiment but not all may be variables whose effect you are examining. It is only these that we need to identify for the next step in our planning.

In our *Cirsium arvense* investigation, the factors we are investigating are the distance travelled by the seed and the wind speed. The wind speed is the variable whose effect we are examining (Fig. 2.3), i.e. we wish to examine the effect of only one variable (wind speed).

---

iv. How many categories do I have? How many observations in each category?

Here we return to the sampling method (Figs 2.2 and 2.5) and decisions made about replication. The term sample can be misleading. For example, you may sample 400 items from one source and then place these into four categories each with 100 items. In this example you would describe your design as having one sample, with four categories, each category having 100 items. An alternative investigation may require you to go to four different locations and at each location sample 100 items. The end result for the design is the same as before in that you will have four categories each with 100 items. The term sample is often used to describe both the sample (e.g. 400 items from one location) and the number of items in each category (i.e. 100). This can be confusing. We therefore prefer to use the term category and numbers in a category when referring to experimental designs and to keep the term 'sample' for referring to sampling methods.

Categories When each observation may fall into only one of two or more mutually exclusive groups, these groups are known are categories

In our experiment on seed dispersal in *Cirsium arvense*, we have five plants in each wind speed category. The furthest distance travelled by the seed will be recorded, i.e. we have five items in each of three categories. One observation was recorded for each item (Fig. 2.5).

v. Therefore, which statistical test(s) seem appropriate?

Chapters 9 and 10 give details about the most common statistical tests that are appropriate for testing hypotheses of the type we have in our experiment. In Chapter 9, all the tests relate to data that are parametric. In Chapter 10, all the tests relate to data sets that are non-parametric. We have already decided that it is most likely that our data will be non-parametric, so we should refer to Chapter 10.

The first section in Chapter 10 sets out to guide you to the most appropriate statistical test. The following is an extract from Chapter 10 and uses a number of criteria to help you choose between different statistical tests. For example, if you had decided in 2.2.7ii–iv that you wish to examine the effect of one variable, where there are two categories or samples and the data are non-parametric and unmatched, then you would, using the table, decide that you should probably use a Mann–Whitney U test to test your hypotheses and so determine whether there is a significant difference between your samples.

Criteria from Chapter 9	Test
You wish to examine the effect of one variable. You are going to compare two categories or samples. The data are unmatched.	Mann–Whitney U test (10.1)
You wish to examine the effect of one variable. You are going to compare two categories or samples. The data are unmatched. The data are measured on a continuous scale and you have more than 30 observations in each sample.	z test for unmatched data (9.1)
You wish to examine the effect of one variable. You are going to compare two categories or samples. The data are matched. You have fewer than 30 pairs of observations.	Wilcoxon's rank paired test (10.2)
You wish to examine the effect of one variable. You are going to compare two samples. The data are matched. You have more than 30 pairs of observations.	z test for matched data (9.3)
You wish to examine the effect of one variable. You are going to compare three or more categories or samples.	One-way ANOVA (Kruskal–Wallis test) (10.3). If the outcome from this test is significant, you may follow it with a multiple comparisons test (10.4)
You wish to examine the effect of more than one variable. You are going to compare two or more categories or samples. You have equal numbers of observations in each category.	Two-way non-parametric ANOVA (10.5). If the outcome from this test is significant you may follow it with a multiple comparisons test (10.6).
You wish to examine the effect of more than one variable, each with two or more categories.	Scheirer–Ray–Hare test (10.7)
You wish to examine the effect of two variables, each with two or more categories. The design is orthogonal. There is only one observation in each category. These may be repeated measures.	Friedman's test (Sokal & Rohlf, 1994; Zar, 2010)

To choose the correct test, you should consider the following. Each statistical test has several requirements that must be met and these details are given at the start of the section relating to that test. The guide given in the extract from Chapter 9 uses a subset of these criteria to take you to the most likely test for your data. It is assumed here that you have non-parametric data.

From the table, it is clear that for our experiment with one variable (wind speed) and three categories (low, medium, and high wind speeds) we should use the Kruskal–Wallis test (Fig. 2.7.2).

The full list of criteria for using a Kruskal–Wallis test are outlined in 10.3. We need to check these, as we hope that they will confirm our choice of test and help us to choose sample sizes. To use a Kruskal–Wallis test, the following must apply:

1) You wish to examine the effect of one variable with three or more categories or samples.

2) You have data that is non-parametric but can be ranked.

3) You must, if there are only three samples, have more than five observations per sample.

4) You do not need equal sample sizes.

5) You may not use this test if you have little variation in a relatively large sample size.

In our experiment, we do have one variable (wind speed) and three categories (low, medium, and high wind speeds). From our background reading, we believe that the data are likely not to be normally distributed and so will be non-parametric. The scale is measured in cm and is an interval scale that can be ranked. From criterion 4, we can see that, although in our design we have the same number of items in each category (a seed from each of five plants), this is not critical. We cannot make an assessment about criterion 5 until we have collected our data so this criterion cannot be used in our planning, but we must return to this when the experiment has been carried out. Finally, if we check against criterion 3, we can also see that, as we have only three categories, we should have more than five observations in each. In our current design, we had only five plants per category. So although most of the criteria for using this test are met, the size of sample proposed in our current design is not large enough for us to be able to use the Kruskal–Wallis test. We must either redesign our experiment or seek out another statistical test.

Many statistical tests of hypotheses have minimum and maximum guides to the numbers of observations in the sample. These may be determined by the range over which the test is most effective. Alternatively, in large samples, it may be that you will obtain no additional useful information from having more observations. Clearly, by checking your design in the manner we have described you can select the most appropriate statistical test and, as part of this process, you will also have checked on

details such as how many observations you need in each category. If your draft design does not meet the criteria, you still have time to amend it before you carry out the experiment.

In our current draft design, the sample size we had chosen arbitrarily (2.2.2) is not going to be adequate. Therefore, we will increase the sample size to 30 and assign ten plants at random to each treatment (Fig. 2.7.2). The minimum number of observations for our experiment has therefore been determined by the criteria for using a Kruskal–Wallis test and the maximum is determined by practicalities and by ensuring that we do not devastate the ecological population at Windmill Hill by our sampling.

---

online resource centre

When planning experiments, it is this step that is probably the hardest. But do not shy away from it. There are lots of examples in the Online Resource Centre to give you practice in choosing the most appropriate statistical test and so enable you to boost your confidence.

vi. Finalizing your aim and objectives

The steps above often lead to a change in your objective(s) and sometimes in your aim. Therefore, you should now finalize them, taking into account the planning stages you have completed.

Aim: To investigate the effect of wind speed on maximum distance seeds are dispersed in *Cirsium arvense* at Windmill Hill.

Objective: To examine the maximum distance travelled by seeds from 30 *Cirsium arvense* sampled from one ecological population in Worcester when placed in a wind tunnel and exposed to low, medium, or high wind speeds that reflect the range and average wind speed in the natural population (Fig. 2.1).

---

vii. Writing your hypothesis

We introduced you to hypotheses in Chapter 1 (1.8). If you are designing an experiment and have got this far in your planning, you should be able to write your hypothesis. More information about writing hypotheses is given in Chapter 6.

In our example, we believe that the data we are collecting will be non-parametric, so our hypothesis will be phrased in terms of population medians (6.1.2).

H_0: There is no difference in the median maximum distance (mm) that seeds from *Cirsium arvense* are dispersed when exposed to low, medium, or high wind speeds (m/s) in a wind tunnel.

H_1: There is a difference in the median maximum distance (mm) that seeds from *Cirsium arvense* are dispersed when exposed to low, medium, or high wind speeds (m/s) in a wind tunnel.

2.2.8 Do I need to take action to ensure that I comply with UK law?

In this chapter, we have considered a number of different types of investigation and the factors you need to think about as you plan your research. This aspect of the topic is covered in detail in Chapter 4. It is not possible to consider all UK legislation within the scope of this book so we have focused on the four areas that our students most often encounter. The decision web (Appendix b) is set up to illustrate the decisions you need to make at this stage in your planning but the list is not exhaustive.

The four areas we consider are health and safety, sampling and access, animal welfare, and working with humans.

Cirsium arvense is not a protected species. The sampling will be carried out on private land managed by the Shropshire Wildlife Trust and therefore permission to sample the plants must be sought. As we are using electrical equipment and handling soil, we will complete a risk assessment (Fig. 2.8).

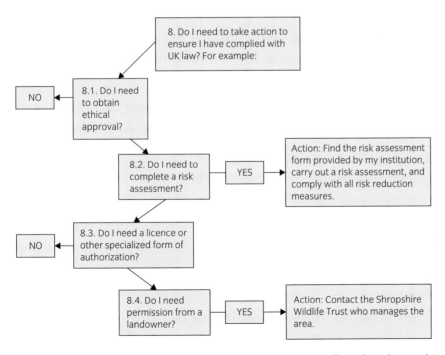

Fig. 2.8. Considering the law when planning the experiment: the effect of wind strength on maximum seed dispersal in the common creeping thistle (*Cirsium arvense*).

2.2.9 Are there any causes of possible bias? Have I made any assumptions?

i. Influencing outcomes and causes of bias

Some particular methods may themselves influence the outcome of an investigation. If this is likely, then you should try to estimate the extent of this effect and take it into account when planning your experiment and when you interpret the results. One of the most obvious examples of this can occur when studying animals, where your own presence and behaviour and the presence of experimental apparatus may influence the outcome. The contamination of samples from your own body or equipment is another example. The problems that can arise from not wishing to influence the investigation are clearly seen in Example 2.2 where the undergraduate needed to ensure that her presence did not temporarily alter her volunteers' behaviour.

In the experiment we have been designing in this chapter, it is difficult to see how we might influence the outcome. This step in our planning is probably not applicable in this instance.

---

ii. Assumptions

Despite all your careful planning, you may still have to make certain assumptions about the experimental system. Make a note of these assumptions as you become aware of them. These assumptions may be testable as a separate experiment. You should always show that you are aware of these assumptions and possible causes of bias when you interpret your results and communicate your findings. In research that involves animals, including humans, one common cause of bias is that of 'self-selection'. Some animals can become 'trap-happy' and choose to be caught, usually because traps contain a source of food. Clearly, such a subset of animals may not represent the population under investigation. Similarly, humans who volunteer to take part in a study may not be representative. This can be exacerbated if a reward is offered as an incentive (4.5.2).

Look through the information you have noted down when 'having a go'. How many assumptions and possible causes of bias can you identify? For our design these include the following:

- The random method employed when collecting plants from the wild did result in a representative sample.
- No bias was introduced by the loss of any seeds from the inflorescence during the sampling.

- The random allocation of plants to the treatments minimized the effect of non-treatment variables.

- The wind speeds used in the experiment were representative of those experienced by the species at Windmill Hill, Worcester.

- The order of testing the plants within the wind tunnel had no effect on the distance travelled by the seeds.

- The seeds were blown along in a manner that is similar to that in the field, including seeds that were blown along the ground (Fig. 2.9).

2.2.10 Will I repeat the investigation?

If an effect is real, then it is reasonable to expect that a repeated investigation, carried out using the same method, will identify the same trend. A scientific finding that has been confirmed in this manner is considered to be 'sound'. Clearly there may be circumstances that do not allow for a repeated experiment. For example, an investigation into the effects of hurricane Ivan on the distribution of manatee along the Florida coast could not be repeated.

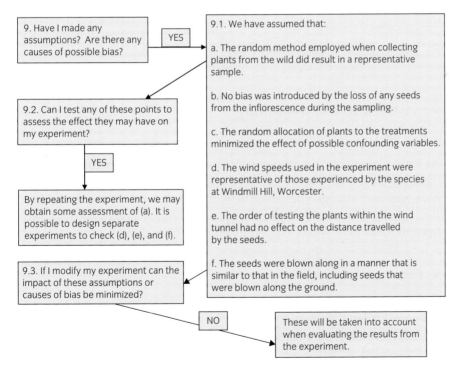

Fig. 2.9. Assumptions and bias when planning the experiment: the effect of wind strength on maximum seed dispersal in the common creeping thistle (*Cirsium arvense*).

One common failing is to confuse a repeated experiment with a replicate. If an experiment is repeated at a different time, this is not a replicate. The data obtained in this way cannot be combined with the data from the earlier experiment without using certain mathematical steps to prove that it can be pooled. We give one example of how this might be done in 7.2.

> In our example, the experiment could be repeated with another 30 plants taken from the same population in a similar manner (Fig. 2.10).

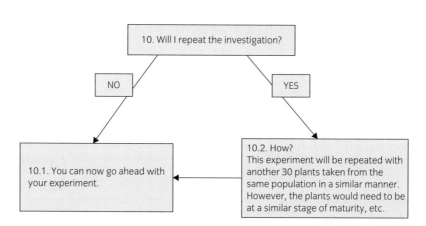
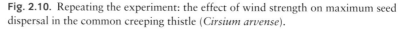

Fig. 2.10. Repeating the experiment: the effect of wind strength on maximum seed dispersal in the common creeping thistle (*Cirsium arvense*).

2.2.11 **Back to the beginning**

i. *Review your planning*

As we explained at the beginning of this chapter, when designing an investigation you usually have to develop a draft and then refine it, because at each step any change you make may affect steps you have already been through. So at this point look through your design again using our checklist and make sure your design is as good as it can be.

ii. *Experimental design and methods*

Our decision web focuses on the design of our experiment, but this does not necessarily mean that you will have a record of all the technical details in your method. As part of your final review of your plan, you can check that all technical details are now included in your design.

The experimental design we came up with is summarized below. How does it compare with your design?

Topic: The effect of wind strength on maximum seed dispersal in *Cirsium arvense*.

Aim: To investigate the effect of wind speed on the maximum distance seeds from *Cirsium arvense* are dispersed, at Windmill Hill, Worcester.

Objective: To examine the maximum distance travelled by the seeds of 30 *Cirsium arvense* sampled from one ecological population in Worcestershire when placed in a wind tunnel and exposed to low, medium, and high wind speeds that reflect the range and average wind speed in the natural population.

Hypotheses

H_0: There is no significant difference in the median maximum distance (mm) that seeds from *Cirsium arvense* are dispersed when exposed to low, medium, or high wind speeds (m/s) in a wind tunnel.

H_1: There is a significant difference in the median maximum distance (mm) that seeds from *Cirsium arvense* are dispersed when exposed to low, medium, or high wind speeds (m/s) in a wind tunnel.

Method

All the apparently mature flowering plants with an apparently undisturbed inflorescence within a specific ecological population at Windmil Hill in Worcestershire were given a number. Thirty numbers were chosen at random using a random number table and these plants were dug up. Care was taken not to dislodge any seeds. Each plant was potted up in 12cm pots in John Innes No. 1 compost.

Three wind speeds (low, medium, and high) that reflected the range of wind speeds experienced by these plants in the natural population were selected. Each plant was given a number. The first ten numbers chosen at random were assigned to the low wind speed treatment, the second ten numbers chosen at random were assigned to the medium wind speed treatment, and the remainder were assigned to the high wind speed treatment. In this way, any variation between the plants in terms of their maturity, inflorescence size, and height should be distributed equally across each of the three treatments.

This numbering system was also used to determine the order in which the plants were tested. Again, using a random number table, a plant was selected and then exposed to its predetermined treatment. A second plant and treatment were selected in the same way until all plants had been tested. This ensured that the order in which the three treatments were tested was also randomized. To minimize the non-treatment effects in the experiment, the temperature and humidity of the test area were consistent throughout the experiment. Each plant was placed at the same point in the wind tunnel. Each plant was exposed to the given wind regimen for 10 minutes and the distance from the centre of the pot to the seed that had travelled the furthest was measured (mm). The hypotheses for this experiment will be tested using a Kruskal–Wallis test. The experiment will be repeated with another 30 plants.

2.3 Managing research

Key points Having planned your experiment, there are a few further practical points to consider including how you manage your time and the space you have to work in. Suggestions are also given here about how to keep track of work in progress.

To be effective as a researcher, either as an undergraduate or a graduate, you must develop good management skills. In research, management falls into at least three areas: managing time, managing space, and managing data.

2.3.1 Time management

Time management is critical. You may have steps in a method that need to be carried out for certain periods of time, or you may need to collect samples or complete preparations all on one day. In addition, most research is carried out within a time frame determined either by your course or by funding. It is therefore important to consider, before you start, how long each part of your investigation is likely to take and to draw up a timetable. A timetable should help you identify critical points in the execution of your investigation or even experimental designs that just cannot be carried out in the time available. The time constraints imposed by seasonality and so on must also be recognized in the planning, as should external factors such as assessment points for other courses. In Example 2.2, the student taking and examining swabs from volunteers' hands collected the samples on one morning, but it then took all afternoon and into the late evening to inoculate the microbial plates. This student's planning was exemplary and she knew from her experience with a pilot experiment how long the work would take. She was therefore able to plan her time effectively and carry out the project, having made proper arrangements in advance to work late.

From our experience, there are two common errors that arise in relation to time management because of inexperience. The first of these is developing complex experimental designs. It is very tempting to try to investigate lots of factors all within one experiment. This can be counterproductive for two reasons. The factors can interact and you may not be able to untangle which factors had which effect. Trying to do so can be very time-consuming and is usually not very successful. Complex designs often need complex statistics. This is fine as long as you have thought this all through before you start and then allowed in your planning for the time it will take to analyse the data correctly. It is usually better to have more objectives to examine an aim than one complex objective. Secondly, research takes longer than you think. A general rule of thumb is to plan your investigation and then double the time you have allowed

for each part. If you are required to write a thesis or report you should also allow yourself twice as much time as you think it should take. This realistic estimate of time should be built into your timetable.

2.3.2 Space management

In research, you are often working in a confined space, such as a laboratory, and in the company of other scientists. This means that your management of your space is very important, allowing you to complete your research without interfering with other people's work. You must be considerate of other users of the areas around where you are working: we consider this further in 4.2 and 4.5.

2.3.3 Data management

When designing and carrying out research, your ability to keep and manage comprehensive records is paramount. For most of us, records are kept in two ways: in a transportable hard-copy format, such as a laboratory or field notebook, and as electronic files.

i. Laboratory or field notebooks

The best approach to recording information in such notebooks is to record more detail rather than less. You should date each day's work, give full details of the method used that day, which may for fieldwork include a site description and details of the weather, and a record of the results you obtained. You may also need to record contacts' details, but you should be careful when making any records that may compromise volunteers' confidentiality (4.5). Finally, add notes about your ideas, why you are taking the research in a particular direction, what the data appear to show you, and how this may relate to other people's studies. All this information will be invaluable, both in tutorials and when communicating your results in a more formal way (Chapter 11). Having more detail rather than less is particularly important for graduates, who may not write formal reports for several months and therefore will be less able to rely on memory to compensate for unclear notes.

EXAMPLE 2.3 The growth rate of rye seedlings

Researchers at an agricultural research station were interested in the natural variation in the rate of growth of rye. The first measurement was made when the rye was at the seedling stage and the leaf length (mm) was recorded for 15 seedlings. In her notebook, the researcher noted down the following:

12.0, 12.0, 11.5, 18.0, 14.0, 11.0, 14.5, 11.5, 10.0, 10.0, 19.5, 19.0, 21.0, 15.5, 14.5.

Q3 Look at Example 2.3. What additional details should the researcher have also recorded?

In Chapter 4, we consider health and safety legislation. One outcome from this is that you will need to produce a risk assessment. This should also be pasted firmly into your field or laboratory notebook. There is no point in finding out how to deal with an emergency if you do not have the relevant information with you when it is needed!

ii. Electronic records

There are two useful practices here that can save enormous heartache from lost or apparently missing files. The first of these is keeping track and the second is backing-up. As your research project develops and you alter electronic files, you need to know which version of a file you are working with. The simplest method to achieve this is to include a table at the top of each file as illustrated in Table 2.1. Using a table like this will prompt you to make sure you keep track of the date the file was last altered and your thoughts as you develop your work. When saved, files do indicate the date last altered, but we have found this 'table' system to be more effective as a file-management tool.

Backing-up electronic files is essential and you should get into the habit of doing this at least every day, so a file-management system (e.g. using a table as Table 2.1) is invaluable. Data storage and collection are subject to several laws, including the Human Rights Act 1995 and the Data Protection Act 1998 (4.5.4 and 4.5.5): you must therefore make sure that you are careful about how all records are stored, where, and for what period of time.

Table 2.1. Information that helps you keep track of electronic files

Title of project
Last revised
Aim
Objectives
Check for accuracy
Check for inadvertent plagiarism
Checked supervisor's feedback
COMMENTS

Summary of Chapter 2

- By working through the first section in this chapter, you will realize that you already know about experimental design and are able to identify the strengths and weaknesses of a design (2.1). We consider how some of the weaknesses may be avoidable and some not, without changing the aim.

- We then consider the nine steps that take you through designing an experiment including: identifying an aim and objective(s) (2.2.1), determining the statistical population (2.2.2), identifying the treatment and confounding variables (2.2.3 and 2.2.4), replication (2.2.5) and the use of controls (2.2.6), the role of statistics when planning experiments (2.2.7), and the importance of compliance with UK law (2.2.8). We introduce you

to a decision web (Appendix b) that may be used when planning an experiment. Some sections (e.g. 2.2.7) draw on subsequent chapters.

- Tools for managing research in relation to the planning stages, time management, space considerations, and data management are also discussed in this chapter (2.3) and developed in Chapters 4, 11, and Appendix a.

- The Online Resource Centre includes interactive exercises that test your understanding of this chapter along with other topics, particularly those considered in Chapters 7–11.

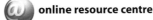 **online resource centre**

Answers to chapter questions

A2 Yes, controls are needed to check that the bacteria are coming only from the participants' hands. The controls in this case need to be a series of plates inoculated with: (i) nothing; (ii) triclosan; (iii) tea-tree oil; (iv) an unused swab; and (v) the solution used for serial dilution.

A3 Clearly not enough information has been recorded. In a few weeks' time, the data will be useless, as not enough

will be reliably known about them. Other information that should be recorded includes units of measure, date, time of recording, details about the experiment, and the name of the investigator (if more than one is involved in the project).

3

Questionnaires, focus groups, and interviews

In a nutshell

Here we give a brief introduction to the use of questionnaires, focus groups, and interviews to gather scientific information. We discuss the design of the questions in relation to the sample size and how the data may be analysed quantitatively.

We have considered the essential steps needed to design most investigations in Chapter 2. However, some information may be gathered using methods such as unstructured and structured interviews, questionnaires, and focus groups. If you are planning to use one of these methods, there are additional points that need to be considered.

Questionnaires, focus groups, and interviews may allow you to collect both quantitative (numerical) data and qualitative (descriptive) information. Qualitative approaches are based on the notion that reality varies for different people in different contexts; therefore, you cannot use a single scale against which you make measurements. An example of this is people's perception of pain. Qualitative information can be summarized but the tools used to do this are outside the scope of this book. In this section, we therefore focus on how these methods may be used effectively to obtain numerical or quantitative data and how these data may be analysed. We have, of necessity, had to assume some prior understanding of

the use of hypothesis testing (Chapter 6) and chi-squared statistical tests (Chapter 7).

Further examples relating to the analysis of quantitative data from questionnaires are included in the Online Resource Centre. It will take about 1 hour to work through this chapter. The answers for these exercises are at the end of the chapter.

online
resource
centre

3.1 What is a questionnaire, interview, or focus group?

Key points Questionnaires, focus groups, and interviews differ in the degree of interaction between volunteers and the researcher.

These three methods for obtaining information differ in terms of the degree of interaction between participants and the interaction between the researcher and the participant.

3.1.1 Questionnaires

Questionnaires are, as the term implies, collections of questions given to all participants. The participant writes down their answer and returns the questionnaire to the researcher. Questionnaires are usually anonymous and confidential (4.5). They require very little contact between the researcher and volunteers and usually there is no interaction between participants. A questionnaire is most useful if you wish to generate quantitative data.

Questionnaire A collection of questions given to all participants in an investigation, for completion without the involvement of the researcher

3.1.2 Focus groups

A **focus group** is a discussion-based interview involving more than two people. A theme or focus is provided by the researcher, who also directs the discussion. These discussions are usually recorded as audio- or videotapes and evaluated later. Most of the output from a focus group will be a record of who said what when. Therefore, less-quantitative data are usually generated by this method.

Focus group A discussion-based interview involving more than two people, managed by the researcher

3.1.3 Interviews

An **interview** is a meeting between the researcher and a participant. Interviews can be structured where the interviewer asks the same series of questions to all the interviewees, or unstructured where the interviewee is left to make their own comments about a topic. Recording the

Interview A directed discussion between the researcher and a participant

results from an interview can be done at the time if the researcher is able to take notes. More often, the interviews are recorded as audio or video images to be evaluated later. Interviews can be used in a similar manner to questionnaires and may also be useful if you wish to generate numerical data.

3.2 **Open and closed questions**

Key points Open and closed questions are phrased in a different way and tend to elicit more (closed questions) or less (open questions) quantitative data.

Closed question Where you give the participant a limited number of answers and they choose one or more or these options

Open question Where the answers are not prescribed

There are two types of questions: open and closed. A closed **question** is one where you give the participant a limited number of answers and they have to choose one or more of these options. An **open** question is where the answers are not prescribed.

For example:

Closed question: Are you warm at present? YES/NO (delete as applicable)

Open question: How warm do you feel at present?

3.2.1 **Closed questions**

Closed questions may be used in questionnaires or structured interviews. They have an advantage over open questions in that the answers can be collated and the frequency of respondents giving a particular answer can be recorded. For example, if we asked 20 people whether they felt warm at present, it may be that 18/20 replied YES and 2/20 replied NO. These data can be presented and summarized using the methods outlined in Chapters 5 and 11. If you are comparing two or more groups of people, then a chi-squared test can be used to compare the relative distribution of the answers between the groups. (We discuss chi-squared tests in Chapter 7.) For example, if we ask students studying in two rooms whether they are warm at present, the outcome from this question can be summarized (Table 3.1).

Table 3.1. Responses by two groups of students to the question: Are you warm at present?

	Answer to question	
	Yes	No
Room 114	18	2
Room 130	14	6

3.2.2 **Open questions**

Open questions may be used in all the methods we are considering in this section, including structured interviews and questionnaires. Open questions lead to descriptive answers or comments. The chief advantage of open questions is that you do not restrict the information you gather. However, this information is much harder to summarize. One approach is to record the number of times a keyword appears within the answers. These frequencies can then be evaluated in the same way as data from closed questions. For example, we asked 20 students how warm they felt at present. Looking through the answers, it was clear that several keywords, such as hot, cold, and warm, were used consistently. By looking at each answer, it is possible to categorize each student's response (Table 3.2).

EXAMPLE 3.1 How useful is this book?

A cohort of 40 students was asked to complete a questionnaire about their experience of using this book. Twenty of these students were studying a degree course in Microbiology and 20 a degree course in Forensic Science. The aim of the investigation was to compare the experience of the two groups of students. The questions were:

1. Have you found this book helpful?   YES/NO (*Delete incorrect answer*)
2. If YES, why?
3. How helpful have you found this book? *Circle one value.* 1 (*least*) 10 (*most*)
   1 2 3 4 5 6 7 8 9 10

Q1 In Example 3.1, which of these three questions are open and which closed?

3.3 **Phrasing questions**

Key points It is very important that you plan your questions carefully so that they obtain the information you are after.

Table 3.2. Responses from 20 undergraduates to the question: 'How warm are you at present?' categorized by one keyword in each answer

	Keywords used to answer question		
	Cold	Warm	Hot
Number of students	2	14	4

Whether you use open or closed questions or a combination of both, all questions need to be phrased so that the meaning is clear. The table below provides a checklist with examples that you can use to check your own questions against.

Fault in the phrasing of your question	Examples of poorly structured questions	Improved structure to questions
Vague	How did you get here?	Give the mode of transport used to travel here, e.g. on foot, bus, train, car, bike, motorbike, etc.
Too few options in a closed question	How regularly do you go swimming: every day/never?	How regularly do you go swimming? Every day/a few times a week/once a week/once a month/a few times a year/never.
Leading, likely to create a biased response	What is your opinion of the terrible and frightening developments in molecular biology?	What is your opinion of the developments made in molecular biology in the last 5 years?
Double negatives	Do you not believe that your degree course is not adequate in preparing you for employment?	Do you believe that your degree course is adequate in preparing you for employment?
Jargon	Are you a member of the HEA?	Are you a member of the Higher Education Academy?
Too many topics	Are you in favour of fox-hunting and the control of deer populations through shooting?	a. Are you in favour of fox-hunting? b. Are you in favour of the control of deer populations through shooting?
Unrealistic	How old were you when you read your first word?	(Delete such questions.)
Status questions	Are you employed/unemployed? ('Employed' is the first answer, implying that this has a higher status.)	These questions are very difficult to improve but you need to be aware of the potential impact that the order of the answers may have on a volunteer. The participant may be more likely to lie and/or be offended.
Insufficient details for analysis	How old are you? 10–20, 21–70	How old are you? 10–19, 20–29, 30–39, 40–49, 50–59, 60–69, 70 or over
Overlap in answers	How old are you? 18–20, 20–22, 22–24	How old are you? 18–19, 20–21, 22–23, 24–25

If your method for collecting information is structured as in a structured interview or questionnaire, you should carry out a pilot study before you begin. This is an invaluable way of checking that the questions really do provide you with the answers you are expecting. In an unstructured interview or a focus group, you will have the opportunity to clarify any questions during the interview/discussion.

3.4 **Your participants**

Key points You need to ensure that all aspects of the interview, questionnaire, or focus group are appropriate with a representative sample of volunteers and that they avoid the creation of biased responses.

For any investigation that involves people, you need to make sure your approach is suitable for your 'audience' (4.5). There are many elements in this and not all will apply to all the methods you might use. Here are some prompts:

- **Language.** Use appropriate language. If necessary define terms and provide explanations of terms or ideas without influencing responses.
- **Accessibility.** Consider the layout of a questionnaire. Is the format and font suitable, for example, for dyslexic participants or for children? Do you need to produce a Braille copy? If you are providing a questionnaire online, can it be read by the standard screen-reader software? Is the room you are using accessible to all participants? Do you need a translator?
- **Environment.** If the location for carrying out this research is important, as in the interviews and focus groups, is the room free from distractions?
- **Time.** How long will the interview, questionnaire, or discussion take? Do your participants have that much time?

3.5 **Sample sizes**

Key points When you test hypotheses from data obtained by questionnaire, interview, or focus groups, the choice of statistical test can indicate your sample size. Sample size is also affected by the number of answers given in both closed and open questions. The most common statistical test used to analyse quantitative data derived from questionnaires, focus groups, and interviews is the chi-squared test.

For most investigations, you will not involve the whole population; instead, you will sample (1.4). Apart from deciding on your sampling strategy (1.4.2), you will also need to decide on the size of your sample. The size of your sample will be determined in part by you satisfying yourself that your sampling strategy will generate a representative sample. In addition, sample size is determined by the way in which the data will be analysed. We consider this last point in particular for closed (3.5.1) and open (3.5.2) questions.

3.5.1 Closed questions

If you wish to compare two or more groups and you are using closed questions, then your sample size is dependent on two criteria: firstly, is your sample representative (1.4.1), and secondly, is your sample size appropriate for the statistical test you intend to use?

As we discussed earlier, if you are planning an experiment to test hypotheses, you need to identify the statistical test you will use before you carry out your investigation, as it is usually this that determines the size of your sample. We illustrated this in 2.2.7 and now apply this process to the answers to question 1 in Example 3.1. A full summary of this process is also given in Appendix c. The statistical test used most often to compare two or more sets of answers to a question is the chi-squared test (Chapter 7). Which chi-squared test you use will depend on the number of answers in the closed questions and the number of groups of people you intend to sample. The following illustrates the process where you have collected replies from two or more groups of volunteers. The same steps can be applied when you intend to gather data from only one group of volunteers.

The responses of 40 students in Example 3.1 to question 1 have been collated (Table 3.3). These two groups of students can only choose one of two possible answers: yes or no. To compare the answers from these two groups of students, you would use a chi-squared test for association with Yates's correction (7.4.2). All statistical tests have a set of criteria that should be met by the data that are to be analysed. If the criteria are not met, then the use of the test is invalidated. The criteria for using a chi-squared test for association with Yates's correction are as follows:

1) You wish to test for an association between two treatment variables.
2) You have data that are organized into two categories for each of the variables.
3) You have data that are counts or frequencies and are not percentages or proportions.
4) You have observations that are independent of each other.
5) You have expected values that are greater than 5.

Table 3.3. A comparison between Microbiology and Forensic Science students in their response to Example 3.3, question 1: Have you found this book helpful?

	Number of respondents for a particular answer	
	Yes	No
Microbiology students	15	5
Forensic Science students	10	10

As we do not expect you to have necessarily read Chapters 5–7 at this point, you will have to believe us when we say that all these criteria are met. The criterion that is most important when determining sample size is the last one. Expected numbers are produced as part of the chi-squared calculation. We show you in Table 3.4 how they are calculated for this example. The expected numbers for this set of data are all greater than 5. So for the results from this question, it appears that our sample size was adequate to ensure that all the criteria for using this statistical test were met.

As you cannot calculate expected values until you have carried out the investigation and collected your data, how then can you determine what sample size to use? The answer is to try out your closed questions on a group of people similar to those who will take part in your study. That way you can judge whether you are likely to get an appropriate distribution of answers.

But what happens when we analyse the answers from question 3 of Example 3.1? This is also a closed question but there are ten possible answers and, as Table 3.5 shows us, not surprisingly our data are more 'spread out' compared with question 1.

Table 3.4. Chi-squared test for association comparing Microbiology and Forensic Science students in their response to question 1 of Example 3.1: 'Have you found this book helpful?'

	Number of respondents for a particular answer		
	Yes	No	Total
Microbiology students observed numbers	15	5	20
Microbiology students expected numbers	$25/40 \times 20 = 12.5$	$15/40 \times 20 = 7.5$	
Forensic Science students observed numbers	10	10	20
Forensic Science students expected numbers	$25/40 \times 20 = 12.5$	$15/40 \times 20 = 7.5$	
Total observed	25	15	40

Table 3.5. A comparison between Microbiology and Forensic Science students in their response to question 3 of Example 3.1: 'How helpful have you found this book?'

	Number of respondents for a particular answer									
	1	2	3	4	5	6	7	8	9	10
Microbiology students	1	0	1	2	2	3	2	4	3	2
Forensic Science students	0	0	0	0	7	5	4	3	1	0

As there are more than two possible answers, we use a chi-squared test for association (without a Yates's correction) to test our hypotheses (7.3). The criteria for using this test are that:

1) You wish to test for an association between two treatment variables.

2) You have data that are organized into more than two categories for at least one of the variables and into two or more categories for the second variable.

3) You have data that are counts or frequencies and are not percentages or proportions.

4) You have observations that are independent of each other.

5) You have expected values that are greater than 5.

Again, you may have to take our word for it that criteria 1–5 are met. Criterion 5 is again the one most relevant in determining our sample size. We have calculated the expected values for these data (Table 3.6) and unlike question 1 (Table 3.4), all the expected values are less than 5. This means that, although this sample size of 40 was adequate to allow analysis of question 1 where there were only two possible answers, it is not large enough to allow us to analyse question 2, where there are ten possible answers. In fact, you would need more than 100 participants in each group to generate expected values that were large enough for the answers given by the two groups of students to be compared.

Therefore, what had been an adequate sample size for question 1 is not adequate for question 3. In general, the more 'possible' answers you include for your closed questions, the larger the sample size needed. Clearly, you need to balance the number of possible answers so that you obtain useful information against the size of sample you will then need to survey to generate data that can be analysed.

Table 3.6. Calculation of expected values for numbers of respondents studying Microbiology or Forensic Science when answering question 3 of Example 3.1: 'How helpful have you found this book?'

	Number of respondents for a particular answer									
	1	2	3	4	5	6	7	8	9	10
Microbiology students observed numbers	1	0	1	2	2	3	2	4	3	2
Microbiology students expected numbers	0.5	0	0.5	1	4.5	4	3	3.5	2	1
Forensic Science students observed numbers	0	0	0	0	7	5	4	3	1	0
Forensic Science students expected numbers	0.5	0	0.5	1	4.5	4	3	3.5	2	1

If you unfortunately find that having carried out an investigation where you planned to use either of the chi-squared tests, you do have expected values less than 5, then you should refer to 7.6.

3.5.2 Open questions

In this book, we only consider how you may extract quantitative information from open questions and analyse this. If you wish to evaluate the qualitative elements of your information, you will need to look to other texts (e.g. Robson, 2002).

Imagine some answers that may have been given in response to question 2 of Example 3.3. These might include:

Person 1:   'The glossary was useful.'
Person 2:   'The explanations are clear, the glossary and boxes are helpful, and I like the illustrations.'
Person 3:   'Having a glossary meant I could check unfamiliar terms.'

If you have a set of open answers like this, it is possible to compile a list of words that appear regularly. Each person's answer can then be checked for the presence of these keywords (Table 3.7).

What can you do with the data? You could total the number of times a keyword or phrase is mentioned. But if one person mentions four keywords and others only one, then the person who includes four keywords is over-represented in your sample and these observations are said to be not independent of each other (9.3). An alternative approach would be to record the first keyword in each answer. Although this resolves the problem of independence, clearly this approach does not then reflect all the information you have. Trying to obtain quantitative data from qualitative answers can be indicative at best.

One of the chi-squared tests may be appropriate to analyse this type of data (Chapter 7). Clearly you again have a problem when trying to determine sample size, as you will not know in advance how many keywords or phrases you will identify from your respondents' answers. Again the solution is to run a pilot test on a similar group of people.

Table 3.7. Record sheet for the evaluation of question 2.

I have found this book helpful because:

	Clear explanation	Glossary	Boxes	Illustrations
Person 1		×		
Person 2	×	×	×	×
Person 3		×		

The general relationship between numbers of possible answers and sample size applies here to the number of keywords and phrases you identify. The more keywords and phrases there are, then in general the larger the sample size required.

3.5.3 Achieving the required sample size

One of the great drawbacks of using these methods to gather information is obtaining a large enough sample size simply because the members of your population are not sufficiently motivated or do not have enough time to take part. You need to bear this in mind when writing your aim and objectives. Will the population identified by the aim provide sufficient numbers of participants to allow you to test any hypotheses you have? However, you must work ethically in your research, so any investigation that includes an element of coercion is not acceptable. Forms of coercion can include asking your family and friends, asking everyone in a lecture or tutorial group, or offering a reward. We discuss this in 4.5.

Summary of Chapter 3

- There are additional points that need to be considered when designing investigations in which you may use questionnaires, focus groups, or interviews to obtain quantitative data, including the need to consider the type of question and how this may determine how the data are analysed, and the sample size.

- The Online Resource Centre includes interactive exercises that test your understanding of this chapter along with other topics, particularly those considered in Chapters 7–10.

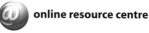 **online resource centre**

Answers to chapter questions

 Question 2 is open. The respondent is left to comment freely in response to the question. Questions 1 and 3 are prescribed and the respondent has to choose from a limited number of options. These are closed questions.

Research, the law, and you 4

In a nutshell

This chapter examines UK law primarily as it relates to undergraduate bioscience research. The way in which law is developed within the UK is outlined. In most respects at present UK law is similar across England, Northern Ireland, Scotland, and Wales with the exception of the laws of trespass. This chapter examines the law as it relates to heath and safety, access, wild species, domesticated animal species, and humans.

Research and the law are intertwined in numerous ways and at all levels from the individual to the institution in which the research is carried out. It is a huge area, and our aim in this chapter cannot be to cover all relevant legislation. Instead, we provide an introduction to the areas that appear to be most relevant to bioscience undergraduates and graduates. Although there are differences between the law in England, Wales, Scotland, and Northern Ireland, on the whole the law that impinges on bioscience research is sufficiently similar that we can cover all four countries at the same time. Where there are notable differences, these are highlighted. The Internet in general is an invaluable and usually reliable resource in relation to the law for any student requiring further details. We provide links to relevant websites in our Online Resource Centre.

online
resource
centre

Clearly, as an undergraduate or graduate student, your institution carries some responsibility for you and what you do. As a result, your institution will make available to you guidelines and information that it is your responsibility to read and respond to. Some of the legal issues we cover in this chapter may therefore be addressed by your institution centrally and through staff such as your supervisor and technical staff. You have responsibilities both to your institution and under the law to comply with legislation whether it is channelled through your institution or by some other means. For example, your institution will have health and safety guidelines in place that put into practice the health and safety legislation. You are required to comply with these guidelines and be aware

of the relevant legislation. However, if you work with wild plants and animals, your institution is less likely to have a central policy; none the less, you are responsible for being aware of the relevant legislation and complying with it.

Unlike other chapters, there are not many exercises here. Instead, we have suggested a number of discussion topics at the end that will allow you to relate real undergraduate project proposals to the areas of law we cover in this chapter. It will take you about one hour to consider these discussion topics in detail.

4.1 About the law

Key points Laws in England, Northern Ireland, Scotland, and Wales are drawn from international treaties and legislation, national common (case) law, legislation, institutional writers, canon law, and custom.

In the UK, the body of law is derived from two sources: international law and national law. We use the term 'law' to mean any legislation or regulation derived from an Act of Parliament (or equivalent national body) or other ratification of international agreements that is potentially binding on the individual and where failure to comply may lead to sanctions.

4.1.1 International law

Many nations voluntarily negotiate agreements at an international level. These agreements, however, are not legally binding until ratified by that country.

An example of this is the Council of Europe where the European Convention on Human Rights and Fundamental Freedoms was developed. This was later ratified within the UK as the Human Rights Act 1998. Another example is the Convention on International Trade in Endangered Species of Wild Flora and Fauna (CITES). Originally agreed in 1975, this was ratified in the UK in 1976 as the Endangered Species (Import and Export) Act.

Other international law derives from the European Community (EC). The 'UK' is a member of the EC so legislation is negotiated by UK representatives on behalf of all countries in the UK. Agreements usually take on one of four formats: Regulations, Directives, Recommendations, and Opinions. Regulations take immediate effect as law in all member states and may subsequently also be included in national law. A Directive is binding but it is not law in a particular country until the national authorities pass an appropriate Act. Recommendations and opinions do not have a binding effect on nation states. An example of EC law is the Habitats Directive 1992, which became incorporated into legislation primarily in

the Conservation (Natural Habitats, &c.) Regulations (as amended) 1994 and the Conservation (Natural Habitats, &c.) Regulations (Northern Ireland) 1995.

4.1.2 National law

The UK comprises four countries—England, Scotland, Northern Ireland, and Wales. Each has its own body of law developed over many centuries. The formation of the UK as a combined political unit at various times has resulted in points at which law for all four countries was made by the one parliament in London. In 1998, governance in relation to many aspects of specific law making was devolved to Scotland, Wales, and Northern Ireland, with the UK Parliament retaining powers to adopt Acts of Parliament for the UK as a whole.

In Northern Ireland, parliament had powers to adopt legislation for Northern Ireland from 1921 to 1972. These powers were suspended in 1972, and until 1998, the Westminster Parliament used direct rule powers to legislate for Northern Ireland and legislation was then adopted by Order of Council rather than by Act of Parliament.

Law developed within the UK is, for the purposes of this book, called national law. In the UK, the body of national law in general comes from five sources: common (case) law, legislation, institutional writers, canon law, and custom. Where national law and the legislative processes differ significantly between countries within the UK, this has been indicated.

i. Common law (case law)

The legal system in England, Northern Ireland, and Wales is based on common law derived from legal decisions made during specific court cases, i.e. we have a doctrine of legal precedent. These cases make rules that must be followed subsequently by all courts of similar or lower standing in the court hierarchy. Common law has been particularly important in shaping the law relating to trespass.

In Scotland, whenever a judge or sheriff pronounces his or her findings, he or she is influencing the development of the law on the point at issue. Like other countries in the UK, this decision has to be followed by all lower courts if the same point is disputed at a future date. However, unlike common (case) law in other UK countries, the Scottish courts seek to discover the principle that justifies a law rather than searching for an example as a precedent.

ii. Legislation

These are laws enshrined in an Act of Parliament, e.g. the Health and Safety Act 1974. The Acts of Parliament are divided into chapters,

sections, and parts. Detailed information is usually included in Schedules at the end of the Act. For example, the lists of species covered by specific parts of the Wildlife and Countryside Act 1981 are included in the Schedules.

Law introduced through an Act of the United Kingdom Parliament or of the parliaments/assemblies in Scotland, Wales, and Northern Ireland may empower other authorities to make further provisions or amendments. The provisions are often known as Orders or Regulations. Principal among those empowered to make secondary legislation are ministers in a particular government office, e.g. the Home Secretary. The law may also empower government departments to license certain otherwise prohibited activities, and in developing the conditions for the licence further statutory controls are introduced. For example, to obtain a licence under the terms of the Animals (Scientific Procedures) Act 1986, an applicant must adhere to the code of practice produced by the Home Office for the housing and care of animals used in scientific procedures.

iii. Local law

Parliament has identified certain national bodies as being able to make laws. These bodies include the Crown, local authorities, and public corporations, such as the Environment Agency, Forestry Commission, Natural England, and the Countryside Council for Wales. These bodies may establish laws within certain remits prescribed by the Statutory Instruments Act 1946 and Local Government Act 1972 and often in response to other specific statutes such as the Wildlife and Countryside Act 1981. Where a public body develops these laws they are usually known as codes of conduct, and where local authorities do so they are known as bye-laws.

iv. Institutional writers

Particularly in Scotland, writers of legal text books may attain a reputation so high that their works acquire authority, and if there is no case law or statute relating to a particular issue, the courts will turn to the views of these writers. The principal of these institutional writers is Sir James Dalrymple, Viscount Stair (1619–95), whose *Institutions* was published in 1681. Other important writers are Baron Hume (1756–1838), Erskine (1695–1768) and Bell (1770–1843).

v. Canon law

This derives from the legal system established by the medieval Roman Church in Western Europe, in turn derived from Roman law. After the

Reformation of 1560, church courts disappeared and their powers, particularly relating to the law on the family, were taken over by the secular courts. Modern law has now largely overshadowed the importance of canon law.

vi. Custom

Custom plays only a very small part as a separate source of law. Many of the older customs have been embedded in Acts of Parliament and judicial decisions in court cases.

vii. Quasi-legislation

Apart from the sources of law outlined above, there are many individuals, and organizations that also develop policies, codes of conduct, etc. These are voluntary codes and do not have the force of law. This can be confusing, especially where such terms as 'code of conduct' are used. You may therefore need to confirm whether a code of conduct is enforceable under the law or not. Failure to follow some recommendations and codes of practice may be used as evidence that a law has been broken, even though these guidelines may not themselves be part of statute law.

viii. Enforcement

The agency responsible for enforcement of legislation varies enormously and is outlined in the original legislation. The police are responsible for enforcing most bye-laws and most statute laws, but there are many other agencies with powers of enforcement. For example, Customs and Excise staff have a major role in enforcing laws relating to imports and exports whereas the Animals (Scientific Procedures) Act 1986 is enforced through an Inspectorate, which reports to the Animals Procedures Committee and to the Home Secretary.

4.2 Health and safety

Key points You are responsible for your own and others' health and safely. To assess the risks that might accrue through your research, you should identify and rate the significance of the hazard, consider the effect of specific activities on the scale of exposure to the hazard, and evaluate the probability of harm and therefore of risk. All risk should be minimized and emergency planning carried out.

You must work in such a way that your safety and that of others is paramount. To do this, you need to identify the hazards of each step of your research protocol and then incorporate practices that minimize the risk to health. At the end of this process, you need to make a judgement as to whether the risk to your health and that of others is sufficiently small that the research may proceed. There are a number of laws that cover health and safety, including the Health and Safety at Work Act 1974 updated by the Management of Health and Safety at Work Regulations 1992, and a number of detailed regulations including the Control of Substances Hazardous to Health Regulations 1988 (as amended) (COSHH), Chemicals (Hazard Information and Packaging for Supply) Regulations 2002 (CHIP), Environmental Protection Acts 1990 and Environmental Act 1995, and the Special Waste Regulations 1996.

Health and safety is something that is usually considered in laboratory work where chemicals or micro-organisms are being handled. However, health and safety must be considered in all types of research from carrying out a questionnaire survey to sampling soil in the field. The method for considering health and safety is a risk assessment. This can be divided into seven steps:

1) **Hazard identification and rating.** What hazards may be present if you carry out your research as you have designed it? In what way may these hazards affect your own or other people's health?

2) **Activity.** How might you become exposed to the hazard and what is the possible scale of exposure?

3) **Probability of harm.** Given the nature of the hazard, the possible effect on health, and the possible scale of exposure, what is the probability of harm?

4) **Minimize the risk.** Decide what precautions are needed to control or reduce this risk.

5) **Risk evaluation.** Is this an acceptable level of risk?

6) **Emergencies.** Be aware of, or prepare procedures to deal with, accidents and emergencies.

7) **Action.** Ensure these precautions are in place and are followed.

4.2.1 Hazard identification and rating

What hazards may be present if you carry out your research as you have designed it? In what way may these hazards affect your own or other people's health? In most student research, the hazards will fall into three broad categories.

i. The environment

Your working environment may contain a number of hazards. This is true of both laboratory work and working in the field. In the laboratory, you should consider physical features including temperature extremes such as working in a cold room or in a greenhouse in the summer. There are often many electrical sources, equipment with moving mechanical parts, or dust in the atmosphere from, for example, grinding bones, sorting soil, or allergen studies.

In the field, you will almost certainly encounter particularly hot or cold days. The environment may have rough terrain and may be subject to flash flooding. Additional hazards can come from living things in the environment that are not strictly part of your research. For example, you may need to consider threats from strangers, bites that may cause skin damage and introduce infections, poisonous or allergenic plants or fungi, etc.

You must also be aware of activities that may be carried out in the environment of your research that you are not involved in but that may be a source of hazards. For example, areas that have been sprayed with pesticides, chemical spills, the use of microbes, the presence of a UV source, etc. Chemical hazards found in the environment are considered further in 4.2.1ii.

One example where the environment may have hazards that need to be considered is seen in Example 7.2: the genetics of flower colour in *Allium schoenoprasum*. In this experiment, the investigator is working in a greenhouse. Common hazards of a greenhouse include broken glass, which may cause cuts; electrical sources in contact with water, which may result in electrocution and death; soil-borne pathogens, which can bring about serious illnesses; and allergens from fungi and plants, which may cause asthma. Most universities and other similar institutions carry out general risk assessments for areas such as laboratories. You may therefore be able to make use of these when carrying out your own risk assessment.

ii. Chemicals

Obtaining, storing, using, and the safe disposal of chemicals is covered primarily by the Dangerous Substances and Explosive Atmospheres Regulations 2002 and by the Control of Substances Hazardous to Health Regulations 2002 (COSHH). The first of these two Acts covers all flammable substances. COSHH covers 'hazardous substances' that are to be used during work, substances generated during work, naturally occurring substances, and biological agents such as bacteria and other micro-organisms. The term 'hazardous substance' is taken to mean

a substance or mixture of substances classified as dangerous to health under the Chemicals (Hazard Information and Packaging for Supply) Regulations 2002 (CHIP). These regulations do not cover all substances, however, and specific regulations cover the safe handling of gases such as helium and the use of pesticides, medicines, cosmetics, lead, and radioactive substances. Under the terms of the Health and Safety at Work Act 1974, these substances must also be considered in hazard identification.

To identify the hazard relating to any substance that you will use in your work, you can refer to the label on the container in which the substance is supplied. Further details can easily be obtained by searching the web using the chemical name or by referring to suppliers' catalogues or through Health and Safety Executive (HSE) publications such as the *Approved Supply List*. Links to useful websites are included in the Online Resource Centre. When carrying out research you may not always know how substances you are using will react and therefore you may not be able to predict what substances will be produced during your work. If you are not sure for any particular protocol, you should refer this matter to your supervisor.

online resource centre

The potential effects of a chemical hazard are described using a number of specific terms. For example, 'caustic' is the chemical burning that occurs on exposed surfaces such as skin, eyes, or internal organs if they become exposed to the substance. The hazard is therefore potentially damaging if you eat/drink it or allow your skin or eyes to come into contact with it (4.2.2i). Other common descriptive terms are flammable, explosive, toxic, harmful, mutagen, carcinogen, corrosive or strongly oxidizing, irritants, radioactive, and harmful to the environment. Links to definitions of these 'risk phrases' are included on many chemical and biochemical hazards websites and are included in suppliers' catalogues.

Q1   Use one of the sources outlined above to identify the hazard(s) associated with concentrated ethanoic acid (glacial acetic acid).

iii. Biological agents

When biological agents are discussed in health and safety, most thought is given to micro-organisms. For example, in the field you may encounter *Leptospirosis* sp., which is a bacterium that can be found in water contaminated with the urine of infected animals. You also need to be aware of organisms that may present a potential hazard, including the common intestinal parasites of dogs and cats (*Toxocara* spp. (ascarids) and *Ancylostoma* spp. (hookworms)), which can be distributed in the environment from animal faeces.

Micro-organisms (bacteria and fungi) present a different challenge from most chemical and environmental hazards. If you are working with microbes in the laboratory, you may not know which of the possible thousands of species you are incubating and the potential virulence and toxicity of a species may vary among strains. It is therefore sensible to consider all micro-organisms as potential pathogens unless the specific nature of the hazard is known. Information about pathogenic micro-organisms can be obtained from a number of sources such as the Advisory Committee on Dangerous Pathogens (ACDP). In general, the hazards associated with micro-organisms are: infection (e.g. botulism, meticillin-resistant *Staphylococcus aureus* (MRSA), the production of toxic substances by the micro-organism *(e.g. Aspergillus flavus* and *Aspergillus parasiticus* are both known to produce aflatoxin), and the potential allergenicity of live or dead airborne micro-organisms leading to the onset of asthma or dermatitis.

Other species including animals, birds, reptiles, and large fish can also be potential biological hazards and should be taken seriously if present or likely to be present in the area you will be working. In the UK, there are few hazards associated with native British flora other than that resulting from an allergenic response to pollen. International students should, however, familiarize themselves with common genera such as *Urtica sp.* and the effects such 'stinging' plants may have.

iv. Rating the hazard

Having identified all the hazards that you may encounter in your work, you need to rate them. The institution you are affiliated with should give guidance on this. In general, however, a major hazard would be one that may cause death or serious injury, a serious hazard may cause injuries or illness resulting in short-term disability, and a slight hazard may cause less significant injuries or illnesses.

 In Example 9.1, the investigator was studying *Littorina* species on the mid to low areas of a rocky shore. What hazards might she encounter, and how would you rate them?

4.2.2 **Activity**

How might you become exposed to the hazard and what is the possible scale of exposure? The way in which you encounter, use, and/or generate hazardous substances will determine the probable extent to which you may become exposed to the hazard. Some substances you handle may be

in a powdered form and so you may risk exposure through inhalation. Some substances can penetrate the skin, whilst others damage the skin.

i. Hazards arising from the activity

Activities can be potentially hazardous in their own right and these need to be recognized as part of your evaluation of health and safety. Examples include where you or others need to move equipment, to sit or stand in awkward or cramped conditions, or to make repetitive movements.

Hazards in some activities will be the result of a particular substance being exposed to a particular treatment, and the potential hazard will differ from either the hazard related to the substance or the hazard relating to the activity alone. For example, agar in its crystalline powder form is believed to present a negligible risk to health. If you then wish to make agar plates, the protocol requires the agar to be placed in a container (usually a glass beaker or bottle) with distilled water and heated to 100°C. The agar needs to be poured when it is still hot, at about 80°C. Splashes from the hot liquid agar could cause serious scalds, which will be hard to treat because the agar sticks to the skin. Therefore, the hazard as a result of the activity is both different from the substance and activity alone and more serious.

ii. Scale of exposure

Having considered in what way you may become exposed to the hazard, you then need to grade this possible exposure. You should be given guidance as to how to grade the likelihood of exposure. However, in general you would assign the hazard to a 'high' category where it is certain harm will occur, 'medium' when harm will often occur, and 'low' where harm will seldom occur. For example, concentrated ethanoic acid is caustic (see Q1/A1) and would be rated as a major serious hazard. You may be provided with 1ml of a very dilute solution of this (e.g. 5%, w/v) of which you will pipette 1μl into a small tube. As you are using small volumes of a solution with a low concentration, the scale of exposure would receive a 'low' rating. (In fact, this dilution of ethanoic acid is commonly found in kitchens as vinegar!)

iii. Storage and disposal

With some potential sources of hazards, such as chemicals and micro-organisms, you need to consider more than just your use of them in an investigation. These substances need to be stored and safely disposed of. Your consideration of hazards and risks needs to extend to the full lifetime of such substances. This is of particular importance, as it is during

these times that other people are likely to become exposed to the potential hazard. Usually, technical staff can advise you about safe storage and disposal, and regulations such as the Special Waste Regulations 1996 and the Dangerous Substances and Explosive Atmospheres Regulations 2002 provide statutory guidance.

4.2.3 Probability of harm

Given the nature of the hazard, the possible effect on health, and the possible scale of exposure, what is the probability of harm?

i. Probability of harm

This, the risk assessment step, is not easy and is usually subjective and qualitative. Most institutions provide guidelines to help you. For supplied substances, the HSE has developed a generic web-based guide to risk assessment called *COSHH Essentials*.

In a risk assessment, you need to make a judgement about each hazard you have identified using the following information:

- The hazard and the nature of its possible effect on health (major, serious, or slight).
- The possible scale of exposure as a result of the way in which the hazard will be encountered or used (high, medium, or low).

The probability of harm is the first of these combined with the second. Therefore, in 4.2.2iii, the major caustic hazard of ethanoic acid will only be present in a form where the risk of exposure is low. In this example, we would grade the overall probability of harm as low.

ii. Variation among individuals

Human beings all differ, and some of this variation can increase or decrease the risks from any one hazard. Points that often need to be considered include height, handedness, sensitivity to allergens, a suppressed immune system, the side effects from medication (such as antihistamines causing drowsiness), and pregnancy. For example, you may not be pregnant, but you may need to alert those who are to a hazard that is of more significance to them than to yourself.

iii. Occupational exposure limits

Some substances have occupational exposure limits. As an undergraduate, you are unlikely to be using such substances, although as a graduate

this is possible, e.g. radioactive substances. There are two types of these occupational exposure limits: the maximum exposure limit (MEL) and the occupational exposure standard (OES). Further information about occupational exposure limits is available from the HSE.

4.2.4 Minimizing the risk

Decide what precautions are needed to control or reduce this risk. It is impossible to cover in this section all possible types of hazard with suggestions as to how you may reduce the risk from each hazard. We have concentrated on the most common, but for each of the hazards you have listed, you must consider how to prevent the hazard or reduce the risk (i, ii) and then re-evaluate the probability of harm (iii).

i. Prevention

No risk to health is really acceptable; therefore, ideally you should prevent exposure to the hazard. Prevention can be achieved by changing the process or activity (e.g. using a different field site) and/or by using a safer alternative (e.g. using a pelleted form of a substance rather than a powder).

ii. Reduction

If prevention is not reasonably practicable, you must adequately control exposure and therefore you need to put in place control measures suitable for the activity and consistent with your risk assessment.

a. Wear personal protection   You may wear and/or provide personal protection appropriate to the nature of the hazard. In the biosciences, this protective clothing most often consists of laboratory coats, general safety glasses or face masks, UV light shields, and gloves known to exclude the particular hazard you may be exposed to (thermal or chemical). In the field, additional items providing personal protection include strong boots, waders, waterproof clothing, sun cream, and insect repellent. It should be noted, however, that the sun cream and insect repellent are substances and as such their use should also be considered under COSHH.

Additional medical protection may be appropriate, for example vaccinations (e.g. tetanus or hepatitis), medication to control allergies, and medical/biological monitoring. If you are ever in doubt or feel unwell, you should always seek medical assistance and draw to the attention of the medical staff the nature of your research and the hazards you have identified.

b. Reduce quantities   You may reduce the hazard by ensuring, for example, that the quantity of substances you are storing, using, or will need to dispose

of are limited. For example, you may keep most of your stock solution in safe storage and only take out the volume required at the time.

c. Control ventilation   You may use controlled ventilation as in a fume hood or laminar flow hood to limit exposure to a hazard. This is common practice where volatile substances are used or produced, and in microbiology.

d. Organize space   The most common method used in risk reduction is to organize your space. Eating and drinking should be prohibited in all laboratories to prevent accidental ingestion of a hazardous substance. In the field, the equivalent practice is to ensure that you are able to wash your hands with clean water before you eat. Other examples of good practice are to keep flammables away from a Bunsen burner and to keep electrically powered equipment away from liquids. For some research, a dedicated area to which access is restricted is appropriate (e.g. microbiology areas and areas where radioactive sources are used and stored). For some substances such as flammables and poisons, special storage should be provided. If you are in doubt your supervisor or technical staff will be able to advise you.

e. Take a break   This is especially important if you are doing a repetitive task or if you are using light sources such as a microscope.

f. Inform   It is essential that you alert other people to potential hazards. With stocks of substances this is usually already done for you by the manufacturers. Solutions or mixtures that you prepare yourself must be similarly labelled with details of the contents including concentrations, the hazard(s), and the date. There are internationally recognizable symbols that are used to indicate certain hazards. Laboratories often carry stocks of sticky labels with the symbols for chemical hazards such as 'explosive' (Fig. 4.1) or 'oxidizing' (Fig. 4.2). These symbols can be stuck onto the containers holding your solutions or substances.

If you will be working in the field, then providing information is important in a number of ways. For example, you may wish to leave markers protruding from the ground to indicate the location of permanent quadrats.

Fig. 4.1. Standard symbol for explosive hazard.

Fig. 4.2. Standard symbol for oxidizing hazard.

If so, you should consider the need to alert other people to their presence so that they will not fall over them. When working in the field, you should ensure that at least one responsible person knows exactly where you are working and when you are expected back. It is also advisable to take a phone so that you can keep in touch.

g. Company It is general practice in most institutions where students may be working independently to require or at least to recommend that the researcher always has someone else with them or near by. In laboratories this can most easily be achieved by agreeing a schedule of work that is compatible with another student also working in that laboratory. For those

working in the field (both urban and rural areas), this may not be so easily arranged but should be achieved if at all possible.

b. Be informed   Part of the process in risk assessment is planning for emergencies. This level of planning is a form of risk reduction. We consider this in 9.2.6.

iii. Re-evaluate the probability of harm

Having put into place control measures to reduce the risk, you can now re-evaluate both the hazard and the nature of its possible effects on health (major, serious, or slight) and the possible scale of exposure as a result of the way in which the hazard will be encountered/used (high, medium, or low). Most control measures work by addressing the second of these. You will now have two measures of the probability of harm, one where there are no control measures (4.2.3) and one where the control measures are in place and effective (4.2.4).

4.2.5 Risk evaluation

Is this an acceptable level of risk? Using these two sets of risk evaluations, you are now in a position to determine whether the level of risk is acceptable. We cannot provide detailed guidance on this as there are so many differing potential hazards. You must therefore seek specific guidance from your supervisor. However, in general, undergraduate work will normally be allowed to continue if the risk, with the controls in place, is low and occasionally moderate and where the controls are not likely to fail, and where if they did the risk would be no more than moderate. Graduates may be allowed to accept a higher risk, but in our experience a severe risk of injury to health is not usually deemed acceptable for any student.

4.2.6 Emergencies

Be aware of, or prepare procedures to deal with, accidents and emergencies. There are two types of emergency that you need to consider before you may finally complete this part of your research preparation. The first point you need to consider is what to do if one or more of your measures introduced to control the hazards fail. The second is what to do if there is a fire or other external emergency that means you must leave your research immediately.

i. Failure of control measures

You must consider here both in what ways these protective measures may fail and what needs to be done in response. For example, if a laminar

flow hood fails, you are more likely to become contaminated with the micro-organisms you are handling. You may therefore need to carry out biological monitoring or be aware of symptoms of any illnesses that may develop as a result. You may also wish to satisfy yourself that the laminar flow hood is appropriately maintained. A second example of such an emergency would be if you spilt a hazardous substance. You will need to know who needs to be informed and how, how quickly anyone else needs to be informed, who can tidy the spill and how, whether the area needs to be evacuated, and if so who is responsible for this. In the field, this preparation can include ensuring that you have emergency phone numbers, additional food, water and clothing, and basic first aid, and to have worked out an emergency escape route.

ii. External emergencies

In preparation for such events as a fire alarm sounding, you should discuss with your supervisor and technical staff what needs to be done to ensure that the hazard(s) in your work continue to be contained and other hazards do not arise. There may be a need, for example, to turn off a Bunsen burner or safely store hazardous substances. As part of their emergency procedures, your institution will have guidelines as to who should carry out such actions: it may not be you.

iii. Information

If your control measures fail or you experience an external emergency, then the risk assessment that you carried out will inform you and those around you as to what action is required. Therefore, it is essential that you keep it with you. For example, if you splash yourself with something, you may urgently need to be able to communicate details about that substance and its hazard to others. Therefore, you should keep a copy of your risk assessment including your emergency plans in your field or laboratory notebook and keep the notebook with you.

4.2.7 Action

Ensure these precautions are in place and are followed. Having carried out your risk assessment and determined that the risk is acceptable, you can then carry out the research. However, you will need to:

- Ensure that the precautions to control the hazard(s) are in place, are followed, and that they are effective
- Ensure that your emergency planning information is with you when you are working and that others in the vicinity are aware of this
- Contact medical assistance immediately if you start to feel unwell, and take your risk assessment details with you.

4.3 **Access and sampling**

Key points Public right of access does not allow you to sample or set up equipment (except in Scotland). In addition, certain habitats or areas are subject to specific protection. Many wild species are protected. You need to familiarize yourself with the level of protection given to the species you may encounter while working on your research and comply with these restrictions. You can apply for a licence for some work on protected species. There are laws that control the movement, import, export, or release of some species.

In this section, we consider the law that relates particularly to the studying or sampling of wildlife. The legislation relating to access and to handling plants and animals is extensive. We cover some of the key areas most often encountered by undergraduates, but have restricted ourselves largely to land-based work. Where there are differences in legislation among different parts of the UK, this is highlighted below.

4.3.1 **Access**

The laws relating to access in England and Wales are fundamentally different from those in Scotland and Northern Ireland. In particular, the concept of trespass is virtually absent from Scottish law and recent Scottish legislation has codified and simplified access rights.

i. Access in England and Wales

The laws relating to access are derived primarily from national law. The statutory regulations are included in a number of Acts, such as the Environmental Protection Act 1990, the Countryside and Rights of Way Act 2000, and the Water Industry Act 1991.

In essence, in England and Wales you may become a trespasser:

- If you enter land without the permission of the landowner
- If you go outside the area that you are allowed on because you have a statutory right or you have permission from the landowner
- If you use land in a way in which you have not been authorized to do so.

Trespass is not usually a criminal offence in that you cannot be prosecuted but you can be sued. This may result in being required to pay compensation. In addition the landowner can use reasonable force to eject a trespasser and obtain an injunction if necessary to prevent you from returning. Trespass can be a criminal offence on Ministry of Defence land, land adjacent to railways, and some nature reserves. In addition to

trespass, you may also commit offences of criminal damage if you put up signs, dig holes, etc., without the permission of the landowner.

a. Public right of way   There are some areas such as registered footpaths where you have a public right of way. However, this only means that you may use the path to travel, usually on foot, from point A to point B. You may therefore be considered to be trespassing if you carry out other activities such as taking samples without the permission of the landowner.

b. Open country   There are some parts of England and Wales that are considered to be 'open country'. There are two main pieces of legislation relating to access in open country: the National Parks and Access to the Countryside Act 1949 and the more recent Countryside and Rights of Way Act 2000 (CRoW). Both of these provide for the possibility of access to 'open country' such as mountains, moors, heath, and down. The way in which these rights are managed varies slightly in that within the first Act details about access are largely set out in an agreement or order made in conjunction with the owner. In the more recent Act, access is determined more by the terms of the Act than by local agreements. Whichever Act covers a particular piece of open ground, normally there will be a right of access on foot for open-air recreation, such as walking, bird-watching, picnicking, running, and climbing, but not usually for driving a vehicle on the land, using boats, hunting, fishing, collecting anything from the area including rocks or plants, camping, or lighting fires. There are variations and exceptions to both Acts; for example, some areas within the designated 'open country' regions in the Countryside and Rights of Way Act 2000 will not be subject to the new access rights. These include buildings and livestock pens, land ploughed or drilled during the previous 12 months to grow crops or planted with trees, quarries and other active mineral workings, land used as a golf course or race course, and land where military bye-laws apply.

c. Waterways   There are a number of different types of waterways: tidal and non-tidal rivers, and streams, canals, natural and man-made lakes, and reservoirs. Ownership and responsibility for the management of these areas varies. However, none of these areas has a general right of way unless they are included within the regions of 'open country' identified in the Countryside and Rights of Way Act 2000. However, many bodies responsible for waterways allow some rights of navigation and access. This permissible access and associated activities will be laid out under agreements, codes of conduct, or bye-laws. Key bodies that may be consulted in relation to waterways are the Environment Agency and British Waterways.

d. Foreshore   The foreshore between the high- and low-water marks is the property of the Crown except where it has been sold or leased, usually to a local authority. Again access is customarily permitted, usually subject to bye-laws. The Countryside and Rights of Way Act 2000 gives the Secretary

of State powers to propose the extension to the right of access to coastal land, but no significant changes have occurred to date. Most beaches above the high-water mark are owned by the local authority and access is subject to bye-laws.

e. Commons and village and town greens   Commons and village and town greens are not necessarily places with a right of public access. For most of this type of land, consent is usually given by the landowner to allow access and some activities. This permission is outlined either in an agreement or in bye-laws under the terms of statutory law such as the Commons Act 1876 and 1899 and the Village and Town Greens Act 1965. Local authorities have in the last 50 years or so clarified the ownership of and public rights of way on commons, etc. Therefore, if you need to contact an owner, the council may be able to help you. Village and town greens may have increased public rights of access if local people have openly used the land for recreation for at least 20 years without the permission of the owner. These increased rights still only relate to access and recreational activities.

f. Other areas   There are some areas where you may feel you can roam at will, e.g. National Trust properties and country parks. Most of the latter are owned by the local authority. Acceptable activities in these areas will also be outlined in, for example, a code of conduct or in bye-laws. Similarly, in woodlands and forests you do not have a general right of access. There may be public rights of way such as footpaths passing through the woodland, or local agreements giving some access. Again, these local agreements usually relate to walking and a few prescribed activities such as bird-watching.

ii. Access in Northern Ireland

In Northern Ireland, the basic premise in relation to trespass, as set out in 4.3.1i. apply equally to Northern Ireland. However, the law in relation to rights of way is different. Basically, Northern Ireland does not have a legally recognized concept of open country, as the Countryside and Rights of Way Act 2000 does not apply to Northern Ireland. Instead, the Access to the Countryside (Northern Ireland) Order 1983 places Northern Ireland's 26 District Councils under a statutory duty to assert, protect, maintain, and record public rights of way. For all access to the countryside, you must therefore identify the landowner and consult with them and obtain permission to carry out any study.

iii. Access in Scotland

The law in Scotland relating to access has always been fundamentally different from that elsewhere in the UK. In Scotland, the simple crossing of a piece of land to get from one point to another has never been considered

trespass; damages for mere trespass are not recoverable, although an interdict could be granted by the courts. The Land Reform (Scotland) Act 2003 enshrined this 'no law of trespass' concept and laid down the formal basis for the Scottish Outdoor Access Code. The Act allows for access to land for the purposes of *'carrying on a relevant educational activity'*. This includes *'(a) furthering the person's understanding of natural or cultural heritage; or (b) enabling or assisting other persons to further their understanding of natural or cultural heritage'*.

Whilst the Scottish Outdoor Access Code is not an authoritative statement of the law, in any dispute the Sheriff will consider whether the guidance in the Code has been disregarded by any of the parties. The Code states that access rights extend to individuals undertaking surveys of the natural or cultural heritage where these surveys have a recreational or educational purpose within the meaning of the legislation. A small survey done by a few individuals is unlikely to cause any problems or concerns, provided that people living or working nearby are not alarmed by your presence. If you are organizing a survey which is extensive over a small area or requires frequent repeat visits, or a survey that will require observation over a few days in the same place, consult the relevant land manager(s) about any concerns they might have and tell them about what you are surveying, for what purpose and for how long. If the survey requires any equipment or instruments to be installed, seek the permission of the relevant land managers.

The main places where access rights do not apply are:

- houses and gardens, and non-residential buildings and associated land
- land in which crops are growing
- land next to a school and used by the school
- sports or playing fields when these are in use and where the exercise of access rights would interfere with such use
- land developed and in use for recreation and where the exercise of access rights would interfere with such use
- golf courses (but you can cross a golf course provided you do not interfere with any games of golf)
- places such as airfields, railways, telecommunication sites, military bases and installations, working quarries and construction sites
- visitor attractions or other places that charge for entry.

4.3.2 Theft

Taking things from the environment, such as stones, wood, and earth, is a form of theft unless you have been authorized to do so by the landowner. Therefore, if you wish to remove leaf litter or soil, for example, you

should ensure you have discussed this first with the landowner. Many species are protected in this regard by additional legislation (4.4.3), but wild plants and fish not covered by the additional legislation are the property of someone (e.g. Theft Act 1968). Therefore, it is illegal, for example, to uproot any wild plant for commercial purposes without authorization from the landowner.

4.3.3 Plants, animals, and other organisms

Currently, the key legislation in England and Wales that provides statutory protection for organisms is the Wildlife and Countryside Act 1981 (with amendments in each country) (the 1981 Act), the Conservation (Natural Habitats, &c.) Regulations 1994 (with amendments in each country), and the Countryside and Rights of Way Act 2000 (with amendments in each country). Some species are subject to additional legislation, for example the Protection of Badgers Act 1992.

In Scotland, the current key legislation is the Nature Conservation (Scotland) Act 2004, which provides updated amendments to the Wildlife and Countryside Act 1981, the Wildlife and Countryside Act (Amendment) 1985, the Conservation (Natural Habitats, &c.) Regulations 1994, and the Protection of Badgers Act 1992.

In Northern Ireland, the legislation relating to the protection of species is primarily contained within Nature Conservation and Amenity Lands (Northern Ireland) Order 1985, the Wildlife (Northern Ireland) Order 1985 (as amended), and the Environment (Northern Ireland) Order 2002.

Badgers are protected in Northern Ireland under the Wildlife (Northern Ireland) Order 1985.

Wildlife protection falls into two categories: protection of specific species and the protection of habitats. In this section, we consider the protection of specific species under three subheadings: birds, animals, and plants. The Schedules in the Wildlife and Countryside Act 1981 and the Wildlife (Northern Ireland)) Order 1985 (as amended), which list the birds, animals, plants, and other species covered by the legislation, are reviewed every five years. For current lists of Scheduled species, we refer you to the websites for the Joint Nature Conservation Committee, Countryside Council for Wales (CCW), the Scottish Natural Heritage website: The Law and You, and the Northern Ireland Environment Agency.

i. Wild birds

a. General protection   All (approximately 500) species of wild birds are covered by current legislation across the UK. Unless you have a licence or are otherwise authorized under the relevant legislation, these acts make it an offence to:

- Kill, injure, take, or possess any wild bird
- Take, damage, or destroy the nest of any wild bird while that nest is in use or being built
- Take or destroy an egg of any wild bird or stop it from hatching
- Possess any live or dead wild bird or any part of a wild bird
- Possess any egg or part of any egg from a wild bird.

There are exceptions to this in that it is not an offence to kill a bird if it has been mortally injured or to take a wild bird to treat and release it if it has been hurt, but only where the bird has been injured in a way that does not contravene the Act (or Order). You may also be exempt if the outcome is '*the incidental result of a lawful operation and could not have been avoided*'. For example, the action is carried out by an authorized person for the purpose of preserving pubic health, public safety, air safety, preventing the spread of disease, or preventing serious damage to livestock, crops, fruit, growing trees, or timber, or otherwise in accordance with a licence granted by the Minister.

b. Enhanced protection   Some species have enhanced protection under Schedule 1 of the Wildlife and Countryside Act 1981, the Wildlife (Northern Ireland) Order 1985 (as amended) and the Nature Conservation (Scotland) Act 2004. For the species listed in these Acts, the penalties for the offences described above are higher and it is an offence to:

- Disturb any wild bird included in Schedule 1 while it is building a nest or is in, on, or near a nest containing eggs or young
- Disturb dependent young of such a bird.

Some of the birds so listed are given this additional protection for the whole year, but for some it covers only the spring and summer (February–August). The exceptions that relate to the general protection of wild birds also apply to this enhanced protection.

There are other regulations relating to game birds, 'pests', and swans. Game birds are covered by Schedule 2 Part 1 of the Wildlife and Countryside Act 1981 and the Wildlife (Northern Ireland) Order 1985 (as amended). This provides regulations concerning the killing of game birds. Some birds such as pigeons may be 'pests' and Schedule 2 Part 2 of the Act/Order provides regulations concerning the control of these birds. Wild swans in England belong to the Crown and, whilst they are subject to the 1981 Act, they are also Crown property unless they have been tamed and are on private waters. In Northern Ireland, swans are included in the Wildlife (Northern Ireland) Order 1985 (as amended).

ii. Wild animals

The Wildlife and Countryside Act 1981 (as amended), Wildlife (Northern Ireland) Order 1985 (as amended), and the Nature Conservation (Scotland) Act 2004 prohibit certain methods of killing and taking wild animals. Further protection, however, is only given to Scheduled species. The species in these Acts at present includes certain mammals, reptiles,

amphibians, fish, butterflies, moths, beetles, hemipteran bugs, crickets, dragonflies, spiders, crustaceans, sea mats, molluscs, annelid worms, sea anemones, and sea horses.

Under the terms of Section 9 of the Wildlife and Countryside Act 1981 (as amended), Schedule 6 of the Nature Conservation (Scotland) Act 2004, and Schedule 5 in the Wildlife (Northern Ireland) Order 1985 (as amended) for the animals listed it can be an offence to:

- Intentionally kill, injure, or take such an animal
- Possess or control a live or dead animal, part or derivative
- Damage or destruct or obstruct access to any structure or place used by the scheduled animal for shelter or protection
- Disturb the animal occupying such a structure or place
- Sell, offer for sale, possess, or transport for the purpose of sale such an animal (live or dead, part or derivative)
- Advertise for buying or selling such things.

You may be exempted from these laws for the same reasons given for wild birds: for example, if you take a Scheduled animal to tend it if it has been injured due to a reason that does not contravene these Acts.

a. Badgers   Badgers are not included in the Wildlife and Countryside Act 1981. Instead, in England and Wales, they are protected under different legislation including the Badgers Act 1973, the Wildlife and Countryside (Amendment) Act 1985, and the Badgers (Further Protection) Act 1991, and consolidated in the Protection of Badgers Act 1992. In Northern Ireland, badgers are protected by the Wildlife (Northern Ireland) Order 1985 (as amended). There are Scottish amendments to the Badgers Act 1992 in the Nature Conservation (Scotland) Act 2004. This combined legislation provides the same type of protection given to the animals on Schedule 5 of the 1981 Act, but with additional offences relating to the use of dogs, cruelty, and enhanced protection relating to reckless disturbance and causing damage to a badger sett.

b. Deer   Deer are also not included in Schedule 5 of the Wildlife and Countryside Act 1981, but in the Deer Act 1963, Deer Act 1980 Deer (Scotland) Act 1996, and in Schedule 10 of the Wildlife (Northern Ireland) Order 1985. These laws specify which species may be killed, when, and how.

c. Fish   Some fish species are included in Schedule 5 of the Wildlife and Countryside Act 1981 and by the Foyle Fisheries Act (Northern Ireland) 1952 (as amended) and the Fisheries Act (Northern Ireland) 1966 (as amended). In addition to this, common law treats fish in non-tidal waters as property. Therefore, you will need the landowner's consent before carrying out any work relating to freshwater fish. In addition, the Salmon and Freshwater

Fisheries Act 1975 and the Salmon and Freshwater Fisheries (Consolidation) (Scotland) Act 2003 prohibit killing freshwater fish by certain methods, including firearms and spears, or using lights. In UK tidal waters and seas, anyone can fish, but you are subject to some statutes and bye-laws.

d. Bats   Bats are protected by a number of different statutes including the Wildlife and Countryside Act 1981, Wildlife (Northern Ireland) Order 1985, the Conservation (Natural Habitats, &c.) Regulations 1994 (as amended), the Wild Mammals (Protection) Act 1996 (as amended), the Countryside and Rights of Way Act 2000, and the Nature Conservation (Scotland) Act 2004. Under this legislation, all bats are listed as 'European protected species' and it is an offence for any person to:

- Deliberately capture, kill, injure, or take a bat
- Possess or control a live or dead bat, any part of a bat, or anything derived from a bat
- Damage, destroy, or obstruct access to any place that a bat uses for shelter or protection
- Deliberately disturb a bat
- Sell, offer, or expose for sale, or possess or transport for the purpose of sale, any live or dead bat, any part of a bat, or anything derived from a bat
- Set and use articles capable of catching, injuring, or killing a bat (for example, a trap or poison), or knowingly cause or permit such an action; this includes sticky traps intended for animals other than bats
- Make a false statement in order to obtain a licence for bat work
- Possess articles capable of being used to commit an offence, or to attempt to commit an offence.

Again, there are exceptions to this legislation, which are similar to those described in relation to wild birds.

iii. Wild plants

In most UK legislation, the working definition of the term 'plants' is wide and includes algae, lichens, and fungi, as well as true plants such as mosses, liverworts, and vascular plants. Protection is provided at three levels: general, enhanced national, and international protection.

a. General protection   All wild plants are covered by legislation relating to property (4.3.2). In addition, all wild plants are given general protection in the Wildlife and Countryside Act 1981 (as amended) and the Wildlife (Northern Ireland) Order 1985, which makes it illegal to uproot or destroy any plant for any reason without the permission of the landowner, but you

may legally pick plant material for the plants not covered under the enhanced protection of, for example, Schedule 8 of the Wildlife and Countryside Act 1981 (as amended).

b. Enhanced protection   Species listed under the Act/Order have enhanced protection. For these plants it is an offence to:

- Intentionally pick, uproot, or in any way destroy such a plant
- Sell, offer for sale, possess, or transport for the purposes of sale any such plant (live, dead, or derivative)
- Advertise for buying or selling such things.

This enhanced protection therefore includes collecting seeds or spores of any of the species listed in Schedule 8 and there is no exception for the owner of the land on which these plants are found. Again, there is an exemption made under the terms of the 1981 Act, which is where plants are damaged or destroyed as an incidental result of a lawful operation and the damage could not reasonably have been avoided.

In addition to the enhanced protection given to some plant species in the Wildlife and Countryside Act 1981, the local authority can place tree preservation orders, for example to protect ancient trees and/or to preserve a certain ambience in urban areas. If a tree is included in a tree preservation order, then it is an offence to cut down, uproot, or wilfully damage or destroy the tree covered by the order without the consent of the planning authority.

4.3.4 Protection in special areas

Within the UK, species are also protected by being within certain designated areas. This protection comes primarily from the Wildlife and Countryside 1981 Act (as amended), the Nature Conservation (Scotland) Act 2004, the Nature Conservation and Amenity Lands (Northern Ireland) Order 1985 (as amended by the Environment (Northern Ireland) Order 2002), and the Council Directives on the Conservation of Natural Habitats and of Wild Fauna and Flora (EC Habitats Directive) and the Wild Birds Directive, which primarily became UK legislation through the Conservation (Natural Habitat, &c.) Regulations 1994 (as amended in each country). The areas we consider in this chapter are those protected because of their biological importance; however, some areas are protected because they are of archaeological or geological value, e.g. limestone pavements.

Areas within England and Wales may be given one of several different types of designation, and the level of protection varies depending on the status assigned to the area. We briefly review the current types of

designation and the protection offered to the species within these areas that is in addition to that outlined in 4.3.3.

i. Habitats and Natura 2000

Under the terms of the Habitats Directive and EC Birds Directive, Member States are required to put forward a number of national sites for consideration as Special Areas of Conservation (SACs) or Special Protection Areas (SPAs). These will form a network of protected areas known as Natura 2000, which aims to protect certain rare or endangered species and habitats. At present, these candidate sites are being identified within the UK by organizations such as the Joint Nature Conservancy Council (JNCC). Once a site is designated, the regulations that might reduce the impact of future use and disturbance to the habitat will be reviewed and if necessary improved.

The amount of legislative protection for habitats is increasing, as seen in the establishment of Natura 2000 sites. In particular, wetlands and woodlands are subject to additional specific legislation. Wetlands of International Importance, especially waterfowl habitat, are covered by the Ramsar Convention (1971), which was ratified in the UK in 1976. Many of the UK's Natura 2000 sites are also Ramsar sites.

In Great Britain, the Forestry Act 1967 originally focused on forests as purely economic enterprises. This has changed with the Wildlife and Countryside (Amendment) Act 1985 where the terms outlined the need to find a balance between commercial forestry, amenity use, and conservation.

ii. Nature reserves

These are areas that have a special conservation interest for a species or habitat or an unusual geological feature.

a. Statutory local and national nature reserves   The National Parks and Access to the Countryside Act 1949 and the Wildlife and Countryside Act 1981 allow Natural England, the Countryside Commission for Wales, and Scottish Natural Heritage to establish a nature reserve. In Northern Ireland, nature reserves are designated under the Nature Conservation and Amenity Lands (Northern Ireland) Order 1985. These can be in private ownership or owned by the council. If bodies such as the JNCC or the Department of the Environment thinks that a nature reserve is very important, then it may be designated as a national nature reserve. Bye-laws are then established for each nature reserve, which may, for example, control access and in other ways protect the species on the nature reserve. A few marine reserves have been established in this way.

b. Non-statutory nature reserves   Under the National Parks and Access to the Countryside Act 1949, local nature reserves may be declared by local authorities after consultation with the relevant statutory nature conservation agency. Other local nature reserves may be run by non-statutory bodies such as the Woodland Trust, Wildlife Trusts, and the Royal Society for the Protection of Birds. Some of these non-statutory nature reserves are run by County Trusts, which come together under the title of the 'Royal Society of Wildlife Trusts'. None of these management bodies can make bye-laws and they are therefore dependent on voluntary codes of conduct and national legislation to provide protection to the species within the nature reserve.

iii. National Parks, Areas of Outstanding Natural Beauty, and Sites of Special Scientific Interest

In England and Wales, the purpose of National Parks is to conserve and enhance landscapes within the countryside while promoting public enjoyment of them and having regard for the social and economic well-being of those living within them. The National Parks and Access to the Countryside Act 1949 established the National Park designation in England and Wales. In addition, the Environment Act 1995 requires relevant authorities to have regard for nature conservation. Special Acts of Parliament may be used to establish statutory authorities for their management (e.g. the Broads Authority was set up through the Norfolk and Suffolk Broads Act 1988).

The National Parks (Scotland) Act 2000 enabled the establishment of National Parks in Scotland. In addition to the two purposes described above, National Parks in Scotland are designated to promote the sustainable use of the natural resources of the area and the sustainable social and economic development of its communities. These purposes have equal weight and are to be pursued collectively unless conservation interests are threatened.

In Northern Ireland, the Amenity Lands (Northern Ireland) Act 1965 and Nature Conservation and Amenity Lands (Northern Ireland) Order 1985 made provision for the designation of National Parks in Northern Ireland. The designation of such an area is currently under discussion.

The primary purpose of the Areas of Outstanding Natural Beauty (AONB) designation is to facilitate the conservation of natural beauty, which by statute includes wildlife, physiographic features, and cultural heritage, as well as the more conventional concepts of landscape and scenery. Account is taken of the need to safeguard agriculture, forestry, and other rural industries, and the economic and social needs of local communities. AONBs have equivalent status to National Parks as far as conservation is concerned. AONBs are designated under the National Parks and Access to the Countryside Act 1949, amended in

the Environment Act 1995. The Countryside and Rights of Way Act 2000 clarifies the procedure and purpose of designating AONBs in England and Wales. Originally designated in Northern Ireland under the Amenity Lands Act (Northern Ireland) 1965, AONBs are now designated under the Nature Conservation and Amenity Lands (Northern Ireland) Order 1985. In Scotland, National Scenic Areas are broadly equivalent to AONBs.

The Sites of Special Scientific Interest (SSSI) (Area of Special Scientific Interest (ASSI) in Northern Ireland) series has developed since 1949 as the national suite of sites providing statutory protection for the best examples of the UK's flora, fauna, or geological or physiographical features. These sites are also used to underpin other national and international nature conservation designations. Most SSSIs are privately owned or managed; others are owned or managed by public bodies or non-government organizations. The SSSI/ASSI designation may extend into intertidal areas out to the jurisdictional limit of local authorities, generally Mean Low Water in England and Northern Ireland and Mean Low Water of Spring tides in Scotland. In Wales, the limit is Mean Low Water for SSSIs notified before 2002, and, for more recent notifications, the limit of Lowest Astronomical Tides (LAT), where the features of interest extend down to LAT. There is no provision for marine SSSIs/ASSIs beyond the low water mark, although boundaries sometimes extend more widely within estuaries and other enclosed waters.

Originally notified under the National Parks and Access to the Countryside Act 1949, SSSIs have been renotified under the Wildlife and Countryside Act 1981. Improved provisions for the protection and management of SSSIs were introduced by the Countryside and Rights of Way Act 2000 (in England and Wales) and the Nature Conservation (Scotland) Act 2004.

ASSIs are notified under the Nature Conservation and Amenity Lands (Northern Ireland) Order 1985. Measures to improve ASSI protection and management are contained in the Environment (Northern Ireland) Order 2002.

Land within these areas remains in the same, usually private, ownership as they were prior to designation. However, owners of the land and authorities such as the planning authority then have a statutory responsibility for these areas and may be limited as to what they are able to do.

iv. Ministry of Defence land and National Trust land

Land used by or owned by the Ministry of Defence is subject to access restrictions (sometimes amounting to prohibitions) and often carries the additional risk of unexploded ordnance. Other land, for example National Trust properties, land owned by the Duchy of Cornwall, the

Windsor Estate, and the Malvern Hills, Worcestershire, whilst not statutory nature reserves, may also be covered by bye-laws, etc., which may
control access and provide additional protection for species within the
area.

4.3.5 Movement, import, export, and control

There are several national statutory controls and EC regulations that control the movement of animals, plants, and other organisms. Controlling
potential pests and diseases requires restrictions on the movement of
materials around the globe and suitable containment if work is carried out
in the UK. For example, the Plant Health (Great Britain) Order 1993 prohibits the import, movement, and keeping of certain plants, plant pests,
and other material, including soil. Some species are listed in the Wildlife
and Countryside Act 1981 Part II of Schedule 9, which prohibits any
person from releasing and allowing to escape into the wild certain plants
such as *Fallopia japonica* (Japanese knotweed). The Act also prohibits
anyone from releasing or allowing to escape any wild animal that is not
ordinarily resident or a common visitor to the UK. Some 'alien' animals
have become established in the wild, but it is not considered to be desirable to add to their number by allowing further escapes/introductions.
These are included in Part I of Schedule 9 of the 1981 Act (as amended).
Other legislation that relates to the control and release of species are
the Ragwort Control Act 2003 and European Directive 2001/18/EC
on the release of genetically modified organisms (GMOs) and the EC
GMO Regulations (Deliberate Release) 2002.

At an international level, the most substantive regulation in relation
to rare and endangered species is the Convention on International Trade
in Endangered Species (CITES) 1975 where 160 Member States have
ratified controls regulating the movement of more than 2,500 animals
and 25,000 plants. This has been extended by the Control of Trade in
Endangered Species (Enforcement) Regulations 1997 (COTES). There
are three levels of protection. Those species listed in Appendix I of the
convention are those that may be threatened with extinction and where
international trade is only allowed in exceptional circumstances and
where the prohibition extends to dead or live individuals or derivatives
such as ivory and furs. The trade in species in Appendix II is monitored
through a licensing system, whilst in the final category (Appendix III)
are those species not threatened on a global level but protected under
national legislation within Member States.

In addition to international and national legislation, carriers within
the UK, such as the Royal Mail, have regulations concerning what
may be carried and the method and appropriate packaging required.

These regulations extend to micro-organisms and other biomaterials such as blood or tissue samples.

4.3.6 Permits and licences

If you wish to work on a species or habitat that is protected by legislation or wish to transport a protected species, etc., it may be possible to obtain a licence or permit that allows you to carry out otherwise proscribed activities on protected species or in protected areas. The bodies responsible for granting these permits or licences are usually listed within the Act of Parliament. Most commonly, these are the Department for Environment, Food and Rural Affairs (Defra), Natural England, the Countryside Commission for Wales (CCW), Scottish Natural Heritage (SNH), and the Northern Ireland Environment Agency. For example, taking photographs of bats is considered to be unlawful disturbance and you therefore need a licence from the relevant body to do so. More unusual arrangements exist for some species; for example, sturgeon and whales in British waters may only be taken under licence from the Crown.

4.4 Animal welfare

Key points The main principles of animal welfare encompassed in UK law are: to provide a suitable environment, provide a suitable diet, to enable the animal to exhibit normal behavioural patterns, to provide suitable housing (with or apart from other animals), and to protect from unnecessary suffering, injury, and disease. Some procedures may require a personal or institutional licence.

Much of the relevant legislation for working with animals in the wild has been considered in 4.3. However, for animal work that involves working with an animal in an institution such as a university or working with a domestic or captive animal, there is additional legislation. At present, there are three main Acts that are most pertinent to this: the Protection of Animals Act 1911, the Animals (Scientific Procedures) Act 1986 (as amended), and the Animal Welfare Act (2006) (Animal Health and Welfare (Scotland) Act 2006). If you are working with agricultural animals there is considerably more legislation that is likely to be relevant, including the Welfare of Animals (Slaughter or Killing) Regulations 1995, the Welfare of Farmed Animals Regulations (England 2000/Northern Ireland 2000/Wales 2001). For more information about working with agricultural animals and for more details in general about animal welfare, we refer you to Radford (2001).

4.4.1 **Protection of Animals Act 1911**

This Act is the current core of much animal protection legislation. The essence of the law is that it is an offence to be cruel to any captive animal, which in this context includes any 'wild' bird, fish, and reptile confined or in captivity, as well as domestic animals. The offence of 'cruelty' can be applied to a wide variety of circumstances and events, and this has been the strength of this law. The notion of what is cruel has changed over time in response to greater scientific understanding and changes in social attitudes. Therefore, what may have been acceptable practice when the bill was first passed may now be considered to contravene the Act.

The definition of cruelty varies but relates primarily to the idea of causing unnecessary suffering. Cruelty in the Act is defined in terms of types of conduct including:

- To cruelly beat, kick, ill-treat, override, overload, torture, infuriate, or terrify any animal; or cause, procure, or, being the owner, permit any animal to be so used

- To wantonly or unreasonably do or omit to do any act, causing unnecessary suffering to any animal; or cause, procure, or, being the owner, permit any such act

- To convey or carry any animal in such a manner or position as to cause it any unnecessary suffering; or cause, procure, or, being the owner, permit any animal to be so conveyed or carried

- To wilfully, without a reasonable cause or excuse, administer any poisonous or injurious drug or substance to any animal; or cause, procure, or, being the owner, permit such administration or wilfully, without any reasonable cause or excuse, cause any such substance to be taken by any animal

- To subject any animal to any operation that is performed without due care or humanity; or cause, procure, or, being the owner, permit any animal to be subjected to such an operation

- Without reasonable cause or excuse, to abandon an animal in circumstances likely to cause it any unnecessary suffering; or cause, procure, or, being the owner, permit it to be abandoned.

Further legislation has introduced amendments and extensions to this Act; for example, the Welfare of Animals (Transport) Order 1997 relates to carrying animals in the course of trade and the Protection of Animals (Anaesthetics) Act 1954 adds legislation referring to anaesthetization of mammals during an operation.

4.4.2 Animals (Scientific Procedures) Act 1986 and Amendment, 1998

The Animals (Scientific Procedures) Act 1986 based on a European Directive (under review) regulates scientific procedures that may cause pain, suffering, distress, or lasting harm to protected animals. Unlike the definition of animals in 4.3 and 4.4.1, in this instance a protected animal includes fetal, larval, and embryonic forms (within certain limits) of any living vertebrate (excluding humans) and the invertebrate *Octopus vulgaris* (common octopus).

There are a number of procedures covered under these regulations including:

- Any experimental or other scientific procedure applied to a protected animal that may have the effect of causing it pain, suffering, distress, or lasting harm
- Anything done for the purpose of, or liable to result in, the birth or hatching of a protected animal if this may have the effect of causing it pain, suffering, distress, or lasting harm
- The administration of an anaesthetic or analgesic to a protected animal, or decerebration or any other such procedure, for the purposes of any experimental or other scientific procedure
- The humane killing by certain methods of a protected animal when this is for experimental or other scientific purposes and carried out in a designated establishment
- The removal of blood or tissues from a live protected animal.

Some procedures are not included in the Act, such as those carried out as part of normal veterinary, agricultural, or animal husbandry practices; ringing or other methods of tagging animals to allow their identification, provided that this causes no more than momentary pain and distress; the humane killing of a protected animal by a method listed in Schedule 1 of the Act; and the administration of materials to animals as part of a medicinal test in accordance with the Medicines Act 1968. However, in these circumstances, the Protection of Animals Act 1911 still applies.

Under the terms of this Act, a system of licensing enables such work to be carried out when the benefits that the work is likely to bring (to humans, other animals, or the environment) outweighs the pain or distress that the animals may experience. Other criteria that have to be met before a licence is granted are that there are no alternatives, the procedure uses the minimum number of animals, involves animals with the lowest degree of neurophysiological sensitivity, and causes the least pain, suffering, distress, or lasting harm. In addition, within the terms of the Act,

certain types of animal must also be obtained from designated breeding or supplying establishments.

The Act includes the arrangements for a three-level licensing system where those carrying out procedures must hold a personal licence, which will not be granted unless they are qualified and suitable. The programme of work must also be authorized in a project licence, and the work must also normally take place at a designated-user establishment. However, in specific circumstances (such as field trials), work can be carried out elsewhere with the Home Secretary's authority. Clearly, there will be considerable variation between institutions in terms of which type of licence, if any, is already held. Therefore, if you are intending to work with such animals, you will need to obtain advice about the existing licences for your institution from your supervisor.

4.4.3 Animal Welfare Act (2006) and Animal Health and Welfare (Scotland) Act 2006

These review and extend legislation relating to animal welfare previously included as a small element in a number of specific Acts, for example the Animal Health Act 1981, the Zoo Licensing Act 1981, the Welfare of Animals at Market Order 1990, and the Welfare of Farmed Animals Regulations England 2000/Wales 2001/Wales 2007, and in the Animals (Scientific Procedures) Act 1986 and Protection of Animals Act 1911.

The Welfare Acts define a protected animal as those not living in the wild, which are commonly domesticated or in some way under the control of humans. The main responsibilities outlined in the Act are to provide a suitable environment, to provide a suitable diet, enable the animal to exhibit normal behavioural patterns, to provide suitable housing (with or apart from other animals), and to protect from unnecessary suffering, injury, and disease.

4.5 Working with humans

Key points   There is an extensive body of law that relates to any research involving humans. Ethical research is based on the principles of non-maleficence (doing no harm) and beneficence (doing good). Where possible, all research should be anonymous. It is important to obtain informed consent. Special consideration should be given to ensure appropriate working in relation to children and vulnerable adults, and to ensure equality.

After the atrocities of World War II, the Nuremberg Code (1947) was developed and was the first internationally recognized code of human ethics.

This code has informed policies relating to work on humans since this date, and has been restated and extended in other international agreements such as the World Medical Association's Helsinki agreements of 1964 and 2000. The Nuremberg Code and subsequent agreements recognize the need during research to minimize harm, to find a balance between the risks to the individual and the benefits to the individual and to society, to recognize the importance of obtaining fully informed and voluntary consent from humans participating in research, and to ensure that the research has real validity.

Within the UK, there is a considerable body of legislation that also impinges on studies involving humans, including the Obscene Publications Act 1964, the Sex Discrimination Act 1975, the Race Relations Act 1976 and the Race Relations (Amendment) Act 2000, the Computer Misuse Act 1990, the Data Protection Act 1998, the Human Rights Act 1998 (as amended), the Freedom of Information Act 2000, the Special Educational Needs and Disability Act 2001, the Disability Discrimination Act 2005 (as in all countries), the Safeguarding Vulnerable Groups Act 2006, the Special Educational Needs (Information) Act 2008, and health and safety legislation (4.2). Relevant Scottish legislation includes the Freedom of Information (Scotland) Act 2002, the Scottish Commission for Human Rights Act 2006, the Protection of Vulnerable Groups (Scotland) Act 2007, and the Adult Support and Protection (Scotland) Act 2007. Apart from this legislation, there are also extensive guidelines that come from professional bodies and codes of practice within your institution. It is impossible for us to cover in detail all this legislation and the guidelines, and therefore in this section we look instead at the general approach taken in research involving humans that is both good science and reflects appropriate practice. Different legislation covers the use of human tissues and we do not consider this here. If you are working in a laboratory using human tissues, there should already be procedures in place that you will need to follow.

Most institutions have some form of ethics scrutiny, which considers research involving human volunteers, so you are unlikely to be left making decisions about ethics in isolation. This is essential in research for several reasons. Firstly, researchers themselves, especially if they are inexperienced, may not foresee circumstances where their subjects may be harmed. Secondly, the researchers have their own ethical and moral principles and these will influence the design of a project. Researchers may also have a vested interest in carrying out this particular research project. For most students, this will relate to their wish to complete and be successful on a degree or higher degree programme, but can also relate to career opportunities and obtaining further funding. Therefore, it is advisable to have an independent scrutiny system in place for all human-based studies. There can be some variation in terms of practice

among different disciplines, so a subject specialist should be a key part of such a review.

4.5.1 Harm, risks, and benefits

There are two basic principles behind an ethical basis for research: these are non-maleficence and beneficence. Non-maleficence is the principle of 'doing no harm' as a consequence of the research. This can be applied both to direct action such as exposing volunteers to a substance and more widely, for example, in the collection and use of confidential information. Therefore, as part of your research planning, you must seek to minimize any harm or potential risks to your volunteers. This may not always be easy, as it can be difficult to know what another individual will be sensitive about. For example, someone may be sensitive about their weight, height, status, etc. This notion of 'do no harm' also covers those who may be affected by the results such as particular groups in society, other researchers, and the institution to which you are affiliated.

Beneficence is the requirement to 'do good'. Again, this is applied broadly across all aspects of working with humans. Both terms (non-maleficence and beneficence) are essential, although at first glance they appear to be two sides of the same coin. However, to 'do good' does not necessarily mean that we do no harm. For example, some studies using questionnaires may produce information that can be used to improve some aspect of human experience, but there may be a potential for unknown and unquantifiable harm in terms of psychological risks to those you collect the information from, especially where questions are personal. Collecting confidential information may also carry the risk of this information becoming accessible to inappropriate individuals. In medical science, novel treatments or the use of a placebo may be 'harmful' to the patient, but the understanding gained about a novel treatment may be beneficial to new patients. Clearly, a balance needs to be struck between the potential for doing harm and the benefits, and this balance needs to be in favour of the volunteer, not the research.

For example, in the study of the effectiveness of tea-tree oil (Example 2.2), the student took hand swabs from staff at a GP's practice. Clearly, if it became known that certain staff had 'dirty' hands, then there may have been adverse consequences for those individuals. Therefore, the student carrying out the research assigned numbers to each sample. In this way, neither she nor the managers of the practice could identify which sample belonged to which person. When the results were known, it was clear that some individuals carried much higher bacterial loads on their hands than others and may therefore have posed an increased risk to patients. The practice managers were able to act on this information in a general way, whilst the confidentiality of the volunteers was maintained.

4.5.2 **Consent**

Another strand in an ethical approach to research is autonomy. Autonomy is being able to make choices and decisions for oneself and by oneself. In practice this means both providing appropriate information about the research to allow volunteers to make an informed decision when giving or withholding their consent and exerting no pressure on a volunteer to take part in the research programme.

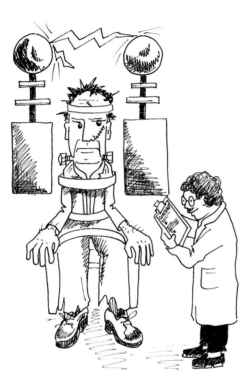

The information you provide to a potential participant should make it clear what you are asking the volunteer to do, why you wish to carry out this research, what you hope to gain from it. There may be some risks, for example if you are examining human physiological responses. These need to be made explicit in the information you provide. You may need to exclude certain people from your research either on the grounds of health and safety or due to your experimental design. If you have exclusion criteria these need to be outlined and the reason for them explained. The name of the person who will carry out the work or is responsible for the project, and the location, may need to be included in the information you provide. However, you should not provide personal contact details. Where external funding has been obtained, it is also ethical to advise any potential volunteers as to who is funding the work. Some volunteers may not wish to take part, for example if your work is funded by a tobacco or

pharmaceutical company. Information about what you are going to do with the data must also be included; for example, how will you ensure confidentiality, to whom will the results be made available, and in what format? Finally, you need to make it clear that the volunteer does not need to take part and may stop/leave at any time.

By asking your volunteers to sign a consent form or in some other way indicate their consent, you are not passing your responsibilities for the well-being of your volunteers to them. You remain responsible for their health and safety and for the ethical conduct of your research.

Information should be provided in a written format that can be taken away and referred to by the volunteer at a later date if they wish. The information should be explained in language that the informant can easily understand. For most work involving humans, you should obtain written consent. If you do have written consent forms, you must consider where you keep this information as it is subject to the Data Protection Act 1998 and is central to your ability to preserve confidentiality.

To obtain honest, informed consent, you cannot directly or indirectly put pressure on (coerce) someone to take part in your research. Most coercion is negative; for example, there may be some sort of penalty for not taking part or not completing the research. But coercion can be quite subtle. For example, a department may expect all students to take part in each others' projects, or you may ask your family or friends to take part, or you may only provide the information about the project just before you intend to carry it out and so allow no time for the potential volunteer to reflect on the information. Coercion can also be positive, for example if you provide a reward. This is not only unethical but can lead to bias in sampling and you may not obtain the representative sample you need for your research (1.4).

There are some circumstances where informed consent may not be obtained. These fall into two areas: either you are working with young people (under 16 years of age) or vulnerable people (for example mentally incapacitated adults or elderly people), or the nature of your investigation is such that you do not want to provide too much information in advance as it may influence the outcome. We consider the first of these in 4.5.3. For the latter, you should first make certain that there is no alternative approach, that the benefits justify not asking for informed consent, that you will not mislead participants, and then if you cannot provide information before the study, this should be done afterwards and the volunteers given the opportunity to withdraw retrospectively.

4.5.3 Special cases: children and vulnerable people

For young people or vulnerable people, such as elderly people and adults with learning difficulties, there are additional codes of practice and legal

requirements in place to protect them, even if the work is indirect, for example observing behaviour at playtime from outside the school premises. If a child is under 16 years old, then you will normally be expected to obtain informed consent from the parents, carers, or guardians. If you are working with children or vulnerable adults in a school or other similar setting, then you should also obtain the consent of the school from an appropriate person such as the head teacher or chair of governors. In addition, when working with young or vulnerable people, you will be required to obtain clearance from the Criminal Records Bureau (CRB). This disclosure scheme is designed to enhance public safety by providing criminal history information on individuals who wish to work with or in the course of their work may come into contact with vulnerable adults or children. This requirement for obtaining a disclosure includes undergraduates and graduates carrying out projects. The institution you are affiliated with should have a well-established procedure for obtaining a CRB disclosure.

Ethically, there are difficulties working with vulnerable people and with whether it is acceptable to ask someone else to provide informed consent. It is therefore even more important to be sure that all elements of your project are ethical. In some areas of research such as within the Health Service, this difficulty is recognized and it is therefore considered to be unethical to conduct research on children where there is no direct benefit to the individual child. If it is feasible, young and vulnerable people should also be asked for their consent, having been given information in an appropriate format.

4.5.4 Equality

In Chapter 1 (1.4), we discussed the notion of a representative sample. This is particularly pertinent when studying humans where it can involve additional planning and preparation to ensure that all potential volunteers can take part in a research project. In addition, some legislation (e.g. the Race Relations Act 1976) makes it an offence to exclude certain groups of people through either practices or attitudes. The Human Rights Act 1998 prohibits discrimination on any grounds, such as gender, race, colour, language, religion, political or other opinion, national or social origin, association with a national minority, property, birth, or other status. The Disability Discrimination Act 1995 requires you to make reasonable adjustments to enable a disabled person to take part in your research. Making a 'reasonable adjustment' can involve the provision of materials that are more accessible by individuals with dyslexia or with restricted sight, or providing an environment more suitable to someone with a physical disability, for example. We touch on some of these points in 3.4. Most higher education establishments have student

support service staff who will be able to provide more specific advice on this matter.

Other rights enshrined within the Human Rights Act 1995 that need to be considered when designing and carrying out research to ensure equality in relation to both inclusion and communication of results are that individuals have a right to freedom of expression; a right to freedom of thought, conscience, and religion; a right to respect for private and family life; and a right to be treated with justice and fairness.

4.5.5 Anonymity, confidentiality, information storage, and dissemination

When working with humans, you may wish to ensure that each volunteer's contribution is either anonymous or confidential. The rationale for doing so is that you may reduce potential harm to the volunteers, and participants may be more willing to give full cooperation if they will not be identified personally. To ensure anonymity no one, including the researcher, must know which data derive from which individual. This can be achieved, but it is difficult to ensure total anonymity where written informed consent is obtained or where personal information or visual records are required. Although anonymity in human research is usually the ideal, more often the research is confidential. Here, only one researcher can identify the individuals within the research, and when reporting, the researcher endeavours to ensure anonymity. To protect confidentiality or anonymity, data collection, storage, and dissemination have to be handled with care and consideration. Legislation that impinges on these activities includes the Human Rights Act 1995 and the Data Protection Act 1998. Under the terms of these Acts, you must consider what data you collect and how, how the data will be stored, who will be able to access the data and through what media, and how long the data is to be stored. The overriding principles are those outlined in earlier sections, including an individual's right to privacy, not to be misrepresented, and not to be harmed.

Confidentiality can be broken inadvertently either by storing information where names are linked to data and this stored data is accessed by a third person or by communicating information in reports, etc., with enough detail for certain individuals to be identified. For example, if you have a relatively small cohort, known to a third person, and you report the data from the only individual over 50 years of age, it will be easy to identify this individual. In undergraduate and graduate work, you may therefore need to remove certain information from reports, such as location, age, or gender, to ensure that individuals cannot be identified. For example, when the undergraduate in Example 2.2 reported on her

findings relating to bacterial loads on the hands of staff in a GP's surgery, she did not include information about the geographical location of the GP's surgery.

There may be some occasions when you may need to reveal a person's identity. This should not be done unless you have received the volunteer's written permission. You may also need to extend this to members of the public if they are included in photographs, video, or film. There is considerable sensitivity at present about recording and storing images of children. If your work requires this, you must discuss it with your supervisor first and ensure that the information you provide to those giving informed consent makes it clear that this is what you intend doing.

4.6 Discussion topics

Consider the following scenarios. What issues arise from these research proposals in relation to the law?

 Q3 An undergraduate keeps stickleback at home. She wishes to use them in her honours year project and investigate their behaviour when presented with different food sources in a number of different environments.

 Q4 An undergraduate wishes to compare the effectiveness of two teaching media (book versus computer) in a primary school. The class will be divided and half will cover a topic from the national science curriculum using a book resource, while the other half will use a computer-based resource. The children will be tested on the topic before and after.

Q5 An undergraduate wishes to test the effect of music on cats in a cattery. Individual cats are observed during times when music is playing and when it is not. Several types of music are tested.

 Q6 A student wishes to follow the colonization of a new pond in a primary school in Worcestershire where the great crested newt is relatively common.

Q7 To investigate the bactericidal properties of *Allium* species, an undergraduate grew a laboratory strain of *Escherichia coli* on a solid agar medium and added small pieces of bulbs from a number of different *Allium* species to the plates.

 Q8 A student wishes to compare the diet of elderly patients in hospital with their normal diet at home.

The answers to these discussion topics are given at the end of this chapter.

Summary of Chapter 4

- The aim of this chapter is to provide a necessarily brief overview of the nature of law and regulations in the UK (4.1).

- The chapter provides an introduction to the law in relation to four key areas; health and safety (4.2), access and working with wildlife (4.3), working with animals in captivity (4.4), and working with humans (4.5).

- The information in this chapter should help you to carry out a risk assessment and to carry out your research safely (4.2).

- You are encouraged to be aware of your responsibilities in relation to research involving wild species and to be able to extend and update your knowledge of this body of law (4.3).

- You are encouraged to be aware of your responsibilities in relation to research involving captive animals and to be able to extend and update your knowledge of this body of law (9.4).

- You are introduced to some of the legislation relating to working with humans. We consider ethical approaches to research and the ideas of non-maleficence and beneficence. These are discussed in the contexts of consent, children and vulnerable people, equality, and data storage and handling (4.5).

- A number of examples from undergraduate honours project proposals are included in the chapter for you to discuss and to challenge your understanding of the topics covered in this chapter (Q3–Q7).

- The Online Resource Centre includes interactive exercises that test your understanding of this chapter with other topics, particularly those considered in Chapters 2–10.

@ online resource centre

Although every effort has been made to ensure that the information in this chapter is accurate, it should not be taken as a definitive statement of the law, nor can responsibility be accepted for any errors or omissions.

Answers to chapter questions

 A1 Corrosive, very harmful if swallowed.

R10 Flammable

R20 Harmful by inhalation

R21 Harmful in contact with skin

R22 Harmful if swallowed

R35 Causes severe burns

The R (risk) codes are used as a shorthand when providing some safety information. They are recognized standard descriptions of different hazards. For example, R10 always means flammable. (There are also S (safety) codes that indicate the safety precautions that should be followed.)

A2 The hazards and ratings include: being swept out to sea and drowning (major), falling over and breaking a bone (serious), and other injury such as a cut (serious slight) or sprain (slight), effects of temperature (hypothermia, sunburn, etc.) (major–slight), waterborne pathogens causing an infection (major–slight) or toxic reaction (major–slight), and being attacked by other people or their dogs (major–slight).

A3 The two main areas relevant to this proposal are firstly, that the student will need to complete a health and safety assessment. Secondly, the work will be subject to the laws outlined in the Protection of Animals Act 1911 and the Animals (Scientific Procedures) Act 1986 (and Amendment 1998).

A4 All the topics covered in 4.5 are pertinent. The student would have to obtain a CRB disclosure. Decisions about consent would need to be made. To reduce harm to the children, the student would have to consider how to protect confidentiality or anonymity, how to ensure that no child's learning was affected, etc. To protect the school from harm, the student would have to consider how the school's anonymity is protected. The student would have to ensure that the 'harm' to the children is outweighed by the benefit. The student should also carry out a risk assessment. Risks might include picking up infections, head lice, etc.

A5 A health and safety assessment would need to be carried out by the student. The student would need to comply with the requirements under the licence given to the cattery. Permission would need to be sought from the cats' owners. The student would be subject to the Protection of Animals Act 1911.

 A6 This project has three elements. Firstly, the student will need to carry out a health and safety assessment, particularly in relation to waterborne pathogens. Secondly, the student will need permission from the school to ensure that they are not trespassing. He or she will need a CRB disclosure as they are working in an area where children may be present. The great crested newt is a European Protected Species. The student may continue to sample from the pond until such time as they have confirmed that great crested newts have colonized the pond. Should this happen, he or she will need to apply for a licence under the terms of the Wildlife and Countryside Act 1981 from English Nature or the Countryside Council for Wales.

A7 One *Allium* is currently listed on Schedule 8 of the Wildlife and Countryside Act and the student should not therefore be using this species in the research. In addition, the student must carry out a health and safety assessment.

A8 The student must consider the topics and review the legislation introduced in 4.5 such as obtaining consent, ethics, and data handling. The student would probably require CRB clearance. In addition, the student must complete a risk assessment for themselves both in relation to working in a hospital and in relation to visiting other people's homes (4.2).

Section 2

Handling your data

What to do with raw data 5

In a nutshell

Knowing the type of measurements you will record is critical when planning your experiment and when summarizing it for presentation. We start in this chapter by explaining the terms that are used to describe your data, the scales they are measured on, and the type of distributions seen in your data. This allows you then to decide whether your data are parametric or non-parametric and to calculate the correct summary statistics.

So far, we have looked at the general terms that you may encounter when reading about or carrying out research relating to experimental design (Chapter 1) and the steps you need to follow when planning an experiment or evaluating other people's research (Chapters 2 and 3). In this chapter, we consider what to do with the data you have collected from your investigations; these ideas are then developed in Chapters 6–11.

As we explained in 2.2.7 and 3.5, the first time to think about the data that your research may generate is while you are planning your experiment. This is critical. Most research uses statistics as a tool to help identify the trends in the data. But each statistical test has certain requirements that need to be met. For example, to use the Mann–Whitney U test (10.1), you need to have between five and 30 observations in each sample. If you have not decided on which statistical test to use before you start your investigation, your sample size may be too small and you will not be able to analyse your data. This is a waste of your time and reflects badly on you as a scientist, as it is clear you have not planned your work properly in the first place. To choose the correct test, you first need to understand terms such as 'qualitative' and 'parametric'. These are explained in this chapter (5.1 and 5.8). You also need to understand about 'distributions' (5.2) and 'transforming data' (5.9). We provide an overview of how to choose the correct statistical test in Appendix c.

The second time you need to think about your raw data is after you have completed your investigation. You will need to identify the trends in your data and to communicate these to other people. When you come to communicate your findings, you may wish to use a figure (11.8.2), a table (11.8.1), or summary statistics (5.3–5.7). In learning how to summarize your data, you will also be introduced to some of the central steps in statistics, which are the calculation of a sum of squares, variance, and standard deviation (see Box 5.1).

If you work through the exercises in this chapter it will take about two hours. The answers for these exercises are at the end of the chapter. All our examples are based on real undergraduate research projects. If these examples are not in your subject area, you will find more in the Online Resource Centre.

online resource centre

5.1 Types of data

Key points When recording observations, you use a particular scale of measurement such as grams (e.g. mass) and centimetres (e.g. height). These scales have important features. They may, for example, be either continuous scales or discontinuous scales. It is these features that primarily determine how data should be summarized and which statistical tests can be used to test a hypothesis and are therefore important to understand.

When carrying out an investigation, you will generate data as a series of observations measured on a particular scale. For example, if you are investigating the change in human body temperature in relation to exercise, the scale of measurement used here is degrees Celsius (°C). There are several terms that are used frequently to describe the scales of measurements used when collecting data. You need to become familiar with these terms: they are essential in helping you decide how to design your investigation and how to communicate your findings. These terms are: qualitative, quantitative, discrete, continuous, rankable, nominal, ordinal, interval, and derived variables.

Qualitative Observations that are assigned to named, descriptive categories that are mutually exclusive and non-numerical. The categories are always discrete

Qualitative refers to information that is not numerical but descriptive. In Chapter 3, we refer to methods such as focus groups and interviews, which may gather qualitative responses, opinions, and thoughts expressed in words. The term 'qualitative' can also be used when collecting numerical data but where the scale of measurement is qualitative. In this case, qualitative data will be numerical observations assigned to named, descriptive categories that are mutually exclusive and non-numerical, for example the number of *Lotus corniculatus* (bird's-foot trefoil) with a yellow or a red keel. The categories yellow and red are qualitative. These scales of measurement are always discrete.

Quantitative refers to the use of numbers. A quantitative scale of measurement is one where the observations are assigned to ordered numerical categories; for example, the height of adult males (cm) is measured on a continuous quantitative scale.

Discrete measurements fall into a series of distinct, mutually exclusive categories and the number of categories is limited. For example, the Royal Horticultural Society's (RHS) scale for recording petal colours (e.g. red, pink, white) is a qualitative discrete scale of measurement. The number of eggs in a clutch is a quantitative discrete scale of measurement. In this context, you cannot have half an egg.

Continuous scales are ones where observations do not fall into a series of distinct categories and may take any value within the scale of measurement. For example, height (cm) can be measured from 0cm upwards with no limit. Some scales, such as percentages and pH, are also continuous, but only within prescribed boundaries. For example, the percentage scale is restricted to 0–100%.

Rankable applies to scales of measurement where the categories can be ranked (put in a consistent order). These can include qualitative, quantitative, discrete, and continuous, and therefore ordinal and interval scales. For example, if a questionnaire included the question, 'How warm are you?', the answers (very hot, hot, warm, cool, cold), although qualitative, can be ranked. Similarly, a numerical but discrete scale, such as the number of eggs in a clutch, can be ranked (e.g. 0, 1, 2, 3, etc.). Continuous data are ranked in two ways: either by numerical order or by category order. For example, if the heights of five men were recorded, the observations could be ranked 166.0cm, 166.5cm, 168.0cm, 168.2cm, and 170.0cm. Where you have many observations recorded on a continuous scale, these can be arranged into categories or classes (Table 5.1). These categories can also be ranked. For some statistical tests observations are ranked and then 'assigned a rank order'. We explain this process in 5.8.2.

Nominal scales of measurement fall into discrete categories with no particular order to these categories. The number and nature of the categories can be either an inherent property of what is being measured or imposed by the investigator. The categories should ensure that every observation in the data set can be classified. For example, the RHS scale used for recording petal colours is an artificial device that provides a method for categorizing

Quantitative Observations that are numerical. Data may be either discrete or continuous

Discrete Observations that fall into a series of distinct categories. The number of categories is limited

Continuous Observations that do not fall into a series of distinct categories and may take any value in the scale of measurement e.g. height (cm)

Rankable Observations that consist of named categories or values that have a consistent order to them

Nominal Observations that fall into discrete categories. The categories have no specified order

Table 5.1. Frequency classes for height (cm) of 87 male students

	Height								
	150.0–154.9	155.0–159.9	160.0–164.9	165.0–169.9	170.0–174.9	175.0–179.9	180.0–184.9	185.0–189.9	190.0–194.9
Frequency	3	4	12	15	19	15	12	4	3

flower colours; these colours have no particular order and cannot therefore be ranked. If a closed question is included in a questionnaire (3.2) where the prescribed answers are 'yes', 'no', or 'don't know', these mutually exclusive categories are an inherent property of what is being measured but there is no rationale for ranking them: they are therefore nominal.

Ordinal Observations that fall into discrete categories where the categories have an order (they can be ranked)

Ordinal scales of measurement are similar to nominal scales in that they are also based on discrete, mutually exclusive categories. However, in this case, the categories can be ranked. The categories can be qualitative or quantitative. In ecology, a common scale that is used when examining species abundance is the ACFOR scale, which comprises the categories 'abundant', 'common', 'frequent', 'occasional', and 'rare'. These are a qualitative measure but have an inherent order, can be ranked and are therefore ordinal. The number of eggs in a clutch also falls into discrete categories. These quantitative categories can also be ranked and are therefore ordinal. In Example 3.1, volunteers were asked to rate aspects of this book using a scale of 1 to 10. This scale is numerical but has discrete categories that are rankable, so it is an ordinal scale.

Interval data Measured on a continuous scale; the data are rankable and it is possible to measure the difference between each observation

Interval data are measured on a continuous and rankable scale, and unlike data measured on an ordinal scale, it is possible to measure the difference between each observation. Examples of interval scales include temperature (°C), distance (m), mass (g), and date.

Derived variables The unit of measure for the observations is the result of a calculation, e.g. ratio, proportion, percentage, rate

Derived variables are scales of measurements that are the result of a calculation; they are therefore quantitative and usually continuous. The four types of derived variables are ratios, proportions, percentages, and rates. Some of these scales, such as percentages, are constrained within certain limits (e.g. 0–100%); others, such as a rate (e.g. km/h), are not.

You will usually find that more than one term can be used to describe any one measurement. For example, if an investigation examines the height of students within a higher education institution, the scale of measurement will be centimetres. This is a quantitative, continuous, rankable, interval scale.

 Q1 Which terms best describe these data:
a. The number of prickles on holly leaves

b. Percentage (%)

c. pH

d. grams

The terms we have considered so far all relate to the scale on which the measurements are made. Understanding which type of scale is being used is important in helping you choose the type of statistical test you should use when you design your experiment and analyse your data (Appendix c),

and which figure or table you could use when communicating your findings (11.8). There are two more terms (parametric and non-parametric) that are also critical. These terms relate to your observations and the characteristics of the distribution of your data. To be able to tell whether your data are parametric or non-parametric, you have to first understand distributions (5.2) and how to calculate a mean and variance (5.4 and 5.5). We therefore consider the terms parametric and non-parametric in 5.8 and explain how to tell whether your data are parametric in Box 5.2.

5.2 Distributions of data

Key points When data are plotted, the shape of the graph outlines the 'distribution'. There are a number of common distributions that are encountered. These arise in part due to the scale of measurement used. The distributions can be described mathematically. The distribution determines whether the data are parametric or not.

In investigations examining the effect of one variable, the data may be plotted, for example as a bar chart or histogram (e.g. Fig. 5.1). These figures can be seen to have a particular shape or **distribution**. Some distributions are well known, such as the normal (5.2.1), binomial (5.2.2), Poisson (5.2.3), and exponential (5.2.4) distributions. The shapes of these distributions have been described mathematically and these equations have been used to demonstrate other relationships, including the tendency of a distribution to have a central point (e.g. a mean, 5.4) and the spread of the data around the central point, such as the variance (5.5).

Distribution The shape seen on a graph when data are plotted

5.2.1 Normal distribution

This is a very important distribution, which can best be explained using an example.

In the study of the height of 87 male students, the raw data are first organized into a frequency table (Table 5.1).

As the data are measured on an interval scale, they can be plotted as a histogram (Fig. 5.1). If you collected more observations, then this increased sample size would allow you to reduce the sizes of the classes you are using and the distribution looks much smoother when plotted (Fig. 5.2). If you could extend this study still further, the distribution becomes even smoother, and more symmetrical and 'bell-shaped' (Fig. 5.3), with a single central peak (unimodal). This is typical of a **normal distribution**.

Normal distribution A symmetrical, 'bell-shaped' distribution with a single central peak (unimodal) which can be described by the Gaussian equation. The data are measured on an interval scale

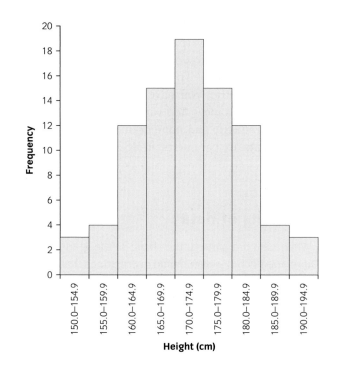

Fig. 5.1. Height of 87 male students.

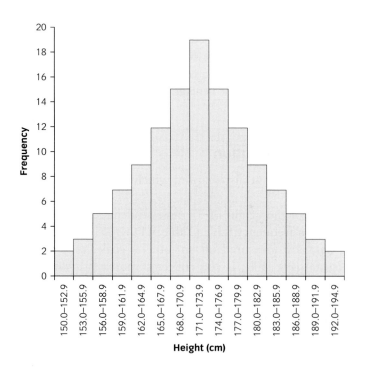

Fig. 5.2. Height of 124 male students.

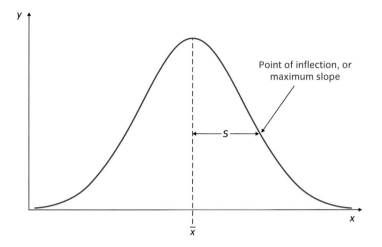

Fig. 5.3. Normal distribution with mean ($\overline{x}$) and standard deviation (s).

If your data are normally distributed like this, then there is a mathematical relationship between the x values and y values. This mathematical description is called the Gaussian equation, where:

$$y = \frac{1}{\sqrt{(2\pi s^2)}} e^{-b}$$

and

$$b = \frac{(x - \overline{x})^2}{2s^2}$$

The terms in this equation may not be familiar to you. The symbols e and π are particular numbers (constants) that occur so often in maths that they have been given letters to refer to them. Their values are 2.72 and 3.14 respectively (rounded to two decimal places). The $\overline{x}$ (mean) and s^2 (variance) are two of the summary statistics that we consider in 5.4 and 5.5. x is any one observation in your sample and y is the y value calculated for any given x value. This equation is considered again in 7.1.3 where we show you how to test whether your data can be described by the Gaussian equation, and Appendix e where we explain the symbols in this equation in more detail. A normal distribution has several features that are widely exploited in statistics, and we refer back to this distribution throughout the rest of the book.

5.2.2 **Binomial distribution**

The second distribution we consider, binomial distribution, is one you would expect to obtain if each item examined can have only one or

Binomial distribution This may be obtained if each item examined can have only one or another state

another state. For example, a seed can either germinate or not germinate, or in a T maze, a rat can only turn either left or right. In humans, the outcome at birth is to be either male or female. The chance of being born male is 1 in 2 and the chance of being born female is 1 in 2.

The chance of a family having only three girls is $1/2 \times 1/2 \times 1/2 = (1/2)^3 = 0.125$. This calculation of probability, the probability of having three girls in a three-child family, can be extended and it is possible to write a mathematical equation that allows you to work out the probability for any particular number of males and females in any family of a particular size using the binomial equation:

$$ y = \frac{n!}{x!(n-x)!} \times p^x \times q^{(n-x)} $$

This equation describes the mathematical relationship between y (the probability of obtaining a particular number of children of one gender in a given number of births) and x (the selected number of children of one gender). If you are not familiar with the terms we have used in this equation (e.g. !), we give more details and some worked examples in Appendix e.

We know that when a child is born, the probability that they will be female (p) is 1/2 and the probability of the child being male (q) is also 1/2. If we examined ten births (n), we can use this information to work out the probability of obtaining any number of girls within those ten births (Table 5.2). For example the probability of obtaining four girls and six boys is 0.205; and the probability of all the children being female is very small, only 0.001.

Table 5.2. The probability of having a given number of female (or male) children in ten births

Number of female (or male) children (x) in ten births	Probability (y)
0	0.001
1	0.010
2	0.044
3	0.117
4	0.205
5	0.246
6	0.205
7	0.117
8	0.044
9	0.010
10	0.001

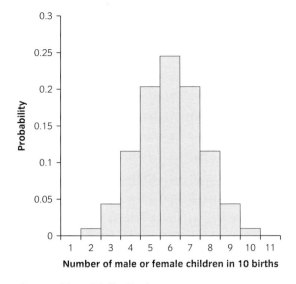

Fig. 5.4. Binomial distribution.

When the data are plotted (Fig. 5.4), you can see that this distribution also has a symmetrical shape with a single (unimodal) high point. If you collect data where you know there can only be two alternative outcomes, where the distribution has this shape and where the data can be described by this mathematical equation, then your data are said to be binomial and they have a binomial distribution. We show you how to check whether your data have a particular distribution in 7.1.3.

5.2.3 Poisson distribution

A **Poisson distribution** is often found where events are randomly distributed in time or space, such as the distribution of cells within a liquid culture or the dispersal of pollen or seed by wind. This distribution is unimodal, but is often very asymmetrical with a protracted tail either to the right (a positive skew) or to the left (a negative skew). Figure 5.5 illustrates four Poisson distributions that vary in their degree of skewness. The Poisson distribution can be described by the equation:

$$y = e^{-\bar{x}} \times \frac{\bar{x}^x}{x!}$$

As we explained when we looked at the Gaussian equation, e is a constant with the approximate value of 2.72 and $\bar{x}$ is the mean of the sample. x is a particular observation in the sample and y is the value for a given x. One Poisson distribution is shown in Table 5.3.

Poisson distribution A unimodal distribution, which is often very asymmetrical with a protracted tail either to the right (a positive skew) or to the left (a negative skew)

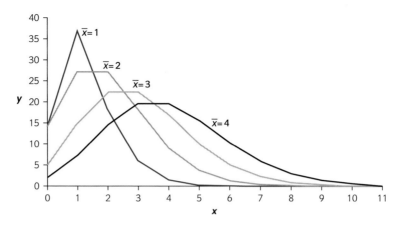

Fig. 5.5. Poisson distribution.

Table 5.3. The distance seeds are dispersed from the canopy edge of one *Taxus baccata* (yew) tree

Distance (m)	Number of seeds
0	13
1	27
2	27
3	18
4	9
5	4
6	1
7	0
8	0
9	0
10	0

We explain in 7.1.3 and 5.5.3 how you may check your data to confirm that they have a Poisson distribution.

5.2.4 Exponential distribution

This distribution can be found in bacterial liquid cultures during their peak growth where the numbers of individuals keep doubling, or if you recorded the increase in the number of copies of DNA during a polymerase chain reaction (PCR) (Fig. 5.6). The mathematical description of an exponential distribution uses mathematical notation relating to integration, which is a mathematical technique beyond the scope of this book. You should refer to other statistical texts for further details.

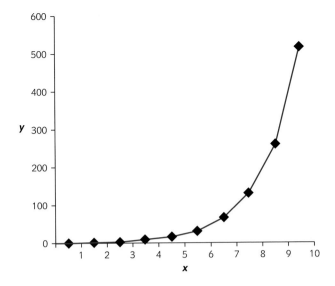

Fig. 5.6. Exponential distribution.

5.3 **Summary statistics**

Key points Data may be summarized to avoid having to present a large body of data in a report and to indicate general trends in the data. Data may also be summarized as a first step when testing hypotheses. The type of summary statistics used depends on the type of data you have. The data are usually summarized using a measure of the central point of the distribution (e.g. mean, median, or mode) and the variation in the data (e.g. standard deviation, variance, or range). The shape of the distribution in terms of skew and kurtosis may also be described.

Summary statistics have two roles in research. In the first, summary statistics are used when communicating your results, to avoid the inclusion of extensive tables of raw data and to allow you to present the main trends in your data more succinctly. The second role of summary statistics is that many of the calculations, particularly the mean and variance, are terms that appear in statistical calculations and it is therefore important to appreciate types of summary statistics, their relative strengths, and how they are calculated.

Summaries of data focus on two things: the central point of the distribution and the variation around this central point. The calculations of the central point can be derived from the mathematical relationship, such as the Gaussian equation (5.2.1) or Poisson equation (5.2.3), and in these instances this central point is called the mean. Where the distribution is not known, then other measures, such as the median and mode, can be used. A similar distinction occurs in relation to estimates of the spread of

Table 5.4. Appropriate summary statistics for different types of data

	Nominal	Ordinal	Interval—unknown distribution	Interval—known distribution
Central Tendency	Mode	Mode Median	Mode Median	Mean
Spread of data (Variation)		Range Interquartile range Percentiles Confidence limits	Range Interquartile range Percentiles Confidence limits	Standard deviation Coefficient of variation Confidence limits

the data about the central point. Some distributions are not symmetrical about the central point and this deviation from symmetry can be indicated by using measures of skewness (5.4.4). The shape of the peak can also vary and a useful measure of this kurtosis is given in 5.4.4.

The summary statistics that we consider in this section can be used in two ways: either you have data where there are no replicates or you have data with replicates. The appropriate summary statistics are indicated in Table 5.4.

In Example 5.1 (Table 5.5), the data for both years are numbers of seeds per umbel, which is a discrete ordinal scale. From Table 5.4, we can see that these data may be summarized using a median, mode, and range. In Table 5.6, we have replicate samples and here we can summarize within each category. If you summarize within a category, for example we may wish to summarize within the category '13 seeds per umbel', then your data within that category will either be ordinal or interval. In this example, the data are ordinal and therefore we may use the median, mode, and range to summarize within the category '13 seeds'.

EXAMPLE 5.1 The number of seeds per umbel in *Allium schoenoprasum*

A single *Allium schoenoprasum* (chive) may produce one inflorescence called an umbel containing many flowers. Each umbel usually therefore produces many seeds. We recorded the number of seeds produced per umbel over two consecutive years (Table 5.5), examining nine umbels in 2004 and eight in 2005.

Table 5.5. The number of seeds per umbel in *Allium schoenoprasum* in 2004 and 2005

Year	Number of seeds per umbel								
2004	13	14	15	15	15	17	19	19	20
2005	13	14	15	15	17	19	19	20	–

In 2004, a number of separate samples were examined and the number of seeds recorded (Table 5.6).

Table 5.6. The number of seeds per umbel in four random samples of *Allium schoenoprasum* in 2004

Sample	Number of seeds per umbel								
	13	14	15	16	17	18	19	20	21
a	1	1	3	0	1	0	2	1	0
b	2	0	3	1	2	0	3	2	4
c	0	3	1	3	1	3	1	0	2
d	2	2	1	2	3	0	0	2	0

The data from Table 5.5 have been reorganized in Table 5.7 so that it is easier to identify the mode and median.

Table 5.7. The number of umbels of *Allium schoenoprasum* producing a certain number of seeds (data recorded in 2004 and 2005)

	Number of seeds							
	13	14	15	16	17	18	19	20
Number of umbels recorded in 2004	1	1	3	0	1	0	2	1
Number of umbels recorded in 2005	1	1	2	0	1	0	2	1

5.4 Estimates of the central tendency

There are three common methods used to measure central tendency: the mode, median, and mean. The mean should not be confused with the average. A mean is a measure of central tendency that can be calculated when the distribution is known. The average is an estimate of the mean using the same method of calculation as the mean for a normal distribution but applied to data whose distribution is not known (interval or ordinal). Although this is an incorrect calculation, it is in common usage.

5.4.1 Mode

The mode is the category that contains the greatest number of observations. If the data are nominal or ordinal, the categories are already specified. For example, the scale of measurement used in Example 5.1, the 'number of seeds', is ordinal with categories (one seed, two seeds, etc.) that can be ranked. When summarized (Table 5.7), it is clear that the mode is 15 seeds in 2004 (i.e. **unimodal**). In 2005, there are two modes: 15 seeds and 19 seeds (i.e. **bimodal**).

Mode One measure of central tendency. When data are organized into categories, the mode is the category that contains the greatest number of observations

Unimodal A distribution that has a single high point or peak

Bimodel A distribution that has two high points or peaks

If the data are measured on an interval scale but the underlying distribution is not known, then you may use the mode as a measure of central tendency but you will first need to organize your data into a frequency table. The modal class is the class with the highest frequency and the mode is the mid-point of this category. Clearly, as you have imposed these categories on the data, there is a degree of artificiality in relation to the mode; grouping the data into other size classes may generate a different mode. You should be aware of this both when choosing the classes for the frequency table and when interpreting your results. (For further comments about choosing size classes, see 11.8.1.)

5.4.2 Median

Median One measure of central tendency, calculated as the middle value of an ordered data set

When data can be ranked (i.e. ordinal or interval data), a simple measure of the central tendency is to take the 'middle' value. This is the **median**.

The median is dependent on the number of observations in the data set (n). When n is an odd number, then the median is the middle value. When n is an even number, then the median is calculated as half the sum of the two middle values. In Table 5.5 for the year 2004 ($n = 9$), the data are already in numerical order and the middle value is 15 seeds per umbel. In 2005 ($n = 8$), here the median is $(15 + 17)/2 = 16$ seeds per umbel.

5.4.3 Mean

Mean The measure of central tendency for interval data where the distribution is known. The sample mean for normally distributed data is calculated as the sum of all observations divided by the number of observations in the data set

The mean is used for interval data where the distribution is known. Here we give the methods for calculating a mean for a normal distribution (described by the Gaussian equation, a binomial distribution, and a Poisson distribution). For other distributions, you will need to refer to other texts and/or computer software.

For any one distribution, a sample mean ($\bar{x}$) or a population mean (μ) may be calculated. The sample mean is used as an estimate of the population mean when not all items in the population have been measured. A value called the confidence interval can be calculated to demonstrate the area around a sample mean in which the population mean will probably fall (5.7).

i. Normal distribution

When calculating a mean for a normal distribution, most often you will have data that are not organized in categories. However, if you have data that are organized into categories a mean may still be calculated using a different method that provides a reasonable estimate of the value.

a. Your data are not grouped in categories   In normally distributed data, the mean is calculated as the sum (Σ) of all observations (x) divided by the number of observations in the data set (n). If you had collected measurements from all items in the population, then the calculation is the same, but you would refer to N rather than n and the mean would be referred to as μ (1.5). If you are not familiar with these terms, you may also wish to refer to Appendix e. The calculation of the mean for a sample is written as:

$$\bar{x} = \frac{\Sigma x}{n}$$

We have demonstrated in Box 5.2 that the data in Example 5.2 appear to be normally distributed and therefore we can calculate a mean assuming that the data are described by a Gaussian equation. For this example, if you add all the x values together (i.e. $\Sigma x = 1 + 5 + 2 + 5 + \ldots + 3 + 4 + 2 + 5$), this equals 254 and there are 50 observations. Therefore, the sample mean is:

$$\bar{x} = \frac{254}{50} = 5.08\text{mm}$$

b. Your data are grouped into categories   Some data may be from a normal distribution but may be grouped. In this case, the mean can be estimated

EXAMPLE 5.2 Length (mm) of two-spot ladybirds (*Adalia bipunctata*)

An investigator was interested in the length of two-spot ladybirds (*Adalia bipunctata*). In an observational investigation, she measured the length (mm) of 50 ladybirds collected at random from a garden (Table 5.8).

Table 5.8. The length (mm) of 50 *Adalia bipunctata* sampled in a garden

Length of *Adalia bipunctata* (mm)

1	5	2	5	7	8	3	6	7	4
4	5	6	4	5	5	7	5	3	5
4	5	1	7	9	2	6	5	6	3
3	6	8	6	4	6	6	8	5	6
7	4	8	9	5	4	3	4	2	5

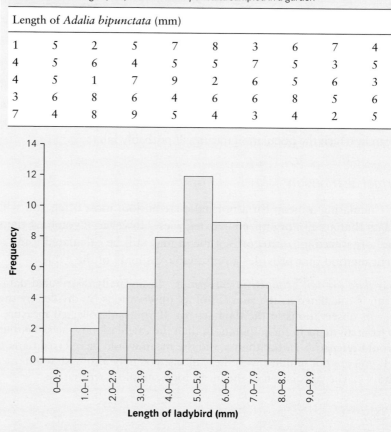

Fig. 5.7. Length of ladybirds, showing a bell-shaped distribution.

in a different way. For this calculation, you need the mid-point for each category (*m*) and the frequency within that category (*f*). Then:

$$\overline{x} = \frac{\Sigma mf}{\Sigma f}$$

Table 5.9 is a frequency table for the data from Example 5.2, where the mid-point for each class and the mid-point multiplied by the frequency (*mf*) are given.

Table 5.9. Frequency table of the data from Example 5.2 on the length of two-spot *Adalia bipunctata* (ladybirds) showing how to calculate a mean for grouped data

	Size classes for length (mm) of *Adalia bipunctata* (ladybird)								
	1.0–1.9	2.0–2.9	3.0–3.9	4.0–4.9	5.0–5.9	6.0–6.9	7.0–7.9	8.0–8.9	9.0–9.9
Mid-point of class (*m*)	1.45	2.45	3.45	4.45	5.45	6.45	7.45	8.45	9.45
Frequency (*f*)	2	3	5	8	12	9	5	4	2
mf	2.9	7.35	17.25	35.6	65.4	58.05	37.25	33.8	18.9

We know that there are 50 observations, so $\Sigma f = 50$. From the table, we can add all the *mf* values together (Σmf) = 276.5. Therefore, $\bar{x} = 276.5/50 = 5.53$. This method will tend to overestimate the mean and therefore should only be used when you do not have the raw data.

 Q2 Calculate the mean and mode using the data from Table 5.1.

ii. Binomial distribution

In 5.2.2, we illustrated the idea of a binomial distribution in relation to the gender of children. The total number of girls you would expect in your sample would be the number of births (*n*) multiplied by the probability that each child was a girl (*p*). If you had *N* families each with *n* children, then you would expect to get $n \times N \times p$ girls. For example if we collected data from six families each with five children, then the total number of girls will be $5 \times 6 \times 1/2$ girls $= 15$ girls in total. The mean for this binomial data will be this total number divided by the number of families (*N*):

$$\bar{x} = \frac{n \times N \times p}{N}$$

The two *N* values cancel each other out, so the mean can be worked out simply as $\mu \approx \bar{x} = np$, and so for six families with five children, $\bar{x} = 5 \times 1/2 = 2.5$ girls.

iii. Poisson distribution

The mean in a Poisson distribution relates both to the skew of the distribution and the variation in the data. The mean for a Poisson distribution is calculated as:

$$\mu \approx \bar{x} \approx \frac{\Sigma xy}{\Sigma y}$$

Table 5.10. Calculating a mean and variance for data with a Poisson distribution: the distance seed dispersed from the canopy edge of one *Taxus baccata* (yew) tree

Distance (m) (x)	Number of seeds (y)	xy	x^2y
0	13	$0 \times 13 = 0$	$0^2 \times 13 = 0$
1	27	$1 \times 27 = 27$	$1^2 \times 27 = 27$
2	27	$2 \times 27 = 54$ etc.	$2^2 \times 27 = 108$ etc.
3	18	54	162
4	9	36	144
5	4	20	100
6	1	6	36
7	0	0	0
	$\Sigma y = 99$	$\Sigma xy = 197$	$\Sigma x^2y = 577$

The terms Σxy and Σy appear in many statistical calculations and therefore we have shown you in detail how these terms are calculated. The steps in this calculation are shown in Table 5.10. in the first three columns. The final column (x^2y) relates to the calculation of a variance for data with a Poisson distribution, which we explain in 5.5 and which can be ignored for now. The mean for this example is:

$$\mu \approx \overline{x} \approx \frac{\Sigma xy}{\Sigma y} = \frac{197}{99} = 1.99\text{m}$$

5.4.4 **Skew and kurtosis**

Skew When the distribution of the plotted observations is not symmetrical

Kurtosis A measurement that indicates how sharp the peak of the central point is

There are two further measures that may usefully be used to describe features of the central tendency. These are skew and kurtosis. We have already seen an example of variation in the degree of skew in a distribution. In data with a Poisson distribution, as the mean decreases the distribution becomes less symmetrical (Fig. 5.5). A measure of this degree of skew is the relationship between the mean, median, and mode. If a distribution is symmetrical, such as the normal distribution, then the mean should equal the mode, which should equal the median. However, this guide to a symmetrical distribution is rarely absolute and you are often left wondering just how much of a difference between values for the mean, median, and mode can be tolerated before you accept that you are not looking at a symmetrical distribution. If there is a doubt in relation to a normal distribution, you can use the additional criteria outlined in Box 5.2 to determine whether the distribution is normal.

Q3   Does the *mean* = median = mode for the data from Example 5.2?

There are several measures of skew of which the most common is:

$$Skew\ (\gamma^3) = \frac{\Sigma(x - \bar{x})^3}{(n-1)s^3}$$

The value for skew for a perfectly symmetrical distribution should be zero. If the distribution has a positive skew (Fig. 5.8), then the value of γ^3 will also be positive, and for a negative skew (Fig 5.9), the value of γ^3 will be negative. An assessment of skew is one of the tests for normality (Box 5.2.) and may also be used as a descriptive statistic with which to compare distributions.

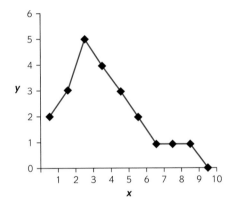

Fig. 5.8. Distribution skewed to the right (positive skew).

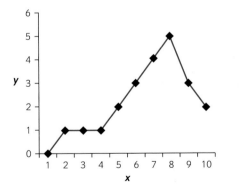

Fig. 5.9. Distribution skewed to the left (negative skew).

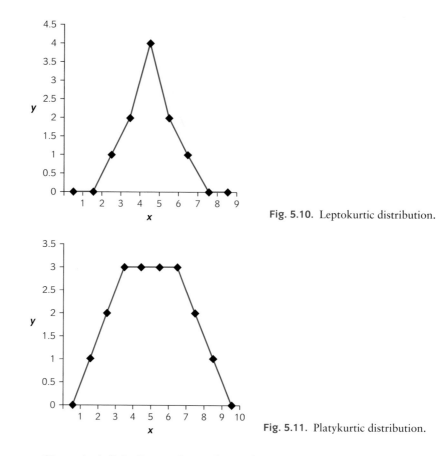

Fig. 5.10. Leptokurtic distribution.

Fig. 5.11. Platykurtic distribution.

Kurtosis (γ^4) indicates how sharp the peak of the central point is. A frequency distribution with a pointed narrow peak is called leptokurtic (Fig. 5.10), a moderate peak is called mesokurtic, and a very flat peak is called platykurtic (Fig. 5.11). Kurtosis can be measured as:

$$kurtosis(\gamma^4) = \frac{\Sigma(x - \bar{x})^4}{(n-1)s^4}$$

Normally distributed data should have a mesokurtic appearance with a kurtosis value of about 3, a leptokurtic distribution will have a kurtosis value of more than 3, and a platykurtic distribution will have a kurtosis value of less than 3.

These two methods for calculating skew and kurtosis assume that the mean and variance have been calculated using the equations outlined for a normal distribution. Therefore, for data with a distribution other than a normal distribution, you may use these methods for indicating skew and kurtosis, having first calculated a mean and standard deviation using the equations for a normal distribution (5.4.3i and 5.5.3i). The seed dispersal around *Taxus baccata* is known to have a Poisson distribution (Table 5.10). If the

mean and standard deviation for this data are calculated using the normal equations, then $\bar{x} = 2.29$, $s = 1.13$, $\gamma^3 = 2.8$, and $\gamma^4 = 6.24$. This supports the notion that this distribution is skewed to the right and is very leptokurtic.

5.5 Estimates of variation

The calculation of a measure of the central point can be used to compare different samples; however, apart from the consideration of skew and kurtosis, it does not tell you much about the shape of the distribution. For nominal data that cannot be ordered, there is no appropriate method for indicating this variation in the data (Table 5.4). For ordinal and interval data, there are a number of measures that can be used to obtain an idea of how the data is distributed around the central point. These (the central point and estimate of variation) together are used to provide a summary of your raw data.

5.5.1 Range

For ordinal and interval data where the underlying distribution is not known,

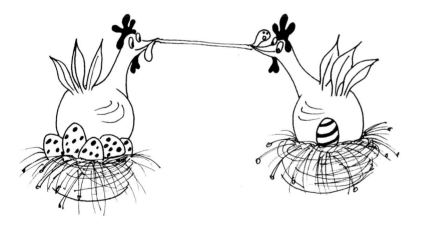

the range is the simplest indication of the spread of data around a central point. The range is the distance between the highest and lowest observations and is therefore calculated by subtracting the value of the lowest observation from the highest observation. For example, the range for the data illustrated in Table 5.5 for both 2004 and 2005 is 20 seeds − 13 seeds = 7 seeds per umbel. The disadvantage of this measure of variation is that it can be vastly altered by the presence of a single 'outlier' and the range does not take into

account the shape of the distribution between the two ends. The range as calculated does not make any allowance for sample size, yet commonly an increase in sample size will tend to increase the range.

5.5.2 Interquartile range and percentiles

Outlier An observation that is noticeably different from all other observations, one that does not follow the apparent trend

Interquartile range A reduced range from the 25th observation to the 75th observation in a data set of 100 observations

The effect of outliers can be reduced by taking a range not from the smallest to the largest observation but at some specified point within the range. Most commonly, the points used are the 25th centile (where 25% of the observations are smaller) and the 75th centile (where 25% of the observations are larger). This is known as the interquartile range. Alternatively, you may select another set of points, such as the 90th centile and 10th centile. These are known as percentiles. The interquartile range is the difference between the observation at the 25% point and the observation at the 75% point. In the data from Table 5.7 for the year 2004, the interquartile range is from 15 seeds to 19 seeds = 4 seeds per umbel. Only 50% of the observations fall within the interquartile range. Within a percentile range of 10–90%, 80% of observations are included. In both these ranges, the impact of outliers is minimized, but neither system conveys much information about the shape of the distribution.

5.5.3 Variance

One of the most useful and frequently used measures of variation in data is the variance. This term may be calculated for interval data where the distribution is known. We illustrate how to calculate the variance for data with a normal, binominal, or Poisson distribution. For other distributions, you will need to find other sources of information. The variance for a sample (s^2) is one measure of the variation in the data. However, the variance is, as the symbol indicates, a squared value and the units of variance are also squared. Therefore, it is common to take the square root of this measure of variation, the standard deviation (s), which has the same units as the original observations. If you have collected observations from every item in your population, then you should use the population equation to calculate the population variance (σ^2) and population standard deviation (σ). Unlike the mean, these calculations for the variance differ depending on whether you are working out the population parameter or the sample statistic. As most biological research works with samples, here we only explain in detail the calculation for samples. The equation for calculating the population variance is given in 1.5.

i. Normal distribution

Normally distributed data can be evaluated using a particular set of statistics called parametric statistics. A common component of these statistics is the calculation of a sum of squares ($SS(x)$), the variance (s^2) and the standard deviation (s). Therefore, we have included these calculations in a box for ease of cross-referencing from future chapters. The box is a format we use throughout the chapters on statistics and is based on two columns: one with the general information about the calculation and a second column with a specific worked example. In this case, we use the data from Example 5.2, the length (mm) of the two-spot ladybird (*Adalia bipunctata*).

a. Your data are not grouped into classes   The **variance** for a sample is calculated as the squared average deviation of an observation from the mean:

$$s^2 = \frac{\Sigma(x - \bar{x})^2}{n-1}$$

It is possible to use this equation as it is. However, there is an alternative form of this equation that provides a very close approximation to the answer. This is the most commonly used equation, especially if you are working the variance out 'by hand'.

The sample variance for normally distributed data (s^2) is:

$$s^2 = \frac{\Sigma x^2 - \dfrac{(\Sigma x)^2}{n}}{n-1}$$

The **standard deviation** for normally distributed data (s) is:

$$s = \sqrt{\frac{\Sigma x^2 - \dfrac{(\Sigma x)^2}{n}}{n-1}}$$

The first part of these equations is called the **sum of squares**. This is a term you will often encounter when using parametric statistics.

The sum of squares ($SS(x)$) is:

$$SS(x) = \Sigma x^2 - \frac{(\Sigma x)^2}{n}$$

We take you through the method for calculating the sums of squares, variance, and standard deviation in Box 5.1. Working out the variance by this method may give you a very slightly different answer to one calculated

Variance (s^2) A measure of the variation in the data, calculated as the squared average deviation of an observation from the mean

Standard deviation A measure of the variation in data where the distribution is known. For data with a normal distribution, it is the square root of the variance and is a value with the same units as the original observations

Sum of squares A common step in the calculation of a variance for parametric data

BOX 5.1 How to calculate a standard deviation and variance for normally distributed (parametric) data

GENERAL DETAILS	EXAMPLE 5.2.
To calculate the sums of squares (SS(x)). Where $$SS(x) = \sum x^2 - \frac{\left(\sum x\right)^2}{n}$$	This calculation is given in full in the Online Resource Centre. All values have been rounded to five decimal places.
a. First add all the observations in a sample together ($\sum x$) and square the total $(\sum x)^2$. Divide this value by n: $\dfrac{(\sum x)^2}{n}$	a. Table 5.8 $$\sum x = 1 + 5 + 2 + \ldots + 4 + 2 + 5 = 254$$ $$(\sum x)^2 = 64516$$ $$n = 50$$ $$\frac{(\sum x)^2}{n} = \frac{64516}{50} = 1290.32$$
b. The next step is to square each of the observations (x^2) and add these squared values together ($\sum x^2$).	b. $\sum x^2 = 1^2 + 5^2 + 2^2 + \ldots + 4^2 + 2^2 + 5^2$ $\quad = 1474$
c. Subtract the value you worked out at step (a) from the value you worked out in step (b).	c. Sum of squares SS(x) $= 1474 - 1290.32 = 183.68$
To calculate the sample variance where $$s^2 = \frac{\sum x^2 - \dfrac{\left(\sum x\right)^2}{n}}{n-1}$$ $$= SS(x)/n-1$$	
d. Divide SS(x) by $n-1$. This is the sample variance (s^2). **Calculate the standard deviation (s) where** $$s = \sqrt{\frac{\sum x^2 - \dfrac{\left(\sum x\right)^2}{n}}{n-1}}$$	d. Variance $s^2 = \dfrac{183.68}{50-1} = 3.74857 \text{mm}^2$
e. The sample standard deviation (s) is the square root of the variance: $s = \sqrt{s^2}$	e. Standard deviation s $$s = \sqrt{3.74857} = 1.93612 \text{mm}$$

Further examples relating to the topic including how to use statistical software are included in the Online Resource Centre.

online resource centre

by a computer or calculator that is programmed with the other equation. However, using the 'estimation' method is accepted as standard practice as the difference between answers is so small.

In many distributions, including the normal distribution and the Poisson distribution, there is a relationship between the mean and the standard deviation. For a normal distribution, this is shown in Table 6.2 and Fig. 6.1. To briefly introduce this relationship, we can consider the height recorded for 87 men (Table 5.1). Clearly, most of the men have an average or near average height. A few are much smaller or much taller. It is possible to use the mathematical description of the distribution to work out how many people you would expect to fall within a particular size class. To do this, you use the standard deviation. The relationships we use most often in normally distributed data are approximately that:

- 68% of the men should have heights in the size range $\bar{x} \pm 1s$

- 95% of the men should have heights in the size range $\bar{x} \pm 1.96s$

- 99% of the men should have heights in the size range $\bar{x} \pm 2.58s$.

This relationship is used to check whether your data are normally distributed. A worked example showing this is included in Box 5.2. We return to this in Chapter 6.

Q4   Using the rye data from Example 2.3, first decide which type of data these are and then summarize the data appropriately.

---

b. Your data are grouped into classes   In most research, you will have access to the full data set, but there may be some occasions when you are only provided with data already summarized in a frequency table. In this case, you may use the following to obtain an estimate of the variance and standard deviation, assuming that the data are normally distributed.

The sample variance is derived by first calculating the Σx^2 from the mid-point of each class (m) and the frequency in each class (f) and where n is the total number of observations:

$$\Sigma x^2 = \frac{n\Sigma m^2 f - (\Sigma mf)^2}{n}$$

The variance is then:

$$s^2 = \frac{\Sigma x^2}{n-1}$$

When this approach is applied to the length of ladybirds (Table 5.9), $\Sigma mf = 276.5$ (as calculated in 5.4.3i(b). Squaring each mid-point (m^2) and multiplying this by its frequency gives a series of values for m^2f. Adding all these values together for this example, $\Sigma m^2f = 1712.725$. Therefore, $\Sigma x^2 = 183.68$, $s^2 = 3.7486$, and $s = 1.93612$. These values for the variance and standard deviation are a very close match to those we calculated in Box 5.1 using the same data but not grouped into classes.

As the mean, variance, and standard deviation for normally distributed data are common components in statistics, calculators and computers invariably include software to allow these terms to be calculated quickly; it is worth becoming familiar with these tools. However, you must make sure that if you have a sample you use sample statistics not population parameters (1.5).

ii. Binomial distribution

In a binomial distribution, the variance is determined by the number of observations (n) and the probability of obtaining each outcome (p, q). Unlike data with a normal distribution, however, the population variance is approximately the same as the sample variance so that $\sigma^2 \approx s^2 = npq$. In our example in 5.2.2, $s^2 = 5 \times 1/2 \times 1/2 = 1.25$ girls; the standard deviation (s) is the square root of the variance (s^2), which for our example is $s = 1.11803$ girls.

iii. Poisson distribution

A measure of both the population and sample variance in a Poisson distribution can be estimated by the following equation:

$$\sigma^2 \approx s^2 \approx \frac{\Sigma x^2 y}{\Sigma y} - \left[\frac{\Sigma \gamma x}{\Sigma y}\right]^2$$

Using the terms $\Sigma x^2 y$, Σy, and Σxy from Table 5.10 and 5.4.3iii, the variance and standard deviation are:

$$s^2 \approx \frac{577}{99} - (197/99)^2 = 1.86858 \text{ and } s = 1.3669$$

In data with a Poisson distribution, the mean should equal the variance. (More detailed proof of this relationship can be found in other statistical texts.) For the data in Table 5.3, the mean distance travelled by the *Taxus baccata* seeds was 1.99m. Using the equation to calculate the variance gives us a value of 1.87m. These estimates are similar to each other although not identical. This means that if you believe you have data with a Poisson distribution, you can initially confirm this by comparing the

online resource centre

mean and variance. The definitive test is to 'fit' the Poisson equation to your data. We show you how to do this in principle in 7.1.3 and in practice in the Online Resource Centre.

5.6 Coefficient of variation

The standard deviation is the most commonly used measure of dispersion. If, however, you wish to compare variation in one set of data with that of another set of data, then we use a relative (scaleless) measure of variation called the coefficient of variation (V). The coefficient of variation is especially useful when data you wish to compare are of different orders of magnitude or if they are measured on different scales. V is determined as a ratio between the mean and the standard deviation expressed as a percentage:

Coefficient of variation (V) The ratio between the mean and the standard deviation expressed as a percentage

$$V = \frac{s}{\bar{x}} \times 100$$

The coefficient of variation can only be used for interval data with a known distribution and where a standard deviation can be calculated. In Chapter 9, we consider a survey carried out at Aberystwyth, Wales, UK, in 2002. Periwinkles from the mid- and lower shore were sampled and the height of their shells (mm) was recorded (Table 9.1). The coefficient of variation for the periwinkles from the lower shore was $(1.89477/7.47) \times 100 = 25.36506\%$ and the coefficient of variation for the periwinkles on the mid-shore was $(2.23011/5.46667) \times 100 = 40.07947\%$. This means that there is relatively more variation in the shell height (mm) of the periwinkles sampled from the mid-shore (40.1%) than in those sampled from the lower shore (25.4%).

5.7 Confidence limits

In most research, you do not collect observations from every item in your statistical population; instead, you collect observations from a sample. A useful measure, therefore, is to know how good a predictor of the population parameter your sample statistic is. We illustrate the principles behind these 'confidence limits' in relation to the means of samples, but confidence limits can be calculated for many population parameters including regression coefficients. For more details, see, for example, Sokal & Rohlf (1994).

Confidence limits The values derived from a sample between which the population mean probably falls and is calculated using the standard error of the mean

The confidence limits for a mean are derived from the standard error of the mean. The idea behind this is that, for a given statistical population, you may take many samples and these can all be described in terms of a mean. If you then used these means as your data, you could calculate a standard deviation of these mean values, thus indicating how much the means from the many samples varied. This standard deviation of the mean is more commonly known as the standard error of the mean. The term standard deviation is usually restricted to the standard deviation you calculate from the items in one sample. The **standard error of the mean** (SEM) is:

$$\text{SEM} = \frac{s}{\sqrt{n}}$$

The distribution of these means will tend to be normal (even if the data in each sample are not normal), and so the standard error of the mean has the same relationship that we introduced in 5.5.3i in that there is a 95% probability that the population mean will fall within the range $\overline{x} \pm 1.96\text{SEM}$ and there is a 99% probability that the population mean will fall within the range $\overline{x} \pm 2.58\text{SEM}$. Most often these confidence limits for means are included as vertical bars around mean values on figures (see Fig. 11.4).

If you have a small sample (less than 30), then the variation in the data due to confounding variables (sampling error) will become relatively large and this can have an undue effect on the relationship between probability and the standard error of the mean. Therefore, a factor that recognizes the effect of the small sample size is included in the calculation of these confidence limits for small samples, so that the confidence limits are defined by the range $\overline{x} \pm (t \times \text{SEM})$. The value for t can be found in Appendix d, Table D6. We explain these tables in Chapter 6. To find a particular t value, you need to choose the probability you wish to use in your confidence limits. A probability of 95% will be indicated by $p = 0.05$; a probability of 99% is indicated as $p = 0.01$. You also need to know the degrees of freedom. For these values of t, the degrees of freedom are $n - 1$. Using your chosen p value and the degrees of freedom, you can identify the value for t, which you then use when calculating the confidence limits. Calculating confidence limits in this way is acceptable for interval data even from non-normal or unknown distributions.

Standard error of the mean (SEM) For a given statistical population, you can take many samples, which can all be described in terms of a mean. Using these means as your data, you can calculate a standard deviation of these mean values, more commonly known as the standard error of the mean

 Q5 What is the 95% confidence interval for the mean from Example 5.2 (Table 5.8) for the length (mm) of two-spot ladybirds (*Adalia bipunctata*)?

5.8 **Parametric and non-parametric data**

It is clear that, where data have a particular distribution, we are able to use this to determine how we manipulate the data. Therefore, for data with a known distribution, such as a normal distribution, we are able to calculate a mean, variance, and standard deviation. Data with a normal distribution should be symmetrical and the mean = median = mode. We have also introduced you to the relationship in normally distributed data between the number of observations in a particular region of the distribution, the probability, and the standard deviation, so that, for example, 95% of all observations should be found in the range $\overline{x} \pm 1.96s$. All this information has been utilized in a branch of statistics called parametric statistics, and normally distributed data are often called parametric data.

Parametric data are measured on an interval scale and so are quantitative, continuous, and rankable. The data have to have a normal distribution e.g. height of men (cm) (Table 5.1). When testing hypotheses (Chapter 6), if you have parametric data you should use parametric statistics (Chapters 8 and 9).

Non-parametric data are data where either the distribution is not normal or more usually the distribution is unknown. When testing hypotheses and you have non-parametric data, you may either attempt to normalize the data by transforming them (5.9) or use non-parametric statistics (Chapters 7–9). Parametric data may also be analysed using non-parametric statistics, but not the other way round.

Parametric data Date that have a normal distribution and are measured on an interval scale and so are quantitative, continuous, and rankable

Non-parametric data Data where either the distribution is not normal or, more usually, the distribution is unknown

5.8.1 **Are my data parametric?**

Some data will have a normal distribution when plotted and can be described by a Gaussian equation (5.2). Data of this sort are called parametric. There is a whole branch of statistics that uses the Gaussian equation as its basis. These are parametric statistics (Chapters 8 and 9). Parametric statistics are more powerful and should always be used where possible. Powerful tests are those that are more effective at detecting a real biological difference when one exists. It is therefore important to be able to check whether your data are normally distributed so that you can then use the parametric statistics.

We have already covered many of the key features of a normal distribution. These are summarized in Box 5.2 for ease of cross-referencing from future chapters.

BOX 5.2 How to check whether your data are normally distributed (parametric)

GENERAL DETAILS	EXAMPLE 5.2
To tell whether your data are probably normally distributed and therefore whether you can use parametric statistics, you should check against the first four criteria at least.	This calculation is given in full in the Online Resource Centre. All values have been rounded to five decimal places for reporting.
a. Are the data measured on an interval scale, such as mm or grams, and therefore are quantitative and continuous? Data that are measured on an ordinal or nominal scale will not be parametric. Derived variables recorded on restricted scales such as percentages, where the scale of measurement only ranges between 0% and 100%, are not usually normally distributed. Some derived variables measured on a continuous scale such as rates (e.g. m/s) may be normally distributed.	a. Yes. The scale is mm, which is an interval scale.
b. Does the distribution appear to be a bell-shaped curve (e.g. Fig. 5.3)? This criterion is often not very convincing as you may have few observations in your sample.	b. Using the frequency table (Table 5.9), the data can be plotted as a histogram (Fig. 5.7) There does appear to be a bell shape to the distribution.
c. Do about 68% of your observations fall within the range $\bar{x} \pm 1s$?	c. For this example, the mean and standard deviation are: $$\bar{x} = \frac{254}{50} = 5.08\text{mm}$$ $s = 1.9361228\text{mm (Box 5.1)}$ Therefore: $$\bar{x} + s = 5.08 + 1.9361228$$ $$= 7.01612\text{mm}$$ $$\bar{x} - s = 5.08 - 1.9361228$$ $$= 3.14388\text{mm}$$ If you examine Table 5.8, it is clear that 34/50 observations in this sample (68%) fall within the range 3.1–7.0mm.
d. Does the mean = median = mode? This can be a difficult criterion to use, as it is not clear how much difference there can be between these values before they indicate that the data are not parametric.	d. The mean has already been calculated ($\bar{x} = 5.08\text{mm}$). Median: Arrange all the values in numerical order. $n = 50$ so the median lies halfway between two values of 5.0mm. Therefore, the median is 5.0mm. Mode: If you examine Table 5.8, it is clear that the most frequent value is 5mm. Therefore, the mean, median, and mode are all very similar.

e. A normal distribution may be mathematically described by the Gaussian equation (5.2.1). Using this equation and your own x values, you can calculate the corresponding y values. These are an 'expected' data set and can be compared with your 'observed' data set by a χ^2 goodness-of-fit test.	e. See 7.1.3. We conclude that the data are normally distributed and are therefore parametric.

Further examples relating to the topic including how to use statistical software are included in the Online Resource Centre.

@ online resource centre

Q6 In Q4, we calculated the summary statistics for data from Example 2.3. Use the first four criteria detailed in Box 5.2 to decide whether the data are parametric.

5.8.2 Ranking non-parametric data

When using parametric statistics (Chapters 8 and 9), you will need to know how to calculate a sum of squares, variance, and standard deviation, and we have explained these steps in Box 5.1. If you are using non-parametric statistics, you will usually need to know how to rank data. This is a central skill for most non-parametric statistical tests (Chapters 7, 8, and 10) and we have provided a step-by-step guide in Box 5.3. To rank data, the values are placed in their logical, usually numerical, order and each observation is given a 'ranking' depending on where in this order they fall. These 'ranking' values are then used in the calculation rather than the observations. Examine the data from Example 2.3 on the growth rate of rye seedlings. These interval data can be organized into a numerical order. (You did this to answer Q6 when finding the median.) Ordinal data may also be ranked. For example, the qualitative ACFOR scale used in ecology can be ordered as shown: A – C – F – O – R. However, nominal data cannot be ranked. For example, if you graded petal colours as red, light red, and spotted red, there is no reason for these colours to be organized in any particular order.

5.9 Transforming data

If your data are normally distributed (Box 5.2), then you are able to use the more powerful and generally more flexible parametric statistics to test your hypotheses. If you have data that are not normally distributed,

BOX 5.3 How to rank data for non-parametric statistics

1. If you have one sample

Sample 1:   1 4 5 5 6 3 4 6 5 5 7 9

i.   Arrange these values in numerical order.

Sample 1:   1 3 4 4 5 5 5 5 6 6 7 9

ii. Assign *possible* ranks.

The first number in order is rank 1. The second number in rank order is 2, etc.

Sample 1:	1	3	4	4	5	5	5	5	6	6	7	9
Rank	1	2	3	4	5	6	7	8	9	10	11	12

iii. Where the data has the same value, these observations all take the average rank.

e.g. In this set of data, there are two 4s. In theory, they should be ranked 3 and 4.

Sample 1:	1	3	4	4	5	5	5	5	6	6	7	9
Rank	1	2	3	4								

But since the observations have the same value, they should have the same rank.
The average rank for the 4s is therefore $\dfrac{3+4}{2} = 3.5$
The ranks are now

Sample 1:	1	3	4	4	5	5	5	5	6	6	7	9
Rank	1	2	3.5	3.5								

This process is repeated for the four 5s and the two 6s.

Rank for 5s $= \dfrac{5+6+7+8}{4} = 6.5$

Rank for 6s $= \dfrac{9+10}{2} = 9.5$

The ranks are now

Sample 1:	1	3	4	4	5	5	5	5	6	6	7	9
Rank	1	2	3.5	3.5	6.5	6.5	6.5	6.5	9.5	9.5	11	12

2. If you have more than one sample.

Many statistical tests require you to combine samples when assigning ranks.

e.g. Sample 1   1 3 5 3 2 2 4

Sample 2   1 1 4 6 7 8 7

i.   Arrange all the observations in numerical order.

e.g. Sample 1	1		2	2	3	3		
Sample 2		1	1				4	

e.g. Sample 1	4	5			
Sample 2			6	7	7	8

ii. Assign ranks. Where values are 'tied', assign the average rank.

e.g. Sample 1	1		2	2	3	3		
Sample 2		1	1					4
Rank	2	2	2	4.5	4.5	6.5	6.5	8.5

e.g. Sample 1	4	5				
Sample 2			6	7	7	8
Rank	8.5	10	11	12.5	12.5	14

iii. When the samples are then analysed, they will have the following ranks.

Sample	1	2	2	3	3	4	5
Rank	2	4.5	4.5	6.5	6.5	8.5	10

Sum of ranks $(\Sigma r_1) = 2 + 4.5 + 4.5 + 6.5 + 6.5 + 8.5 + 10 = 42.5$

Sample	2	1	1	4	6	7	7	8
Rank		2	2	8.5	11	12.5	12.5	14

Sum of ranks $(\Sigma r_2) = 2 + 2 + 8.5 + 11 + 12.5 + 12.5 + 14 = 62.5$

These sums of ranks are often used in non-parametric statistical tests, e.g. Mann–Whitney U test (10.1).

you should always consider **transforming** them. This process of transformation squeezes and/or stretches the scale you used when making your measurements so that the distribution takes on the appearance of a bell-shaped curve. Transformations are only appropriate for ordinal or interval data. Selecting which type of transformation to use depends firstly on whether you have observations for one variable or more than one, and secondly on the shape of the actual distribution seen in your original (untransformed) data. The presence of zeros in your data can have an undue influence in some transformations. Therefore, it is common practice, if you have zeros, to add one to all observations before transforming the data.

When transforming data, you first take each observation and transform it (Table 5.11). Next, you should check to confirm that the data have been normalized (Box 5.2.) before using the transformed values in your parametric calculations. When reporting the results from most tests of hypotheses for one variable, you will not need to take further action in that the test statistic (this term is explained in 6.2.1) is usually independent of the original scale of measurement. You report the original untransformed data and the results from the analysis on the transformed data (Chapter 11).

Transforming two variables is not necessarily complex; however, interpreting the results from the analysis of transformed data following, for example, a regression analysis, can be complex and is outside the scope of this book. We therefore suggest that if you are considering the transformation of data where you have two treatment variables, you should refer to other texts.

When you are considering transforming the data from one treatment variable, the first step is to plot your data and examine the original distribution. If the data are not normally distributed, then the distribution will have one or more of the following features: the data are very spread out and the distribution may be platykurtic (Fig. 5.11); there is little variation and the distribution may appear to be leptokurtic (Fig. 5.10); the distribution is skewed to the right (Fig. 5.8); the distribution is skewed to the left (Fig. 5.9). Table 5.11 indicates which transformation is most likely to be effective at normalizing your data. Most statistical calculators and computer software will carry out these transformations.

Transforming data A mathematical calculation that has the effect of squeezing and/or stretching the scale you used when making your measurements so that the distribution takes on an alternative shape, usually a normal distribution

Table 5.11. How do I transform my data: one variable?

Type of distribution	Other features	Transformation
There is a lot of variation in the data so that the distribution appears to be spread out and may appear to be platykurtic.	The original measurements are on a scale where the values are greater than 10.	Take the $\log_{10}$ of each observation.
	Some of the original values are 0.	Add 1 to each observation and then take the $\log_{10}$ of each observation.
The variation is limited and the distribution is therefore narrow and may appear to be leptokurtic.	The observations are recorded as a percentage and there is a wide spread of values.	Divide each value by 100. Take the square root of each observation. Then find the angle whose sine equals each value or 'inverse sine' ($\sin^{-1}$).
	None of the original measurements is 0.	Take the square root of each observation.
	Some of the original values are 0.	Add 1 to each observation and then take the square root of each observation.
	The observations are recorded as percentages calculated from counts with a common denominator (e.g. 1/50, 5/50) and most values lie between 0% and 20%.	Take the square root of each observation.
	The observations are recorded as percentages calculated from counts with a common denominator (e.g. 45/50, 49/50) and most values lie between 80% and 100%.	Subtract each observation from 100 and then take the square root of each of these values.
	The data is recorded as a proportion.	Take the square root of each observation. Then find the angle whose sine equals each value or 'inverse sine' ($\sin^{-1}$).
The distribution is skewed to the right (positive).	There are a number of suitable transformations. The one to use depends on the degree of skewness. The following is in the order of little skew (try a) to considerable skew (try c or d). If you have zeros in your original data, you should add 1 to each observation before transforming the data.	a. Take the square root of each observation. b. Take the natural logarithm (ln) of each observation. c. Divide 1 by each observation (reciprocal, $1/x$). d. Take the square root of an observation and then divide 1 by this value (reciprocal square root $1/\sqrt{x}$).
The distribution is skewed to the left (negative).	Again, there are a number of transformations that may be used depending on the degree of skew from least (try a) to considerable (try b, etc.).	a. Square each observation. b. Cube each observation, etc.

Summary of Chapter 5

- When designing an investigation or choosing a statistical test, you will need to be familiar with the terms used to describe scales of measurement. These terms are quantitative, qualitative, discrete, continuous, rankable, nominal, ordinal, interval, and derived variable (5.1).

- Observations have a particular shape or distribution when plotted. Some of these distributions can be described mathematically including the normal distribution, Poisson distribution, and binomial distribution (5.2).

- Distributions may be described in terms of the central point (mean, median, mode, skew, and kurtosis) and spread of data around the central point (range, interquartile range, percentile, variance, and standard deviation). The methods used depend on whether the data are measured on a nominal, ordinal, or interval scale (5.3, 5.4, 5.5, 5.7).

- The coefficient of variation can be used to compare the relative amount of variation in different data (5.6).

- Data may be normally distributed and these data are also known as parametric data (Box 5.2). If data are not normally distributed or the distribution is unknown, then the data are non-parametric (5.8 and 5.9 and developed in Chapters 7–10).

- Non-parametric data may be normalized by transforming them. The choice of transformation depends on the original shape of the distribution and on how many variables need to be transformed (5.9).

- When using parametric statistics, you often need to know how to calculate the sums of squares, variance, and standard deviation (Box 5.1). If you are using non-parametric statistics, you will need to know how to rank data (Box 5.3).

- The Online Resource Centre includes interactive exercises that test your understanding of this chapter with other topics, particularly those considered in Chapters 7–11.

 online resource centre

Answers to chapter questions

 A1
1) Quantitative, discrete, rankable, ordinal.

2) Quantitative, continuous, rankable, derived variable.

3) Quantitative, continuous, rankable, derived variable.

4) Quantitative, continuous, rankable, interval.

A2 $\Sigma mf = 15003.705$, $n = 87$, $\bar{x} = 172.45638$cm. The mode $= 172.45$cm. See the Online Resource Centre for further details.

 online resource centre

A3 In 5.4.3i, we calculated the mean $(\bar{x}) = 5.08$mm. To calculate the median, arrange all the values in numerical order. The mid-position lies halfway between two values of 5.0mm. Therefore, the median is 5.0mm. If you examine Table 5.9, it is clear that the most frequent value is 5.0mm. This is the mode. Therefore, the mean almost equals the median and the mode.

A4 The data are measured on an interval scale (cm), but there are few observations and it will be difficult to establish the underlying distribution. The data have been organized into a frequency table in Chapter 11 (Table 11.4). This indicates that the data are skewed and are unlikely to be normally distributed. Therefore, the data can be summarized using the median, mode, and a range.

The data can be written in numerical order:

10.0 (×2) 11.0 11.5 (×2) 12.0(×2) 14.0 14.5(×2) 15.5 18.0 19.0 19.5 21.0

There are 15 observations in total; therefore, the median is the eighth observation, which is 14.0cm.

If the raw data are considered, there are four values (10.0cm, 11.5cm, 12.0cm, and 14.5cm) that occur twice and are therefore modes.

If the data are organized into a frequency table (Table 11.4), the modal class is 10.0–11.9cm, with mid-point 10.95cm.

The range is between 10cm and 21cm = 11cm.

The interquartile range is 11.5 – 18.0cm = 6.5cm.

A5 There are 50 observations in this sample; therefore, we can calculate the confidence limits as $\overline{x} \pm 1.96\text{SEM}$ where $\text{SEM} = 1.93612/\sqrt{50} = 0.27381$ and the 95% confidence limits are 4.5–5.6mm.

A6
a. Yes. The observations are measured on an interval scale.

b. No. When plotted, the data does not have a bell shape to the distribution.

c. No. The mean (14.26cm) is approximately the same as the median (14.0cm), but these differ from the mid-point of the modal class (10.95cm).

d. No. The range determined by $\overline{x} \pm s$ is 10.6–17.9mm. Only 60% of observations fall within this range.

Using these criteria, the data (Example 2.3) do not appear to be parametric. You could confirm this using criterion e in Box 5.2. If you would like to see the calculations in full, they are given in the Online Resource Centre.

An introduction to hypothesis testing

6

In a nutshell

In this chapter, we consider the principles behind hypothesis testing and introduce you to a number of key terms such as sampling error, p values, and degrees of freedom. In the chapters on statistical testing (Chapters 7–10), we include a number of boxes with examples of the method for a particular test. These boxes all follow the same five steps: (i) what are the hypotheses to be tested?; (ii) work out the calculated value of the test statistic; (iii) find the critical value of the test statistic; (iv) the rule; and (v) what does this mean in real terms? We explain these steps here.

In this book so far, we have thought about general principles relating to the planning of an experiment (Chapters 1–3) and introduced you to the ways in which you might start to handle raw data (Chapter 4). What we have not done is shown you how to answer the questions your experiment may have set out to examine. This is what hypothesis testing is all about. There are generally five steps that you follow in all hypothesis testing:

1) What are the hypotheses to be tested?

2) Work out the calculated value of the test statistic.

3) Find the critical value of the test statistic.

4) The rule.

5) What does this mean in real terms?

These are the headings we use throughout our worked examples for all statistical tests in this book. We explain the underlying principles for each of these steps here. If you work through this chapter it should only take one hour to read and complete all the exercises. The answers for these exercises are at the end of the chapter. All our examples are based on real undergraduate research projects. If these examples are not in your subject area, you will find more in the Online Resource Centre.

online
resource
centre

6.1 **What are the hypotheses to be tested?**

Key points In an experiment, you may test one of three types of hypothesis: (i) Do the data fit expected values? (ii) Is there an association between two or more variables? (iii) Do samples come from the same or different populations? Hypothesis testing seeks to discriminate between a null and an alternative hypothesis. In the null hypothesis, the variation is said to be due to sampling error, whilst in the alternative hypothesis, the variation in observations is due to the factors under investigation.

In this chapter, we are thinking about experiments that are designed to answer a specific question; for example, is there a difference in mineral content between organic and non-organic potatoes? Some investigations, often those producing baseline data, or a preliminary investigation, will not set out to test a hypothesis. For example, a survey of the flora and fauna of a wood is not 'asking' a specific question: you will be finding out what is there. The data from this type of investigation are usually evaluated using analysis based on information-theoretic models (1.8). They may generate hypotheses at the end of the evaluation, but they do not start out with one. When you are carrying out an experiment where you do wish to test a hypothesis, your first step is to decide which type of hypothesis you are testing.

6.1.1 **Types of hypothesis**

There are three types of hypothesis you may use:

- Do the data fit expected values?
- Is there an association between two or more variables?
- Do samples come from the same or different populations?

i. Do the data fit expected values?

There are some occasions when we have reasons to make a mathematical prediction about our observations and we wish to compare our observations with our expectations. This can arise when:

- You can reasonably argue that all samples or observations should have the same value; for example, if you sampled the numbers of beetles falling into different-coloured pitfall traps, you might expect that there should be the same number of beetles collected in each trap
- You have carried out a genetic cross or sampled from a population and have reason to expect a particular segregation ratio, population

genetic ratio, or sex ratio; for example, you may use the Hardy–Weinberg theorem to predict the allele frequencies in the F_1 generation of a population of *Drosophila melanogaster* exposed to a known selection pressure

• You wish to confirm that your data have a particular distribution, such as a normal or a Poisson distribution (5.2).

The term expectation can be used in two ways, only one of which is correct. You may have a personal hunch ('expectation') about the likely outcome, but this is not a mathematical prediction, also called an 'expectation'. When we talk about expectations or expected values, it is the second of these that we mean.

> **Expected values** Values calculated using a mathematical or biological model against which the observations can be compared

Chapter 7 introduces you to chi-squared tests and *G*-tests, which can be used to test this type of hypothesis.

ii. Is there an association between two or more variables?

Two types of analysis may be carried out in relation to a possible association.

> **Association** Where one variable changes in a similar manner to another variable for unknown reasons

a. The presence or absence of an association   Typical examples might be: Is there an association between blood group and eye colour in people living in north Wales? Is there an association between temperature and the growth rate of bacterial colonies in liquid culture?

When you wish to test for the presence of an association, there are two groups of tests that may be useful: a chi-squared test or a *G*-test (Chapter 7), and a correlation or regression (Chapter 8). There is some overlap between these two groups of tests, so you should look at both chapters before deciding which test is the most appropriate.

b. The nature of the association   Some statistical analyses also allow an association to be described in a mathematical way: for example, in fitting a straight line to data and confirming the linear relationship between flower number and bulbil number in *Allium ampeloprasum* var. *porrum* (leek).

If you wish to examine the nature of an association in this way, then go to Chapter 8 and look at regression analyses.

iii. Do samples come from the same or a different population?

The word 'population' used here refers to a '**statistical population**'. Usually, you will not collect data from all items in the population. Instead, the population will be represented by a sample. Many experiments are

> **Statistical population** All the individual items that are the subject of your research. A statistical population may be the same as an ecological or genetic population but not necessarily

designed so that you can compare samples from two or more populations. In some cases, these will be samples from different ecological or genetic populations. In other cases, you may wish to compare different treatments. For example, you may wish to examine the effectiveness of different bactericide treatments on a number of bacteria, or you may wish to compare phosphate concentrations in water samples collected above and below an outfall.

There are many tests that address this type of hypothesis. Your choice depends on the numbers of observations in each category or sample, the number of categories or samples, whether the data are parametric or not, and the simplicity or complexity of the experimental design. Examples are shown in the table below.

General criteria	Parametric	Non-parametric
One variable Two samples Unmatched data	Chapter 9 t-test z-test	Chapter 10 Mann–Whitney U test
One variable Two samples Matched pairs of data	Chapter 9 t-test and z-test for matched pairs	Chapter 10 Wilcoxon's matched pairs test
One or more variables One or more samples With or without replicates	Chapter 9 Parametric analysis of variance (ANOVA) and Tukey's test	Chapter 10 Non-parametric analysis of variance (ANOVA), Kruskal–Wallis test, Sheirer–Ray–Hare test Chapter 7 Chi-squared test for heterogeneity

To help you choose which of these tests is appropriate, you should check the details in the relevant chapters. We have also included examples of choosing statistical tests and how these inform your experimental design in 2.2.7 and 3.5, with further examples in the Online Resource Centre and a review in Appendix c.

online resource centre

....

Q1 If you designed an experiment to find out whether there is a difference in mineral content between organic and non-organic potatoes, which type of hypothesis would you be testing?

....

Hypotheses may be either general or specific. For example, a **general hypothesis** might be 'There is no difference between the means of samples A, B, and C'. Clearly any outcome from the hypothesis testing may tell you if there is probably a difference, but will not tell you anything more about

General hypothesis This is phrased in a general way; for example, there is no difference between the samples

that difference. Some hypothesis testing is more **specific**. For example, when using a chi-squared or *G* goodness-of-fit test, you compare your observations to a specific ratio determined by biological principles, such as Mendelian laws of segregation or the Gaussian equation describing a normal distribution (7.1). The test of hypotheses here is more specific, asking whether your observed values are significantly different from this *a priori* ratio. Similarly, in simple linear regression (8.5), you may test the significance of an association between two variables by comparing the observed data with that predicted by a regression equation. In Chapters 9 and 10, there are statistical tests that allow you to make other specific comparisons. Tukey's test (9.6) allows you to compare pairs of means, in experiments with more than two samples, enabling you to gain a more detailed insight into the trends in your data. A similar test is considered in Chapter 10 for non-parametric data.

Specific hypothesis Some statistical tests allow you to ask more discerning (specific) questions. Specific tests of hypotheses are usually one-tailed tests

a priori Before the event

6.1.2 **How to write hypotheses**

A **hypothesis** is the formal phrasing of your objective (1.1 and 1.8). You may have general or specific hypotheses. A hypothesis when written correctly should have a number of features, as the following checklist shows.

Hypothesis The formal phrasing of each objective, including details relating to the experiment and the way in which the data will be tested statistically

1) Hypotheses come in pairs, the null hypothesis (H_0) and an alternate hypothesis (H_1). The null hypothesis (H_0) is the negative hypothesis. The null hypothesis assumes that any variation in the data is due to sampling error. We explain what is meant by sampling error in 6.1.3. The alternate hypothesis (H_1) is worded in such a way that if H_0 is true, then H_1 must be false. The alternate hypothesis indicates that the variation in the data is probably due to the factor(s) you are investigating in your experiment.

2) For a general hypothesis, the phrasing indicates the type of hypothesis being tested, i.e. 'expectation', 'association', or 'difference', and/or uses a term relating to the test itself, e.g. 'heterogeneity', or 'correlation'. A specific hypothesis includes details of the particular trend you believe is present in the data.

3) The hypothesis being tested relates to the statistical population under investigation rather than the sample (1.3, 1.4).

4) Hypotheses usually indicate the statistic being examined. For example, most parametric statistical tests of hypotheses are written in terms of the means of the data. When non-parametric tests of hypotheses are used, these refer to the medians.

5) Hypotheses include details of the experiment, such as the number of samples or items, the variable(s), and the species being studied.

EXAMPLE 6.1 Germination of *Allium schoenoprasum* on three soil types

A researcher was interested in the effect of soil types on seed germination in the chive (*Allium schoenoprasum*). She sowed 100 seeds on each of three soil types and after two weeks recorded the number of seeds that had germinated. She expected that the number of seeds germinating on the three soils would be the same. In this instance, the hypotheses would be written as:

H_0: There is no difference in the median numbers of seeds from *Allium schoenoprasum* germinating on the three soil types compared with the expected value.

H_1: There is a difference in the median numbers of seeds of *Allium schoenoprasum* germinating on the three soil types compared with the expected value.

We can compare these hypotheses with our checklist.

Checklist	Example 6.1
1. Hypotheses come in pairs: the null hypothesis (H_0) and an alternate hypothesis (H_1).	Yes, there is a null and alternative hypothesis. The null hypothesis (H_0) is the negative hypothesis: 'There is no difference ...'. The alternate hypothesis (H_1) is worded in such a way that, if H_0 is true, then H_1 must be false: 'There *is* a difference ...'.
2. The phrasing of a hypothesis indicates the type of hypothesis being tested, i.e. 'expectation', 'association', or 'difference', and/or uses a term relating to the test itself, e.g. heterogeneity, correlation.	H_0: '... compared with the *expected value*.'
3. The hypothesis being tested refers to the statistical population under investigation rather than the sample.	In this example, the statistical population is the seed produced by *Allium schoenoprasum*.
4. Hypotheses usually indicate the statistic being examined.	'... difference in *the median numbers* ...'
5. Hypotheses include details of the experiment.	'... seeds of *Allium schoenoprasum* germinating on the three soil types ...'

 EXAMPLE 7.1 The distribution of holly leaf miners on *Ilex aquifolia* (from Chapter 7)

As part of a student project, the number of holly leaf miners was recorded at three heights on a holly tree. The student assumed that the holly leaf miners were distributed at random and so should be found in equal numbers in each part of the tree.

Write suitable hypotheses for this investigation.

6.1.3 **Why hypotheses come in pairs**

To understand the process of hypothesis testing, you need to understand what is meant by the term sampling error. When you record a series of observations these will vary. Sampling error is a measure of the variation in your data that is due either to chance and/or to confounding variables (1.7).

i. Sampling error and chance

One cause of sampling error in your data is 'chance'. We can illustrate this best with an example.

EXAMPLE 6.2 Sampling error in a study of *Cepaea nemoralis* in Jack's Wood

As part of a study of the ecological genetics of *Cepaea nemoralis*, the shell patterns of the snails in a wood were recorded. Every snail in the area (i.e. the whole population) was examined and their shell patterns recorded. Two days later, three samples were collected at random and their shell patterns were also recorded (Table 6.1).

Table 6.1. Percentage of *Cepaea nemoralis* with particular shell patterns in Jack's Wood

	Cepaea nemoralis with a particular shell pattern and colour (%)			
	Banded Yellow	No bands Yellow	Banded Pink	No bands Pink
Sample 1	90	0	5	5
Sample 2	20	5	30	45
Sample 3	0	50	50	0
Population	30	10	30	30

Clearly the samples differ from the population results and from each other. This variation is said to be due to sampling error. By chance, the samples collected did not reflect the actual population frequencies of shell pattern in this species. Two important effects of sampling error are firstly that the samples may not reflect the statistical population values, and secondly that the sampling error can lead to variation between samples by chance.

ii. Sampling error and experiments

Sampling error can also occur in your data as a result of factors that are not part of your experiment (i.e. confounding variables), but which also affect the characteristic you are measuring.

In Chapter 2, we designed an experiment to allow us to examine the maximum distance a seed was dispersed in a wind tunnel in relation to wind speed. We identified a number of confounding variables (2.2.4), including the effects of digging up the plants, the maturity of the seeds and inflorescence, the height of the inflorescence, the temperature and humidity of the wind tunnel, and the location of the plants during testing. All these factors could influence how far the seeds were dispersed. For example, when the plants were dug up, this may affect the plants' turgor, which in turn might affect the adherence of the seed to the inflorescence; if the seeds were not mature, their weight may vary, and the pappus on the seed might not be so effective at providing lift, etc. All these factors can influence the distance a seed might travel and therefore introduce a level of variation into the experiment that has nothing to do with wind speed. These all contribute to the sampling error, which is why you need to take steps when designing your investigation to minimize and/or equalize their impact on your observations (2.2.4).

iii. What does the term 'sampling error' mean?

Sampling error The variation in a set of data that is due to the effects of confounding variables and/or chance

The term '**sampling error**' can therefore mean one of three things:

- Variation between samples collected from a single population that has occurred by chance
- The variation between the population value and sample value that has arisen by chance
- Some variation in the experimental observations that may be due to factors not being investigated by the researcher.

When you collect your data, the variation between the observations will be due in part to one or more of these causes of sampling error. Thus, in an experiment, you have variation in your data due to the factor(s) you are investigating (treatments) and due to sampling error.

iv. Sampling error and hypotheses

This explains one aspect of hypotheses. They come in pairs: a null hypothesis and the alternate hypothesis. The null hypothesis describes the variation seen in the data in terms of sampling error. The alternate hypothesis describes the variation seen in your data in terms of the variable(s) you are investigating.

EXAMPLE 6.3 Attitudes to human cloning

An undergraduate investigated attitudes of students to different forms of human cloning. One question asked students taking a Biology degree and students taking a Health Studies degree if they agreed with somatic cell human cloning. Her hypotheses were:

H_0: There is no association between the median numbers of Biology and Health Studies students in their response to this question.

H_1: There is an association between the median numbers of Biology and Health Studies students in their response to this question.

In this example, the factor under investigation is the education of the students. Are the students taking one course likely to give a different response compared with students on the other course? However, any variation in the data may have arisen due to sampling error. Perhaps the students on the Biology course all happen to be mature students and this influences their opinion. This is a confounding variable and therefore a contributor to the sampling error.

v. Hypothesis testing

In Example 6.1, the hypotheses were:

H_0: There is no difference in the median percentage germination of *Allium schoenoprasum* seeds on the three soil types.

H_1: There is a difference in the median percentage germination of *Allium schoenoprasum* seeds on the three soil types.

The null hypothesis states that there is 'no difference'. In other words, there should be little or no variation between the percentage germination of the seeds sown on the three soil types other than that due to sampling error. If we work out how much the three samples differ from each other, it may be very little, in which case we can probably agree with our null hypothesis. If the three samples are very different from each other, then we can probably say that the null hypothesis should be rejected. But how

much variation between our three samples do we need before we can say that we should probably reject our null hypothesis?

Hypothesis testing allows us to examine our data and to use them to choose one of the two hypotheses: is the variation in our data due to sampling error (H_0) or is the variation in our data due to the factor(s) we are investigating (treatment) (H_1).

6.2 Working out the calculated value of the test statistic

Key points The calculated value of a test statistic is derived from your observations and is calculated using an equation derived from a specific distribution such as a normal distribution or the t distribution.

6.2.1 What is a calculated value or test statistic?

To be able to choose between the null and alternate hypotheses, the observations from an experiment are 'summarized' to produce a single value, which we refer to as a 'calculated value' or 'test statistic'.

i. The distributions

To work out a calculated value, you use a mathematical relationship that is derived from a particular distribution. This is similar to the way in which it is possible to calculate the single value for a variance from a number of observations for data that have a normal distribution, and to calculate this variance you use an equation based on characteristics of the normal distribution.

In Box 7.1, we show you how to calculate the single $\chi^2_{calculated}$ value from the observations in Example 7.1, and in Box 9.3, we show you how to work out the $t_{calculated}$ value for Example 9.2. These use mathematical equations that are derived from characteristics of the chi-squared and t distribution, respectively.

ii. Criteria for statistical tests

This use of a number of different distributions and their mathematical characteristics is why when choosing a statistical test you first check your data to see whether they meet the criteria that are important for a given distribution (e.g. parametric data meet the criteria for using equations based on a normal distribution, data that meet the criteria of chi-squared can use equations based on the chi-squared distribution, etc).

If you look at any statistical test in Chapters 7–10, you can see that each has a set of criteria. For example, if you wish to use a chi-squared goodness-of-fit test (7.1.1), you:

1) Wish to compare your observed values with those predicted by an *a priori* expectation.

2) Have one treatment variable.

3) Have only one sample.

4) Have data that fall into more than two categories.

5) Have data that are counts or frequencies and are not percentages or proportions.

6) Have observations that are independent.

7) Have expected values that are more than five.

If the data from your investigation satisfy the criteria, then the data may be examined in relation to a particular distribution, in this case the chi-squared distribution.

There are many examples in subsequent chapters in which we show you how to work out a calculated value. It will help your understanding if you look briefly at several of these now (for example Box 7.1 and Box 8.1) before you continue reading about hypothesis testing.

iii. Name of the test statistic

The test statistic is usually given a letter or symbol that relates to the name of the distribution or statistical test you are using. If you were using a Mann–Whitney U test (Box 10.1), the test statistic is U. If you are using a chi-squared test, then the test statistic is χ^2 (e.g. Box 7.5).

6.3 Finding the critical value of the test statistic

Key points The critical values can be located in Appendix d or are provided as an outcome when using statistical software. There are a number of parameters that you may need before you can select the correct critical value. These parameters include the degrees of freedom, the p value, and whether you are carrying out a one- or two-tailed test. For an explanation of the *power of a test*, you should refer to the glossary.

To find a critical value, you first need to understand what a critical value is (6.3.1). Most critical values are calculated for you, and there are many possibilities. If the values are not calculated, you will need to interpolate your value from those provided (6.3.6). If the value is provided,

to find the correct critical value you need to select a p value (6.3.2) and to know whether you are testing two tails or one tail of the distribution (6.3.3). You may also need other values, such as the degrees of freedom (6.3.4).

6.3.1 What is the critical value?

The critical value is also found from the mathematical relationships that are derived from the distribution you have chosen to use to test your hypotheses. It is the boundary mark against which you compare the calculated value. In most statistical tests, if your calculated value is greater than your critical value, then you may reject the null hypothesis (e.g. F-test, t-test, χ^2 test). In the Mann–Whitney U test (10.1), if the calculated value is smaller than the critical value, then you may reject the null hypothesis. Critical values have been calculated for you from the distribution you are using to test your hypotheses. Statistical tables with these critical values are included at the end of this book (Appendix d).

There is a plethora of critical values derived from any particular test distribution. You have to choose one of these critical values and to do this you will usually need a probability value (p) and a measure of sample size, which is most often in terms of the degrees of freedom, and you will need to decide whether you wish to use a one- or two-tailed test. We explain these three criteria in this section.

6.3.2 p values

Look at the critical values in the statistical table in Appendix d for the Spearman's rank test and at the excerpt in Table 6.2.

Table 6.2. Excerpt from the critical values for a Spearman rank test

	Critical values for a two-tailed* test for probability (p) values 0.10–0.01			
	0.10	0.05	0.02	0.01
$n = 8$	0.643	0.738	0.833	0.881
$n = 9$	0.600	0.683	0.783	0.833
$n = 10$ (etc.)	0.564	0.648	0.745	0.794

*The terms one-tailed and two-tailed are explained in 6.3.3.

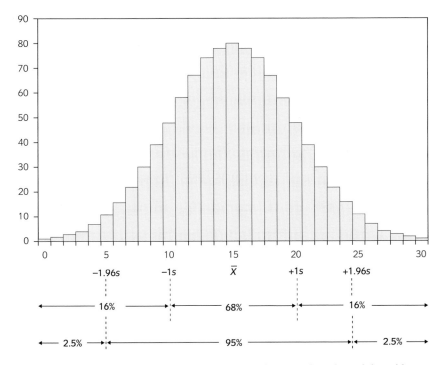

Fig. 6.1. Normal distribution ($n = 1000$, $\bar{x} = 15$, and $s = 5$). The coloured dotted lines demarcate the number of observations that would be expected to fall into the centre of the distribution and the two tails, when the range is defined by $\bar{x} \pm s$. The black dotted lines demarcate the number of observations that would be expected to fall into the centre of the distribution and the two tails, when the range is defined by $\bar{x} \pm 1.96s$.

The first column is n, the number of pairs of observations. The remaining columns are titled 'probability' and there are several values. What is this probability all about?

i. Distributions and p values

As an example, let us first consider one large sample of 1,000 observations measured on a continuous scale with units from 0 to 30. The frequency distribution of this sample is shown in Fig. 6.1. These data have a normal distribution and can therefore be summarized using the mean and standard deviation (Box 5.1).

In this first sample, you can see that most observations are central values. There are far fewer observations at the more extreme ends (e.g. 0–5 units and 25–30 units).

In fact, as these data are normally distributed, we would expect that about 68% of observations would have a value that falls between the mean ($\bar{x}$) minus 1 standard deviation (s) and the mean ($\bar{x}$) plus 1 standard

Table 6.3. Attributes of a normal distribution in relation to the sizes of the tails from $\bar{x} \pm 1.00s$ to $\bar{x} \pm 3.29s$

	Size of tails			
a. Size of the tails are determined by …	$\bar{x} \pm 1.00s$	$\bar{x} \pm 1.96s$	$\bar{x} \pm 2.58s$	$\bar{x} \pm 3.29s$
b. Number of observations in central range (%)	68	95	99	99.9
c. Number of observations in the two tails combined (%)	32	5	1	0.1
d. Number of observations in each tail (%)	16	2.5	0.5	0.05
e. Probability of any one observation falling within the central zone	$p = 0.68$	$p = 0.95$	$p = 0.99$	$p = 0.999$
f. Probability of any observation falling in either one of the two tails	$p = 0.32$	$p = 0.05$	$p = 0.01$	$p = 0.001$
g. Number of times on average where you might expect to be wrong	1/3	1/20	1/100	1/1000

deviation (s) (Table 6.3, rows a and b). We have looked at this before, in 5.2 and 5.5.

This relationship can be turned on its head and you could instead say that you would expect to find 32% of your observations being values that fall into one or both of the two, more extreme, ends of the distribution, i.e. the tails (Fig. 6.1). In this example, you would expect to find half of the 32% of observations in the tail at the left-hand end of the distribution and half of the 32% of observations in the right-hand tail. So each tail would have 16% of your observations and the central part would have 68% of your observations (Table 6.3, rows c and d).

You can rephrase this in yet another way. If you had 100 observations and you took one at random, then you would expect that the probability of this one observation having a value that falls in the range mean ±1 standard deviation is 68/100 or $p = 0.68$; the probability that this one observation will have a value that falls outside the range mean ±1 standard deviation is 32/100 or $p = 0.32$ (Table 6.1, rows d–f). We are now thinking about probabilities. Hypothesis testing uses this idea to allow you to select the level of probability at which you choose whether to reject or not reject the null hypothesis.

 Q3   Examine some of the data from Table 9.2. The mean ($\bar{x}$) and the standard deviation (s) of the periwinkles from the lower shore are $\bar{x}_1 = 5.89231$ and $s_1 = 1.4186$; $n = 13$.

a. How many observations fall within the range $\bar{x} \pm s$, i.e. 4.5–7.3 mm?

b. How many observations fall outside this range?

Now divide the data so that 95% of your observations are included in the central block and only 5% of the observations are included in the tails.

Extract from Table 9.2. Shell height in periwinkles from the lower shore at Porthcawl, 2002

Shell height (mm) for sample of periwinkles on the lower shore ($n = 13$)			
4.0	5.0	6.0	8.4
4.8	5.0	6.2	
5.0	5.5	7.7	
5.0	5.6	8.4	

c. How many observations would be included in the central block and how many would be in the tails?

d. What is the probability that the next periwinkle you measured also falls within these tails?

ii. Hypotheses and p values

We have described the use of p values for a single set of data in 6.3.2i (Fig. 6.1). What, then, if we had a second sample and we wished to test the hypothesis that both samples came from the same population, i.e. there is no difference between the population means for the two samples? If we plotted the data from our second sample on top of the first sample (Fig. 6.2), we would expect there to be an overlap if they came from the same population. We would need to make some allowance for sampling error, so we would not expect a perfect match, but we would expect a reasonable match. If the samples were from different populations, we would expect the overlap between them would be less. If the two samples were very different, then you would reach a point where the observations in sample 2 would only fall into one of the extreme ends of the first samples' distribution. The point at which this lack of overlap is considered to be sufficiently significant as to indicate that the samples do not come from the same population is given in terms of p values.

iii. Which p value?

You therefore have to decide where the cut-off point will be. How small an overlap should there be between samples for you to reject the null hypothesis and to decide that the two samples probably come from different populations? Usually we use a p value of 0.05 or less. A probability of $p = 0.05$ is an arbitrary cut-off point that has been adopted in biological sciences. In other disciplines, such as medical science, a lower p value may be used. With a p value of 0.05, we would expect that 1/20 times the decision we come to will be incorrect (Table 6.3, row f). Making wrong decisions in hypothesis testing in this way is called a Type I error.

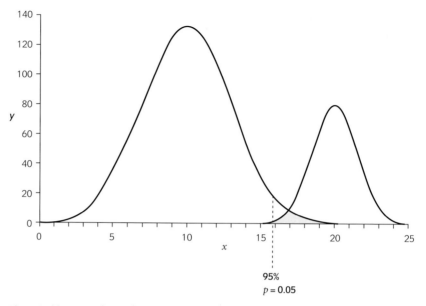

Fig. 6.2. Two samples with a normal distribution ($n = 1000$, $x = 10$, $s = 3$; $n = 300$, $\bar{x} = 20$, $s = 1.5$). The two distributions overlap in the upper tail of the larger distribution.

Power of a test An indicator of how effective a statistical test is in detecting a difference resulting from the variable(s) under examination

We explain more about Type I and Type II errors in the glossary under 'Power of a test'.

If when testing your hypotheses you find that you do reject the null hypothesis at $p = 0.05$, you should see how small a p value you can adopt and still reject the null hypothesis. For example, if you can reject the null hypothesis at $p = 0.001$, then you can expect your decision probably to be incorrect only 1/1000 times (Table 6.3, row f). At this level of stringency, you could carry out 1000 experiments and expect probably to draw the wrong conclusion in only one of them. Therefore, making a decision in hypothesis testing at a low p value increases the likelihood that the decision is correct (e.g. see Boxes 7.1 and 8.1). You may think, then, that a p value of 0 is the ideal. However, no hypothesis testing will set $p = 0.0$ as a cut-off point for rejecting or not rejecting the null hypothesis, because our experimental systems are never without some inaccuracies and you can never be absolutely sure that the decision you have made is correct. Where you do use a p value of less than 0.05, you may wish to indicate this either by giving the exact p value, such as $p = 0.01$, or by using the 'less than' symbol, i.e. $p < 0.05$. More details about these symbols are given in Appendix e.

iv. Critical values and p

The process of hypothesis testing starts, then, with you choosing a particular distribution e.g. chi-squared, F, t, z, etc. This distribution will

have particular mathematical features that you use to summarize all your observations and determine a calculated value. The distribution you have chosen is then used to determine the critical value. This value is also dependent on the p value you have chosen for your decision-making. Therefore, there is a series of critical values for any one distribution for a range of p values. These are given in the statistical tables at the end of the book (Appendix d). You should start with the critical value for $p = 0.05$. We have given examples of choosing critical values throughout Chapters 7–10 (e.g. Box 7.1).

When using statistical software such as those we cover in the Online Resource Centre, the computer will show you the exact p value as one of the outcomes from the analysis. This allows you to be more exact in your reporting of significant results.

online resource centre

6.3.3 **One-tailed and two-tailed tests**

In the table of critical values for the Spearman's rank test, you will see the terms 'one-tailed' and 'two-tailed' tests. These terms are most easily explained in terms of a normal distribution, but apply to any statistics that assume any underlying distribution.

If you were carrying out hypothesis testing in which you were comparing two or more samples, the statistical test would provide an indication of the overlap between the samples and allow you to determine whether there is sufficient overlap between the samples to indicate that they come from the same statistical population (Fig. 6.2).

You will remember that there is a mathematical relationship between the mean, standard deviation, and number of observations (6.3.2i) for a normal distribution so that 95% of your observations should fall within the range of the mean $\pm 1.96s$ (Table 6.1). This means that 5% of your observations will fall outside this range. These extreme values may either fall evenly, 2.5% at each end (Fig. 6.3), or 5% in one or other of the tails only (Fig. 6.4).

Imagine that you had two sets of data. In one set, you had the weights of 250 adult female mice. In the other data set, you had another ten observations, but the gender and age of these mice were unknown. You wish to test the hypothesis H_0: there is no difference between the mean weights (g) of the female mice and the unknown mice.

In the absence of any information about these ten mice, the statistical test used would check to see whether the data overlapped at the top or the bottom end of the distribution. This is a **two-tailed test** and is the most common form of statistical test.

However, you may know that, if the mice do not belong to the adult female population, then they will be adult males and belong to the adult male population, which has a different weight distribution with a higher mean.

Two-tailed test Where the hypothesis testing is against both ends of a distribution. Most general tests of hypotheses are usually two-tailed

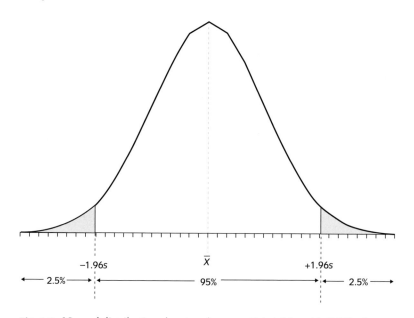

Fig. 6.3. Normal distribution showing the range $\bar{x} \pm 1.96s$, with 2.5% of observations in each tail excluded from this range.

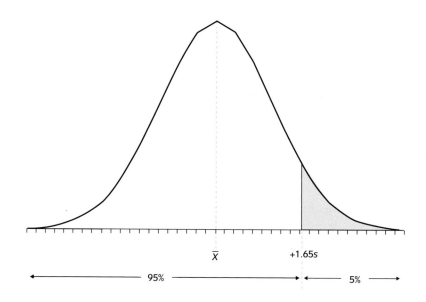

Fig. 6.4. Normal distribution showing the range $\bar{x} \pm 1.96s$, with 5% of observations in the upper tail excluded from this range.

You could then use a one-tailed test that checks for an overlap against only the top end of the distribution (Fig. 6.2). Some statistical tests are by their nature one-tailed tests, e.g. when testing specific hypotheses using a non-parametric analysis of variance. However, most of the time, we do not know enough about our investigative system to be able to predict which tail our second data set should fall into. Thus, you should normally use a two-tailed test.

One-tailed test When the hypothesis testing is such that only one end of a distribution is being considered. One-tailed tests are most often encountered when testing specific hypotheses

 Q4 Using the Spearman's rank critical values table at $p = 0.05$, what are the critical values (r_s) at $n = 10$ for a one-tailed and a two-tailed test? Which value is the greater?

6.3.4 **Degrees of freedom**

Turn to the F_{max} table (D7) in Appendix d. There are three F tables, so you need to select the correct one. In the F_{max} table, the critical value is found using a (the number of samples or treatments) and v (the degrees of freedom). What are these 'degrees of freedom'?

Example 9.1 has a frequently encountered experimental design where there are two categories (periwinkles from the lower shore and periwinkles from the mid-shore), and for each category there are several observations. We illustrate this design in Table 6.4 for $n = 3$ observations.

If you were to fill in Table 6.4 for category 1, then you could choose any values for observations 1 and 2, but once these are chosen, the third observation can only be one value. This final value is already determined and there is no 'freedom' for this final value. The same would be true for the observations for category 2. You only have freedom to choose $n - 1$ observations; the final observation is then fixed. It is this constraint within the experimental design that is recognized by the degrees of freedom. For this design, the degrees of freedom for category 1 (periwinkles

Table 6.4. The impact of experimental design on the degrees of freedom based on Example 9.1: The evolution of *Littorina littoralis* at Aberystwyth, 2002

Observations	Periwinkles from lower shore	Periwinkles from mid-shore
1		
2		
3		
Total	25	35

from the lower shore) are $v_1 = n_1 - 1 = 3 - 1 = 2$ and the same is true for treatment 2 (periwinkles from the mid-shore) as n_2 also equals 3.

Examine Table 6.5. This represents one experimental design where there are two variables and for each variable there are two categories. (This design can be seen in Example 7.5 where the numbers of two snail species were recorded in a hedgerow and woodland.) You can see that, once the grand total is fixed and you have put in one row total, the remaining row total can only be one number. (The missing woodland row total is 12.) The same is true of the columns. The table could be extended to three rows and two columns and again one row total is fixed once the grand total and two row totals are determined. For this experimental design, this constraint is recognized by the degrees of freedom being calculated as (number of rows − 1) (number of columns − 1).

Our final illustration of degrees of freedom is from 7.1.3. Here an additional constraint is recognized in that a mathematical formula has been used to determine values that were then used in the statistical analysis. In this experimental design, there is one variable with several categories (Table 6.6). Usually for this experimental design, the degrees of freedom would normally be the number of categories $(a) − 1$. However, because a mathematical formula in which the population parameter μ has been estimated using the sample statistic $(\bar{x})$, a further degree of freedom is lost and $v = a − 2$ (see Box 7.2.).

Degrees of freedom (v) A measure that reflects the number of observations in your calculation and the experimental design

The **degrees of freedom**, therefore, are a measure that reflects the number of observations in your calculation and the experimental design.

Table 6.5. The impact of experimental design on the degrees of freedom based on Example 7.5: Frequency of *Cepaea nemoralis* and *Cepaea hortensis* in a hedge and wood

	Species of snail		
	Cepaea nemoralis	*Cepaea hortensis*	Total
Hedgerow			148
Woodland			
Total	97		160

Table 6.6. The impact on the degrees of freedom of the use of a Gaussian equation to determine expected values in a chi-squared test

	Variable				
	Treatment 1	Treatment 2	Treatment 3	Treatment 4	Total
Observation					106

You will be told for each statistical test how to calculate the degrees of freedom. It is not the same for all tests.

6.3.5 More than two criteria

There are a number of critical values that can only be found using three criteria. If you look now at the Mann–Whitney U test table in Appendix d, you will see that rows are labelled n_1 and columns are labelled n_2 (Table D12). n_1 and n_2 are the numbers of observations for sample 1 and sample 2, respectively. The third criterion for choosing the critical value for the U distribution is p. When there are more than two criteria for choosing a critical value, as in this case, only one table is usually included for one p value. We have included the set of critical values where $p = 0.05$. Thus, all the critical values in this table are at $p = 0.05$. If you wish to find the critical values for another p value for this test, you will need to refer to other sources.

 Q5 In a Mann–Whitney U test table of critical values, if $n_1 = 7$ and $n_2 = 10$, what is the critical value of U at $p = 0.05$?

Similarly, look at the F tables, either the F table that is used before a t- or z-test, or the F table that is used in a parametric analysis of variance test. Both these require a p value and two degrees of freedom, so again each table of critical values is for a single p value.

 Q6 If the degrees of freedom are $v_1 = 5$ and $v_2 = 10$, what are the critical F values if you were:

a. Carrying out an F-test for homogeneity of variances before a t-test at $p = 0.05$?

b. Carrying out an F-test in a parametric analysis of variance, $p = 0.05$?

Are they the same?
What does this tell you?

6.3.6 Interpolation

The final point we wish to make about finding critical values is relevant to several tables, including the F table for a parametric analysis of variance (Table D8 in Appendix d). A small excerpt is shown in Table 6.7.

To locate a particular $F_{critical}$ value, you use the two degrees of freedom, where the v_1 column intersects the v_2 row. However, it is not possible

Table 6.7. Excerpt from the table of critical values for an *F*-test for a parametric analysis of variance when $p = 0.05$

v_2	v_1		
	8	10	20
28	2.29	2.19	1.94
29	2.28	2.18	1.93
30	2.27	2.16	1.88

in this book to give all possible *F* values for all combinations of degrees of freedom. Clearly, if you needed to find the critical value at $v_1 = 13$, $v_2 = 29$, you will need to estimate it from the values provided. This process is called **interpolation**. It assumes that each missing degree of freedom increases the critical value in a regular way between the two known values.

For example, to find $F_{critical}$ at $v_1 = 13$, $v_2 = 29$, the nearest $F_{critical}$ values on the *F* table are:

Interpolation The estimation of a value when only nearby values are known, one greater than the required value and the other less than the required value

v_2	v_1						
	10	11	12	13	14	15 ...	20
29	2.18						1.93

You therefore need to fill in the gaps between $v_1 = 10$ and $v_1 = 20$.

First, take the difference between 2.18 and 1.93 = 0.25. The number of increments between 2.18 and 1.93 is ten. Divide the difference between the known $F_{critical}$ numbers by the number of increments, 0.25/10 = 0.025. Keep subtracting this number from the largest known $F_{critical}$ value until you have filled in the gaps:

v_2	v_1						
	10	11	12	13	14	15 ...	20
29	2.18	2.16	2.13	2.11	2.08	2.06	1.93

Therefore, the critical value for *F* when $v_1 = 13$, $v_2 = 29$ is 2.11.

6.4 **The rule**

Key points The rule is specific for each statistical test and describes the decision-making process when the critical and calculated values are compared.

This is the point at which the choice is made between the null and alternate hypotheses. The rule tells you about the relationship between your calculated test statistic and your critical test statistic. For example, if you were carrying out a chi-squared test for association, the rule is 'If $\chi^2_{\text{calculated}}$ is *greater* than χ^2_{critical}, you may reject the null hypothesis' (see Box 7.4). Most tests have a rule that sounds similar to this. There are exceptions, however. For example, if you were carrying out a Mann–Whitney U test, then the rule is 'If $U_{\text{calculated}}$ is less than U_{critical}, then you may reject the null hypothesis' (see Box 10.1). The Mann–Whitney rule is the opposite to the majority of rules. We therefore include each rule in all our general and worked examples relating to all the statistical tests in this book.

Applying these rules allows you finally to choose between your two hypotheses and to decide whether to reject or accept the null hypothesis. You can then say, given a certain level of probability, whether there is or is not a significant difference, association, correlation, etc. between your data sets.

6.5 **What does this mean in real terms?**

Key points The final step in hypothesis testing should always be to refer the outcome of the calculations back to your detailed experimental hypotheses.

The final step in hypothesis testing is the most important and yet most often forgotten. This is to refer back to your experiment. A common failing in hypothesis testing is for the first four steps to be completed and for a student to conclude that there is a significant difference. In what? What does this mean? Refer back to your hypotheses to conclude your hypothesis testing. We have included this step in all our examples of hypothesis testing to encourage you to not leave this out.

For example, in the experiment to examine the effect of soil type on seed germination (Example 6.1), a chi-squared goodness-of-fit test (7.1) was used to analyse the data. The outcome may be to reject the null hypothesis ($\chi^2_{\text{calculated}} = 9.00$, $p < 0.05$). But it is not enough to stop there. You should refer back to the alternate hypothesis and add details to this statement about the outcome. For this example, we might conclude that there is a significant difference ($\chi^2_{\text{calculated}} = 9.00$, $p < 0.05$) in the median number of *Allium schoenoprasum* seeds germinating on the three soil types.

Summary of Chapter 6

- You are introduced to the principles surrounding hypothesis testing and to the format used in Chapters 7–10 when a number of statistical tests are considered.

- There are three types of hypothesis: Do the data match an expected ratio? Is there an association between two or more variables? Do samples come from the same or different populations? (6.1.1 and developed in Chapters 7–10).

- Hypotheses may be general or specific. General hypotheses test questions such as, 'Is there a difference between the means of samples 1, 2, and 3?' Specific hypotheses may test questions such as, 'Is the mean of sample 1 significantly greater than the mean of sample 2, which is greater than the mean of sample 3?' (6.1.1 and developed in Chapters 7–10).

- Hypotheses have a number of features. Hypotheses come in pairs: the null and alternative hypotheses. They test the population parameter, indicate the type of data being analysed, indicate the type of hypothesis being tested, and give some details about the experimental design (6.1.2).

- The null hypothesis describes the variation in the data in terms of sampling error. The alternative hypothesis describes the variation in terms of the treatment variable (6.1.3).

- To enable you to decide whether you may reject the null hypothesis, you work out a calculated test statistic derived from a particular distribution (6.2). This is compared with a critical value also derived from the same distribution (6.3) and determined by the chosen level of stringency (e.g. $p = 0.05$) and other factors, such as the degrees of freedom (6.3.4).

- For each statistical test, there is a rule that determines the comparison you make between the calculated test statistic and the critical value of the test statistic. This rule differs between statistical tests (6.4 and developed in Chapters 5–8).

- When concluding your hypothesis testing, you must refer back to your hypotheses and your original objective (6.5 and developed in Chapters 7–11).

- The Online Resource Centre includes interactive exercises that test your understanding of this chapter with other topics, particularly those considered in Chapters 7–11.

@ online resource centre

Answers to chapter questions

A1 You do not start out with any expectation nor are you looking for an association, but you do wish to compare two treatments (organic and non-organic). Therefore, you will be testing hypothesis iii.

A2 This is an example that we use later in the book. The hypotheses are given in Box 9.1. Were you correct? If not, what did you miss out?

A3
a. 9/13
b. 4/13 (4.0, 7.7, 8.4, 8.4)
c. 12, 1 ($13 \times 95/100 = 12.35 \approx 12$, $13 \times 5/100 = 0.65 \approx 1$)
d. $p = 0.05$

A4 One-tailed test $n = 10$, $p = 0.05$, $r_s = 0.564$

Two-tailed test $n = 10$, $p = 0.05$, $r_s = 0.648$

The value for the two-tailed test is higher. This reflects the difference between the cut-off points. In a two-tailed test where $p = 0.05$, the cut-off points are 2.5% at each end of the distribution. In a one-tailed test where $p = 0.05$, the cut-off point will be 5% at one end only.

A5 $U = 14$ at $p = 0.05$, $n_1 = 7$, $n_2 = 10$

A6 a. $F_{critical} = 4.24$

b. $F_{critical} = 3.33$

No, they are not the same. This tells you that you don't want to get these two tables muddled up!

7

Hypothesis testing: do my data fit an expected ratio?

In a nutshell

In this chapter, we consider statistical tests and experimental designs that will be suitable for testing the hypothesis 'Do my data fit an expected ratio?' The statistical tests most often used to test this hypothesis are the chi-squared and *G* goodness-of-fit tests. We explain more about this type of hypothesis, how to carry out the necessary calculations, and how to resolve some of the problems you may encounter. We also consider a second group of chi-squared and *G*-tests (test for association). These relate to the hypotheses considered in Chapter 8. However, as these tests are very similar to the goodness-of-fit tests, we have included them here.

This chapter tells you how you may analyse data from one or more samples where each observation can be assigned to one of two or more discrete categories and allows you to test either the hypothesis 'Do my data fit an expected ratio?' or the hypothesis 'Is there an association between two or more variables?' The categories can be derived directly from the categories inherent in a nominal or ordinal variable, for example flower colour categorized as purple or white (Example 7.2), or imposed on an interval scale of measurement as classes, for example the distribution of holly leaf miners on a holly tree (Example 7.1). The data in these categories or classes are counts, i.e. the numbers of items in a category. The tests that are introduced here are the chi-squared test and the *G*-test.

In all these tests, including those designed to test the hypothesis 'Is there an association?', the observed values are compared with calculated 'expected' values. The term 'expected' can be used in three ways:

Expected values The values generated in response to a mathematical or biological model such as the Gaussian equation or a genetic segregation ratio

1) For some experiments, you may have a reason for expecting certain outcomes from your investigation. This is often referred to as having an *a priori* expectation and arises when you carry out a test of the hypothesis 'There is an expectation.'

2) You can use your own (observed) data to generate 'expected' values given certain rules relating to the statistical test and experimental design to allow you to test the hypothesis 'There is an association.'

3) You may have your own expectations (or predictions) for your experiment. We considered this in Chapter 6 (6.1.1) and return to this topic in 11.6, as a change in the approach to science in schools has led to an increase amongst undergraduates in putting forward such predictions. You must never confuse this, your own personal preference or 'expectation', which is usually for significant outcomes, with the idea of 'expected' values used in hypothesis testing.

The tests we are looking at in this chapter include those where you may have an *a priori* expectation (1) and those where you may not (2). (3) can lead to bias in carrying out an experiment and evaluating the results and should therefore be avoided.

We will look at three forms of the chi-squared test: the goodness-of-fit test, a test for heterogeneity, and the test for association. With a chi-squared test, you use an equation based on the chi-squared distribution to work out 'expected' values. These are then compared with the ones you recorded in your experiment. The three versions of the chi-squared test differ in how the 'expected' values are calculated. It is therefore important to be sure you have selected the right form of chi-squared test for your type of data and hypothesis.

The G-test can be used in the same way as the chi-squared test. It has some advantages over the chi-squared test, especially when numbers are large. The chi-squared test tends to be used most frequently, so we have only included two forms of the G-test: the goodness-of-fit test and an $r \times c$ test for association (7.5).

Worked examples are given for each test. If this is the first time you have used this test, you should work through these examples and the questions and then check your answers before using the test on your own data. If your answer differs considerably from that given, you should check your calculation by going to the Online Resource Centre, where we have included both the full calculation and more examples. If you work through the questions in this chapter it should take you about three hours to complete. The answers for these exercises are at the end of the chapter. All our examples are based on real undergraduate research projects. If these examples are not in your subject area, you will find more in the Online Resource Centre.

online
resource
centre

How to choose the correct test

Your choice is determined by the number of variables and number of categories for each variable. This table directs you to the most likely statistical test. If you are not sure, then go to the specific sections indicated

and look at the examples to see whether they are similar to the work you are planning. Each test has additional criteria that need to be met. These are given in the sections indicated.

You wish to examine the effect of one variable and have only one sample. You have an *a priori* reason for expecting certain outcomes from your investigation. The variable has more than two categories. The data are counts or frequencies.	Chi-squared goodness-of-fit test (7.1) or G goodness-of-fit test (7.5.1)
You wish to examine the effect of one variable and have only one sample. You have an *a priori* reason for expecting certain outcomes from your investigation. The variable has only two categories. The data are counts or frequencies.	Chi-squared goodness-of-fit test with Yates' correction (7.4.1) or G goodness-of-fit test (7.5.1)
You wish to examine the effect of one variable with two or more categories. You have more than two samples in your data set and wish to know whether the samples are similar or different from each other. The data are counts or frequencies.	Chi-squared test for heterogeneity (7.2)
You do not have an *a priori* expectation. You have two variables. At least one of these variables has more than two categories. You wish to test for an association between the variables. The data are counts or frequencies.	Chi-squared test for association (7.3) or G-test for association (7.5.2)
You do not have an *a priori* expectation. You have two variables. Both variables have only two categories. You wish to test for an association between the variables. The data are counts or frequencies.	Chi-squared test for association with Yates' correction (7.4.2) or G-test for association (7.5.2)

One of the criteria for using a chi-squared test is that the expected values are greater than 5. Having calculated your expected values, if you find any that are less than 5, you should refer to 7.6.

7.1 Chi-squared goodness-of-fit test

Key points In some experiments, you may have a reason for expecting your data to be explained by an equation or particular ratio. You may analyse data to examine whether your data fit this *a priori* expectation. If your data are counts or frequencies, one of the most common statistical tests used to test this type of hypothesis is the chi-squared goodness-of-fit test.

There are three types of investigations where you may use a goodness-of-fit test:

1) You can reasonably argue that all samples should have the same value.

This can arise when your sampling is very restricted, e.g. you may record the remains of shells from *Cepaea nemoralis* and *Cepaea hortensis* at the site where a thrush has been breaking them open (an anvil). You might hypothesize that there is no statistically significant difference between the numbers of each species, i.e. you would expect to see one *Cepaea nemoralis* shell for every one *Cepaea hortensis* shell at this thrush anvil. But beware. This sample may not be representative of thrush anvils as a whole. To examine a broader hypothesis, you would need to collect more samples and use a different statistical test.

2) You wish to confirm that your data have a particular distribution, such as a normal or a Poisson distribution.

   In this instance, your expected values are calculated using a particular formula that describes the distribution, such as the Gaussian equation (5.2.1).

3) You have carried out a genetic cross or sampled in a population and have reason to expect a particular segregation ratio, sex ratio, allele frequency, etc.

   In each of these examples, you will be able to use known underlying principles, such as Mendelian genetics or the Hardy–Weinberg theorem, to determine the 'expected' values.

A chi-squared goodness-of-fit test is introduced in 7.1.1 in relation to an experimental design where the expected values are determined by a ratio resulting from 'random' events (Investigation type 1). A second example is considered in 7.1.3 where this test is also used to confirm that data have a normal distribution. (Investigation type 2). In 7.2 and 7.4.1, examples of testing an expectation arising from genetic crosses are shown (Investigation type 3).

A goodness-of-fit test allows you to compare the data you have collected in an investigation with that predicted by an *a priori* expectation and to

calculate how probable the match is. As in all chi-squared tests, there are several common steps to be followed. The first of these is to arrange your data into a contingency table.

EXAMPLE 7.1 The distribution of holly leaf miners on *Ilex aquifolia*

As part of a student project, the number of holly leaf miners was recorded on a holly tree. The heights of the holly leaf miners within the tree were recorded (Table 7.1). If the distribution was random you would expect equal numbers at each height.

Table 7.1. Contingency table for Example 7.1: The distribution of holly leaf miners on a single *Ilex aquifolia* tree

	Height on holly tree (m)			Total number of holly leaf miners
	0.00–1.99	2.00–3.99	4.00–5.99	
Observed number of holly leaf miners	131	38	2	171

Table 7.1 is a contingency table for the data from Example 7.1. A contingency table can be the same as a frequency table, as in this example. At this stage, you will only have 'observed' values. The observed data have been organized into classes: 0.00–1.99m, 2.00–3.99m, and 4.00–5.99m. There is no ambiguity about the categories: they are discrete. Categories are also known as classes.

7.1.1 Using this test

To use this test you:

1) Wish to compare your observed values to those predicted by an *a priori* expectation.
2) Have one treatment variable.
3) Have only one sample.
4) Have data that fall into more than two discrete categories.
5) Have data that are counts or frequencies and are not percentages or proportions.
6) Have observations that are independent.
7) Have expected values that are greater than 5.

The data from Example 7.1 meet these criteria. The *a priori* expectation is that the holly leaf miners should, by chance, be present in equal numbers at the different tree heights. We wish to examine the effect of one variable

Table 7.2. Expected values for a goodness-of-fit chi-squared test using the data from Example 7.1: The distribution of holly leaf miners on a single *Ilex aquifolia* tree

Number of holly leaf miners	Height on holly tree (m)			Total number of holly leaf miners
	0.00–1.99	**2.00–3.99**	**4.00–5.99**	
Observed	131	38	2	171
Expected	57	57	57	171

(height on tree) and there is only one sample. The observations fall into three discrete categories (0.00–1.99m, 2.00–3.99m, and 4.00–5.99m) and the scale of measurement (m) is an interval scale. Each holly leaf miner was recorded only once; therefore, each observation is independent of all other observations in the sample. The expected values are only worked out as part of the calculation, so cannot easily be checked in advance. If you look at Table 7.2 on the 'expected' row, these values are greater than 5. So this criterion is also met. If the criterion is not met, then you should refer to 7.6.

7.1.2 The general calculation

Having organized your observed values in a contingency table and checked that the criteria for using this test are met, you can now proceed with the test. In Box 7.1, we show you the general calculation and a specific example of a chi-squared goodness-of-fit test. In addition, you will need to refer to the contingency calculation table (Table 7.2). Where steps have been abbreviated, the full calculation is included in the Online Resource Centre.

online resource centre

BOX 7.1 How to calculate a chi-squared goodness-of-fit test

GENERAL DETAILS	EXAMPLE 7.1
	This calculation is given in full in the Online Resource Centre. All values have been rounded to five decimal places.
1. Hypotheses to be tested H_0: There is no difference between the expected and observed values. H_1: There is a difference between the expected and observed values.	1. Hypotheses to be tested H_0: There is no difference between the numbers of holly leaf miners found at various heights (m) on the tree compared with those expected. H_1: There is a difference between the numbers of holly leaf miners found at various heights (m) on the tree compared with those expected.

BOX 7.1 Continued

2. Have the criteria for using this test been met?	2. Have the criteria for using this test been met? Yes (7.1.1).
3. How to work out expected values Expected values are calculated using the numerical formula or ratio that you are expecting your data to conform to.	3. How to work out expected values We are expecting a random distribution of the holly leaf miners throughout the tree. This means that we should find equal numbers in each height category, i.e. the ratio we are using is 1:1:1. The number of holly leaf miners we would expect at each level will be 171/3 = 57 (Table 7.2).
4. How to work out $\chi^2_{calculated}$ $$\chi^2_{calculated} = \sum \frac{(observed - expected)^2}{expected}$$	4. How to work out $\chi^2_{calculated}$ $$\chi^2_{calculated} = \frac{(131-57)^2}{57} + \frac{(38-57)^2}{57} + \frac{(2-57)^2}{57}$$ $$= 96.07018 + 6.33333 + 53.07018$$ $$= 155.47369$$
5. How to find $\chi^2_{critical}$ See Appendix d, Table D1. To find the critical value of χ^2, you need to know the degrees of freedom (v). In this goodness-of-fit test, the degrees of freedom are the number of categories (a) – 1. Use a chi-squared table of critical values to locate the value at $p = 0.05$.	5. How to find $\chi^2_{critical}$ The categories in this example are 0.00–1.99m, 2.00–3.99m, and 4.00–6.99m, and therefore $v = 3 - 1 = 2$ and $\chi^2_{critical}$ is 5.99.
6. The rule If $\chi^2_{calculated}$ is greater than $\chi^2_{critical}$, you may reject the null hypothesis.	6. The rule $\chi^2_{calculated}$ (155.47) is greater than $\chi^2_{critical}$ (5.99) at $p = 0.05$ and therefore we reject the null hypothesis. In fact, at $p = 0.001$, the $\chi^2_{critical}$ value is 13.82. Therefore, we can reject the null hypothesis at this higher level of significance.
7. What does this mean in real terms?	7. What does this mean in real terms? There is a highly significant difference ($\chi^2_{calculated} = 155.47$, $p < 0.001$) between the numbers of holly leaf miners found at the various levels on the tree compared with those expected, such that the holly leaf miners are not found in equal numbers at all heights.

Further examples relating to the topic including how to use statistical software are included in the Online Resource Centre.

online resource centre

Q1 In an investigation into the visual responses of beetles, a number of coloured pitfall traps were placed at random in grassland. After 24 hours, the pitfall traps were collected and the numbers of beetles recorded. The results for each trap were red: 20 beetles; yellow: 34 beetles; white: 10 beetles; and black: 40 beetles. What is your *a priori* expectation? Calculate the expected values.

7.1.3 How to check whether your data have a normal distribution using the chi-squared goodness-of-fit test

In Chapter 5, we discussed distributions and in particular the normal distribution. Data with a normal distribution can be analysed using parametric statistics and it is therefore important to be able to check whether your data are normally distributed. Section 5.8 and Box 5.2 explained a number of ways to check your data to see whether they are normally distributed. The best support for deciding whether your data are normally distributed is to calculate expected values using the Gaussian equation and then use a chi-squared goodness-of-fit test to check that your observed data are not statistically significant from that predicted by the equation. To do this requires some extra steps in addition to those described in Box 7.1 and a change to how you work out the degrees of freedom, so we have included a worked example here. The observed values we will use come from Example 5.2: Length (mm) of two-spot ladybirds (*Adalia bipunctata*).

Although we only include an example relating to the normal distribution, the principles are the same for checking against any known distribution, such as the Poisson distribution. All you need to know is the equation that describes the distribution you are interested in. You then follow the same steps using this other equation.

online resource centre

i. Arrange your data into a contingency table

The contingency data for this example are the same as the frequency table in Chapter 5 (Table 5.9). We reproduce this again here in an amended form so that the whole of this process is represented here for convenience.

ii. Do the data meet all the criteria for using a chi-squared goodness-of-fit test?

For this example, the criteria are met. The *a priori* expectation is determined by the Gaussian equation. There is one treatment variable (length). This is measured on an interval scale (mm), which can be

Table 5.9. Frequency table of length of two-spot *Adalia bipunctata* (ladybirds)

	Size classes for length (mm) of *Adalia bipunctata* (ladybird)								
	1.0–1.9	2.0–2.9	3.0–3.9	4.0–4.9	5.0–5.9	6.0–6.9	7.0–7.9	8.0–8.9	9.0–9.9
Mid-point of class (m)	1.45	2.45	3.45	4.45	5.45	6.45	7.45	8.45	9.45
Frequency (f)	2	3	5	8	12	9	5	4	2

arranged into discrete categories. The values within each category are the number of observations falling within that category. There is one sample. Each ladybird was only measured once, so the data are independent. We take you through the steps for calculating the expected numbers in the next section, but in Table 7.3, you can see that not all the expected values are greater than 5. Therefore, this last criterion is not met. We explain how to overcome this problem of small numbers in general in 7.6 and specifically for this example in Box 7.2.

BOX 7.2 To check whether your data are normally distributed using a chi-squared goodness-of-fit test

GENERAL DETAILS	EXAMPLE 5.2
	This calculation is given in full in the Online Resource Centre. All values have been rounded to five decimal places.
1. Hypotheses to be tested H_0: There is no difference between the expected and observed values. H_1: There is a difference between the expected and observed values.	1. Hypotheses to be tested H_0: There is no difference between the observed lengths of ladybirds (mm) compared with that expected if the data are normally distributed. H_1: There is a difference between the observed lengths of ladybirds (mm) compared with that expected if the data are normally distributed.
2. Have all the criteria for using this test been met?	2. Have all the criteria for using this test been met? Most of the criteria are met (7.1.3ii). However, some of the expected values are less than 5 (Table 7.3). We explain how to deal with this in the next step.
3. How to work out expected values Expected values are calculated using the numerical formula you are expecting your data to conform to.	3. How to work out expected values See 7.1.3iii and Table 7.3. It is clear that five of the nine expected values are less than 5. To overcome this, we will add together the expected values for the lower two classes (1.7768 + 4.0951 = 5.8719) and the expected values in the upper two classes (2.2649 + 0.8066 = 3.0717). These expected values will be compared with observed values that have been combined in the same way.

BOX 7.2 Continued

4. How to work out $\chi^2_{calculated}$ $$\chi^2_{calculated} = \Sigma \left[\frac{(observed - expected)^2}{expected} \right]$$	4. How to work out $\chi^2_{calculated}$ $$\chi^2_{calculated} = \frac{(5-5.8719)^2}{5.8719} + \frac{(5-7.2283)^2}{7.2283} + \cdots$$ $$+ \frac{(6-3.0717)^2}{3.0717} = 4.40299$$
5. How to find $\chi^2_{critical}$ See Appendix d, Table D1. To find the critical value of χ^2 at p = 0.05, you need to know the degrees of freedom (v). Usually for a chi-squared goodness-of-fit test, the degrees of freedom would be the number of categories (a) – 1. However, because a mathematical formula in which the population value of μ has been estimated using the sample mean ($\bar{x}$), a further degree of freedom is lost. Therefore, for this chi-squared goodness-of-fit test, $v = a - 2$.	5. How to find $\chi^2_{critical}$ As we combined the lower two classes and the upper two classes, there were only seven classes in this calculation, so $v = 7 - 2 = 5$. When $p = 0.05$, $v = 7$, then $\chi^2_{critical}$ is 14.07.
6. The rule If $\chi^2_{calculated}$ is greater than $\chi^2_{critical}$, you may reject the null hypothesis.	6. The rule $\chi^2_{calculated}$ (4.40) is less than $\chi^2_{critical}$ (14.07) at $p = 0.05$ and therefore we do not reject the null hypothesis.
7. What does this mean in real terms?	7. What does this mean in real terms? There is no significant difference ($\chi^2_{calculated}$ = 4.40, $p = 0.05$) between the observed lengths of ladybirds (mm) compared with that expected if the data are normally distributed. The data can be said to be normally distributed.

Further examples relating to the topic including how to use statistical software are included in the Online Resource Centre.

online resource centre

---

iii. Calculate the expected values

Like the previous example for the chi-squared goodness-of-fit test, you use your *a priori* expectation to determine the expected values. In this case, we use the Gaussian equation, which is the mathematical equation that describes the normal distribution, where:

$$y = \frac{1}{\sqrt{2\pi s^2}} e^{-b}$$

and

$$h = \frac{(x - \bar{x})^2}{2s^2}$$

The terms in this equation are explained in 5.2.1 and Appendix e. The symbols e and π have values of 2.72 and 3.14, respectively. The $\bar{x}$ (mean) and s^2 (variance) were calculated in Box 5.2 and are $\bar{x} = 5.08$mm, $s^2 = 3.74857$mm^2. x is any one observation in your sample. Thus, all these terms have known values apart from y. These y values can be calculated for each x in your sample as follows.

Choose an x value that fits the lowest category of your data. As we have classes, we use the mid-point of that class, so the first $x = 1.45$. By including this x value and all the other known values, you can now work out y. Don't be put off, even though it looks complicated. Break the calculation down into smaller steps.

The first part of the equation is:

$$\frac{1}{\sqrt{(2\pi s^2)}} = \frac{1}{\sqrt{2 \times 3.14 \times 3.74857}}$$

$$= \frac{1}{\sqrt{23.5529}} = \frac{1}{4.8531} = 0.20605$$

The second part of the equation involves the exponential term e. First work out the value for h:

$$h = \frac{(x - \bar{x})^2}{2s^2} = \frac{(1.45 - 5.08)^2}{2 \times 3.74857} = \frac{13.1769}{7.49714} = 1.75759$$

Use your calculator function buttons to find $e^{-1.75759} = 0.17246$
So

$$y = 0.20605 \times 0.17246 = 0.03554$$

This y value is worked out as a proportion and our final step is to calculate expected numbers from these proportions. As the sample size for our ladybird example is 50, then the expected number for y when $x = 1.45$ is $50 \times 0.03554 = 1.77678$.

Repeat this calculation for all mid-point values from 1.45 to 9.45. These are your expected values (Table 7.3) and you can now carry on with the chi-squared goodness-of-fit test (Box 7.2).

Table 7.3. Contingency table for Example 5.2: The length of two-spot *Adalia bipunctata* (ladybirds) with expected numbers calculated from the Gaussian equation

	Size classes for length (mm) of *Adalia bipunctata* (ladybird)								
	1.0–1.9	2.0–2.9	3.0–3.9	4.0–4.9	5.0–5.9	6.0–6.9	7.0–7.9	8.0–8.9	9.0–9.9
Mid-point of class (*m*)	1.45	2.45	3.45	4.45	5.45	6.45	7.45	8.45	9.45
Observed number of ladybirds	2	3	5	8	12	9	5	4	2
Expected number of ladybirds	1.7768	4.0951	7.2283	9.7714	10.1162	8.0209	4.8705	2.2649	0.8066

7.2 Heterogeneity in a goodness-of-fit test

Key points In some experiments, you may have a reason for expecting your data to be explained by an equation or particular ratio. You may analyse data to examine if your data fit this *a priori* expectation. The goodness-of-fit test (7.1) outlines how to analyse this type of hypothesis when you only have one sample. Here we explain how you may compare samples within a goodness-of-fit test.

There are some circumstances where you have several samples and wish to know whether they can be pooled. For example, you may have carried out an investigation into the genetic inheritance of flower colour where you had several pairs of parent plants. Crossing within these pairs would produce F_1 offspring. If you kept the seed from each cross separate from the others, grew these F_1 plants up, and then crossed these, you could collect the results from several different lineages. These separate lineages are known as accessions.

EXAMPLE 7.2 The genetics of flower colour in *Allium schoenoprasum*

Crosses were carried out in three pairs of plants of *Allium schoenoprasum*. In each cross, one parent had purple flowers and the other was white flowered. Seeds from each cross were collected and grown on. The offspring from each cross were kept as separate accessions. These F_1 plants were all purple flowering and thought to be heterozygous for a single gene that controlled the pigmentation in the flowers, and purple was believed to be the dominant allele. These F_1 plants were then crossed among themselves within an accession and the seed from these crosses grown up and the flower colours recorded (Table 7.4). If the genetic theory is correct, then the F_2 plants should occur in the ratio of three purple-flowering plants:one white-flowering plant.

Table 7.4. Flower colour exhibited by F_2 *Allium schoenoprasum* in four accessions

	Flower colour in the F_2 generation		Total number of plants flowering in the F_2
	Purple	White	
Accession 1	127	41	168
Accession 2	123	39	162
Accession 3	107	53	160
Accession 4	130	42	172
Total	487	175	662

Table 7.5. Contingency table with expected values for a chi-squared test for heterogeneity for Example 7.2: Flower colour exhibited by F_2 *Allium schoenoprasum* in four accessions

		Flower colour in the F_2 generation		Total number of plants flowering in the F_2	χ^2 from the goodness-of-fit test for each accession and for the total values
		Purple	White		
Accession 1	Observed	127	41	168	0.03175 (NS)
	Expected	126.0	42.0		
Accession 2	Observed	123	39	162	0.07407 (NS)
	Expected	121.5	40.5		
Accession 3	Observed	107	53	160	5.63333 ($0.05 > p > 0.01$)
	Expected	120.0	40.0		
Accession 4	Observed	130	42	172	0.03101 (NS)
	Expected	129.0	43.0		
Total	Observed	487	175	662	0.72709 (NS)
	Expected	496.5	165.5		

NS, not significant.

In examining Table 7.4, it appears that the total values (487 purple and 175 white) are reasonably close to a 3:1 ratio. When a goodness-of-fit chi-squared test is carried out on just these total values, there is no significant difference ($\chi^2 = 0.73$, $p < 0.05$) between the observed values and that expected for a three purple:one white ratio (Table 7.5). But should you combine the data from your different samples in this way? You can get some idea by carrying out a goodness-of-fit chi-squared test on each accession separately (Table 7.5). These analyses confirm that the results from accessions 1, 2, and 4 also conform to a 3:1 ratio at $p = 0.05$.

Accession 3, however, is different and the test there indicates that the observed values depart significantly from the predicted 3:1 ratio. Does this difference between accession 3 and accessions 1, 2, and 4 mean that the data should not be pooled? In these circumstances, you may use the chi-squared test to see whether the data are statistically heterogeneous (different).

7.2.1 **Using this test**

To use this test you:

1) Wish to test for heterogeneity between samples.

2) Have one treatment variable.

3) Have two or more samples.

4) Have data that fall into discrete categories.

5) Have data that are counts or frequencies and are not percentages or proportions.

6) Have observations that are independent of each other.

7) Do not have expected values of less than 5.

We have checked the data from Example 7.2 against the criteria for using this test and all the criteria are met. We do wish to test for heterogeneity between samples. There is one treatment variable (flower colour) and four samples (accessions). The observations fall into two discrete categories (purple and white) and the number of plants has been recorded. Each plant has been counted only once. All the expected values are greater than 5 (Table 7.5). If this were not the case, you should refer to 7.6.

7.2.2 **The calculation**

At this point, you have constructed a contingency table, checked the criteria for using this test, and wish to compare your samples to see whether they are heterogeneous or can be pooled. You can now proceed with the chi-squared test for heterogeneity (Box 7.3). Where steps have been abbreviated, the full calculation is included in the Online Resource Centre.

 online resource centre

BOX 7.3 How to calculate a chi-squared test for heterogeneity

GENERAL DETAILS	EXAMPLE 7.2
	This calculation is given in full in the Online Resource Centre. All values have been rounded to five decimal places.
1. Hypotheses to be tested H_0: There is no heterogeneity between the samples. H_1: There is heterogeneity between the samples.	1. Hypotheses to be tested H_0: There is no heterogeneity between the F_2 accessions of *Allium schoenoprasum*. H_1: There is heterogeneity between the F_2 accessions of *Allium schoenoprasum*.

BOX 7.3 Continued

2. Have all the criteria for using this test been met?	2. Have all the criteria for using this test been met?
	Yes (7.2.1).

3. How to work out expected values	3. How to work out expected values
These are calculated for each accession and for the total values. The expected values are calculated using the numerical formula you are expecting your data to conform to.	We are expecting three purple-flowering plants:one white-flowering plant. As the grand total number of plants examined was 662, then 3/4 should be purple flowering, i.e. 496.5, and 1/4 plants should be white flowering i.e. 165.5. Using the same ratio, we also calculate the expected values for each accession (sample) (Table 7.5).

4. How to work out $\chi^2_{calculated}$

i. First calculate chi-squared values for each separate sample and for the total, where:

$$\chi^2_{calculated} = \Sigma \left[\frac{(observed - expected)^2}{expected} \right]$$

It is at this point that the process differs from that described for the goodness-of-fit test for one sample and a few further calculations are required.

4. How to work out $\chi^2_{calculated}$

i. e.g. Accession 1.

$$\chi^2 = \frac{(127-126)^2}{126} + \frac{(41-42)^2}{42}$$
$$= 0.00794 + 0.02381$$
$$= 0.03175$$

$$\text{Total } \chi^2 = \frac{(487-496.5)^2}{496.5} + \frac{(175-165.5)^2}{165.5}$$
$$= 0.18177 + 0.54532$$
$$= 0.72709$$

The other values are shown in Table 7.5.

ii. 'Summed' chi-squared. First sum all the chi-squared values calculated for each accession.

ii. 'Summed' chi-squared. In our current example, this 'summed' value is: 0.03175 + 0.07407 + 5.63333 + 0.03101 = 5.77016

iii. 'Deviation' chi-squared. This is the chi-squared value found when examining the total values.

iii. 'Deviation' chi-squared = 0.72709

iv. The 'heterogeneity' chi-squared is found by subtracting the 'deviation' value from the 'summed' value. This is $\chi^2_{calculated}$.

iv. $\chi^2_{calculated} = 5.77016 - 0.72709 = 5.04307$

5. How to find $\chi^2_{critical}$

See Appendix d, Table D1. First, calculate the degrees of freedom (v). Again, there are several steps.

i. v for the 'deviation' value is the number of categories (columns, a) – 1.

ii. v for the 'summed' chi-squared value is the sum of the degrees of freedom from each of the rows (samples) excluding the total.

5. How to find $\chi^2_{critical}$

i. There are two categories or columns (purple and white), so $v = 2 - 1 = 1$.

ii. There are four accessions (rows), each with 1 degree of freedom. Therefore 'summed' $v = 1 + 1 + 1 + 1 = 4$.

BOX 7.3 Continued	
iii. v for the 'heterogeneity' value is the value from (ii) – the value from (i). The critical value is found at $p = 0.05$ and the heterogeneity degrees of freedom from (iii).	iii. For this example, the 'heterogeneity' v is $4 - 1 = 3$. The critical value to test for heterogeneity in the data is found in the chi-squared table at $v = 3$, $p = 0.05$ and is $\chi^2_{critical} = 7.81$.
6. The rule If $\chi^2_{calculated}$ is greater than $\chi^2_{critical}$, you may reject the null hypothesis. If your data are heterogeneous, refer to Chapter 10.	6. The rule $\chi^2_{calculated}$ (5.04) is less than $\chi^2_{critical}$ (7.81) at $p = 0.05$ and therefore we do not reject the null hypothesis.
7. What does this mean in real terms?	7. What does this mean in real terms? There is no statistically significant heterogeneity ($\chi^2_{calculated}$ 5.04, $p = 0.05$) between the accessions of *Allium schoenoprasum* and it is therefore reasonable to sum the data across all accessions and use a goodness-of-fit chi-squared test on the totals.

Further examples relating to the topic, including how to use statistical software, are included in the Online Resource Centre.

online resource centre

7.3 Chi-squared test for association

Key points In some investigations you may wish to examine a possible association between two variables. This is a type of hypothesis testing we consider in more detail in Chapter 8. However, we have included the chi-squared test for association here as it is similar in many respects to the chi-squared goodness-of-fit test.

A chi-squared test for association is used when you have categorical data for two variables and you wish to examine the possibility of an association between these two variables. The term 'association' has a very specific meaning. Before proceeding with this test, you should read the first paragraphs of Chapter 8 to ensure you understand how this term is being used.

As in all chi-squared tests, the data are arranged in a contingency table. If your data have more than two discrete categories, either for variable 1 (columns) and/or for variable 2 (rows), you should use the method in Box 7.4. This method is often referred to as a generalized or $r \times c$ test. If your contingency table has only two categories for variable 1 and two categories for variable 2, you should refer to 7.4.1.

EXAMPLE 7.3 Shell colour in *Cepaea nemoralis* in coastal and hedgerow habitats

An investigation was carried out into the frequency of banding and colour patterns in snail shells and habitat. Four patterns were observed in the two populations studied: pink with bands, pink with no bands, yellow with bands, and yellow with no bands. These results are shown in Table 7.6. The association that is under investigation is therefore between the variables 'shell pattern' and 'habitat'.

Table 7.6. Shell colour in *Cepaea nemoralis* in coastal and hedgerow habitats

Habitat	Shell pattern and colour in C. *nemoralis*				
	Banded Yellow	No bands Yellow	Banded Pink	No bands Pink	Total
Coastal observed	10	19	5	16	50
Hedgerow observed	17	8	19	11	55
Total	27	27	24	27	105

7.3.1 Using this test

To use this test you:

1) Wish to test for an association between two treatment variables.
2) Have data that are organized into more than two categories for at least one of the variables. (If you have only two categories for both variables go to 7.4.1.)
3) Have data that are counts or frequencies and are not percentages or proportions.
4) Have observations that are independent of each other.
5) Have expected values that are greater than 5.

Table 7.7. Contingency table with expected values for a chi-squared test for association using data from Example 7.3: Shell colour in *Cepaea nemoralis* in coastal and hedgerow habitats

Habitat	Shell pattern and colour in C. *nemoralis*				
	Banded Yellow	No bands Yellow	Banded Pink	No bands Pink	Total
Coastal observed	10	19	5	16	50
Coastal expected	27/105 × 50 = 12.85714	27/105 × 50 = 12.85714	24/105 × 50 = 11.42857	27/105 × 50 = 12.85714	
Hedgerow observed	17	8	19	11	55
Hedgerow expected	27/105 × 55 = 14.14206	27/105 × 55 = 14.14206	24/105 × 55 = 12.57143	27/105 × 55 = 14.14206	
Total	27	27	24	27	105

For the data from Example 7.3, we do wish to test for an association between two variables (shell pattern and habitat). Shell pattern has four categories (pink with bands, pink with no bands, yellow with bands, and yellow with no bands) and the habitat has two categories (coastal and hedgerow). The data are numbers of snails in each category. When the expected values are calculated (Box 7.4, Table 7.7) they are greater than 5, so all the criteria for using this test are met.

7.3.2 The calculation

This is the calculation for an *r* × *c* chi-squared test. At this point, you will have arranged your data in a contingency table and have checked the criteria for using this test. You can now proceed to test for an association between the two variables (Box 7.4 and Table 7.7). Where steps have been abbreviated, the full calculation is included in the Online Resource Centre.

online
resource
centre

BOX 7.4 How to calculate an $r \times c$ chi-squared test for association

GENERAL DETAILS	EXAMPLE 7.3
	This calculation is given in full in the Online Resource Centre. All values have been rounded to five decimal places.
1. Hypotheses to be tested H_0: There is no association between the two variables. H_1: There is an association between the two variables.	1. Hypotheses to be tested H_0: There is no association between the distribution of shell patterns observed and the habitat (coastal and hedgerow) of *Cepaea nemoralis*. H_1: There is an association between the distribution of shell patterns and the habitat (coastal and hedgerow) of *Cepaea nemoralis*.
2. Have all the criteria for using this test been met?	2. Have all the criteria for using this test been met? Yes (7.3.1).
3. How to work out expected values In chi-squared tests for association, you have no *a priori* expectation against which to compare your observed data. Instead, the expected values are calculated from the totals of each column and row.	3. How to work out expected values Look first at the expected value in row 1, column 1 in Table 7.7. To calculate this expected value, take the column total for banded yellow (27) ÷ grand total (105) × row total for coastal snails (50) = 12.85714 This calculation is repeated for each row × column combination.
4. How to work out $\chi^2_{calculated}$ The rest of the procedure is the same as that described for the goodness-of-fit chi-squared test, where: $$\chi^2_{calculated} = \sum \left[\frac{(observed - expected)^2}{expected} \right]$$	4. How to work out $\chi^2_{calculated}$ $$\chi^2 = \frac{(10 - 12.85714)^2}{12.85714} + \frac{(19 - 12.85714)^2}{12.85714}$$ $$+ \frac{(5 - 11.42857)^2}{11.42857} + \frac{(16 - 12.85714)^2}{12.85714}$$ $$+ \frac{(17 - 14.14206)^2}{14.14206} + \frac{(8 - 14.14206)^2}{14.14206}$$ $$+ \frac{(19 - 12.57143)^2}{12.57143} + \frac{(11 - 14.14206)^2}{14.14206}$$ $$\chi^2_{calculated} = 15.18472$$
5. How to find $\chi^2_{critical}$ See Appendix d, Table D1. You now need to calculate the degrees of freedom before looking up the critical value. In this case, v is the (number of rows – 1) × (number of columns – 1). The critical value is found in the statistical table at $p = 0.05$ and the degrees of freedom just calculated.	5. How to find $\chi^2_{critical}$ In our example, there are two rows (coastal and hedgerow). Do not include your 'expected' rows as these are part of your calculation. There are four columns (banded yellow, no bands yellow, banded pink, no bands pink). Therefore: $v = (2 - 1) \times (4 - 1) = 1 \times 3 = 3$. The critical value where $v = 3$, $p = 0.05$, is $\chi^2_{critical} = 7.81$.

BOX 7.4 Continued	
6. The rule	**6. The rule**
If $\chi^2_{\text{calculated}}$ is greater than χ^2_{critical} you may reject the null hypothesis.	$\chi^2_{\text{calculated}}$ (15.18) is greater than χ^2_{critical} (7.81) at $p = 0.05$ and therefore we reject the null hypothesis.
7. What does this mean in real terms?	**7. What does this mean in real terms?**
	There is a significant association ($p = 0.05$) between the distribution of shell patterns and habitat (coastal and hedgerow) of *Cepaea nemoralis*. In fact, at $p = 0.01$, $\chi^2_{\text{critical}} = 11.34$ and $p = 0.001$, $\chi^2_{\text{critical}} = 16.27$. Therefore, you may reject the null hypothesis at $p = 0.01$ but not at $p = 0.001$. This can be written as: there is a highly significant association ($\chi^2_{\text{calculated}} = 15.18, 0.01 > p > 0.001$) between the distribution of shell patterns and the habitat of *Cepaea nemoralis*.

Further examples relating to the topic including how to use statistical software are included in the Online Resource Centre.

online resource centre

Q2 Two groups of students were asked to rate on a scale of 1 (not at all) to 10 (outstanding) how helpful they had found this book (Table 7.8). The investigators wished to test whether there was an association between the answers and the subject studied. Calculate appropriate expected values.

Table 7.8. The numbers of Microbiology and Forensic Science students giving particular responses to the question: 'How helpful have you found this book?

	Answers on a scale of 1 (not at all) to 10 (outstanding)									
	1	2	3	4	5	6	7	8	9	10
Number of Microbiology students	1	0	1	2	2	3	3	4	3	2
Number of Forensic Science students	0	0	0	0	7	5	4	3	1	0

7.4. Chi-squared test with one degree of freedom

Key points When contingency tables are small, the degrees of freedom are 1. A chi-squared test may be adjusted to allow for a resulting overestimation of the calculated value using Yates' correction. This may not always be needed (7.6).

In any chi-squared test when there is only one degree of freedom, the $\chi^2_{calculated}$ value is too high if calculated in the ways described earlier in this chapter. These calculations therefore may be modified using Yates' correction. Yates' correction reduces the $\chi^2_{calculated}$ value by subtracting 0.5 from the numerator as shown:

$$\chi^2_{calculated} = \Sigma \left[\frac{(|observed - expected| - 0.5)^2}{expected} \right]$$

Absolute A number without its sign is called the absolute value and is indicated as |number|

The symbols | | indicate that the **absolute** value is used, so that when you take the expected value from the observed value, it does not matter if the outcome is negative: you can ignore this sign. For example, if the observed value was 10 and the expected value was 15, then |observed − expected| = |10 − 15| = 5, not = −5. The negative sign is ignored. For further examples of using absolute values, see Appendix e.

There are two occasions when you may come across a chi-squared test with one degree of freedom: in a goodness-of-fit test where the variable has only two categories or in an $r \times c$ test for association where there are only two rows and two columns. The methods for applying Yates' correction in these two cases are explained in this section. Further details about goodness-of-fit tests and tests for association are included earlier in this chapter (7.1 and 7.3).

7.4.1 Chi-squared goodness-of-fit test when there is one degree of freedom

In the earlier examples relating to a goodness-of-fit test, we considered the distribution of holly leaf miners on a single holly tree (Box 7.1) and we checked to see whether the distribution of a set of data was normal (Box 7.2). Another application of a goodness-of-fit test is when you have a genetic theory that is tested by carrying out a controlled cross. The observed results are then compared with those predicted by your theory. We used an example like this in 7.2 where there were the results from several accessions. Here we only have the results from a single accession and one degree of freedom.

EXAMPLE 7.4 The genetics of flower colour in *Allium schoenoprasum*

A cross was carried out between two plants of *Allium schoenoprasum*. As in Example 7.2, the flower colours were purple and white. The genetic model proposed for the inheritance of the flower colour means that the F_2 plants should occur in the ratio three purple-flowering plants:one white-flowering plant (Table 7.9).

Table 7.9. Flower colour in the F_2 of a single *Allium schoenoprasum* cross

	Flower colour in the F_2		Total number of plants flowering in the F_2
	Purple	White	
Number of plants	131	37	168

Table 7.10. Contingency table with expected values calculated for a goodness-of-fit test using a Yates's correction for Example 7.4: Flower colour in *Allium schoenoprasum* from a single cross

	Flower colour in the F_2		Total number of plants flowering in the F_2
	Purple	White	
Observed	131	37	168
Expected (Box 7.5)	126	42	168

A contingency table for the data relating to this example is shown in Table 7.9. If you analysed this data using a chi-squared goodness-of-fit test, then the degrees of freedom will be the 'number of categories' $(a) - 1 = 2 - 1 = 1$; therefore, you should use Yates' correction as shown in Box 7.5. The criteria for using this modified version of the chi-squared goodness-of-fit test are the same as those described in 7.1.1 apart from the number of categories. For this modified test, you will have one variable with only two categories.

 Q3 Are all the criteria for using this modified chi-squared goodness-of-fit test met for the data from Example 7.4?

Having organized your observed values into a contingency table and checked that the criteria for using this test are met, you can now proceed with the test (Box 7.5 and Table 7.10).

BOX 7.5 How to calculate a chi-squared goodness-of-fit test when there is one degree of freedom	
GENERAL DETAILS	**EXAMPLE 7.4**
	This calculation is given in full in the Online Resource Centre. All values have been rounded to five decimal places.
1. Hypotheses to be tested H_0: There is no difference between the expected and observed values.	1. Hypotheses to be tested H_0: There is no difference between the observed numbers of plants with purple or white flowers in the F_2 of *Allium schoenoprasum* plants and that expected from the a priori expectation of three purple-flowering plants:one white-flowering plant.

BOX 7.5 Continued

H_1: There is a difference between the expected and observed values.	H_1: There is a difference between the observed numbers of plants with purple or white flowers in the F_2 of *Allium schoenoprasum* plants and that expected from the a priori expectation of three purple-flowering plants:one white-flowering plant.
2. Have all the criteria for using this test been met?	2. Have all the criteria for using this test been met? Yes (7.4.1, answer to Q3).
3. How to work out expected values Expected values are calculated using the numerical formula you are expecting your data to conform to.	3. How to work out expected values We are expecting three purple-flowering plants:one white-flowering plant. As the total number of plants examined was 168, then 3/4 should be purple flowering, i.e. 126, and 1/4 plants should be white flowering, i.e. 42 (Table 7.10).
4. How to work out $\chi^2_{calculated}$ $$\chi^2_{calculated} = \sum \left[\frac{(|observed - expected| - 0.5)^2}{expected} \right]$$	4. How to work out $\chi^2_{calculated}$ $$\chi^2_{calculated} = \frac{(|131-126| - 0.5)^2}{126} + \frac{(|37-42| - 0.5)^2}{42}$$ $$= 0.64286$$
5. How to find $\chi^2_{critical}$ See Appendix d, Table D1. To find the critical value of χ^2 at $p = 0.05$, you need to know the degrees of freedom (ν). In this goodness-of-fit test, the degrees of freedom are the number of categories (a) – 1.	5. How to find $\chi^2_{critical}$ The categories in this example are 'purple' and 'white' and therefore: $\nu = 2 - 1 = 1$. When $p = 0.05$ and $\nu = 1$, then $\chi^2_{critical}$ is 3.84.
6. The rule If $\chi^2_{calculated}$ is greater than $\chi^2_{critical}$, you may reject the null hypothesis.	6. The rule $\chi^2_{calculated}$ (0.64) is less than $\chi^2_{critical}$ (3.84) at $p = 0.05$ and therefore we do not reject the null hypothesis.
7. What does this mean in real terms?	7. What does this mean in real terms? The flower colours in the F_2 generation of *Allium schoenoprasum* plants do not differ significantly ($\chi^2_{calculated} = 0.64$, $p = 0.05$) from the predicted ratio of three purple:one white. This indicates that the genetic model is probably correct.

Further examples relating to the topic including how to use statistical software are included in the Online Resource Centre.

@ **online resource centre**

7.4.2 Chi-squared test for association when there is only one degree of freedom

Some experiments are designed to test for an association between two variables, both of which have discrete categories. In our first example (Example 7.3), we carried out a test for association between the numbers of snails with particular shell patterns and their habitat. There were four categories of shell patterns (banded yellow, no bands yellow, banded pink, no bands pink) and two habitats (coastal and hedgerow). But there are many occasions when our two variables will each only have two categories. This is often referred to as a 2 × 2 chi-squared test for association.

EXAMPLE 7.5 Frequency of *Cepaea nemoralis* and *Cepaea hortensis* in a woodland and a hedgerow

A survey of two species of snail was carried out at two habitats: a woodland and a hedgerow. The numbers of snails at each location were recorded (Table 7.11).

Table 7.11. The distribution of *Cepaea nemoralis* and *Cepaea hortensis* in a woodland and a hedgerow

	Species of snail		Total
	C. *nemoralis*	C. *hortensis*	
Hedgerow observed	89	59	148
Woodland observed	16	8	24
Total	105	67	172

In the $r \times c$ chi-squared test for association, the degrees of freedom are $(r - 1)(c - 1)$ (Box 7.4). If we apply this to Example 7.5, the degrees of freedom will therefore be $(2 - 1)(2 - 1) = 1$, so Yates' correction may be applied. The criteria for using this modified version of the chi-squared test for association is similar to that described in 7.3.1 apart from the number of categories. For this modified test, you will have two variables, each with only two categories.

....

 Q4   Are all the criteria for using this modified chi-squared test for association met for Example 7.5?

....

Having organized your observed values in a contingency table and checked that the criteria for using this test are met, you can now proceed with the modified test (Box 7.6, Table 7.12).

Table 7.12. Contingency table with expected values calculated for a 2 × 2 chi-squared test for association using Yates's correction for Example 7.5: The distribution of *Cepaea nemoralis* and *Cepaea hortensis* in two habitats

| | Species of snail | | Total |
	C. nemoralis	*C. hortensis*	
Hedgerow observed	89	59	148
Hedgerow expected	90.34884	57.65116	
Woodland observed	16	8	24
Woodland expected	14.65116	9.34884	
Total	105	67	172

BOX 7.6 How to calculate a 2 × 2 chi-squared test for association

GENERAL DETAILS	EXAMPLE 7.5
	This calculation is given in full in the Online Resource Centre. All values have been rounded to five decimal places.
1. Hypotheses to be tested H_0: There is no association between the two variables. H_1: There is an association between the two variables.	**1. Hypotheses to be tested** H_0: There is no association between the two snail species (*C. nemoralis* and *C. hortensis*) and habitat (hedgerow and woodland). H_1: There is an association between the two snail species (*C. nemoralis* and *C. hortensis*) and habitat (hedgerow and woodland).
2. Have all the criteria for using this test been met?	**2. Have all the criteria for using this test been met?** Yes (see answer to Q4).
3. How to work out expected values In chi-squared tests for association, you have no *a priori* expectation against which to compare your observed data. Instead, the expected values are calculated from the totals of each column. The totals for rows and columns are first calculated and the expected values are then determined.	**3. How to work out expected values** Look first at the expected value in row 1, column 1 in Table 7.12. To calculate this expected value, take the column total for *C. nemoralis* (105) ÷ grand total (172) × row total for hedgerow snails (148) = 90.34884. This calculation is repeated for each row × column combination as shown in Table 7.12.
4. How to work out $\chi^2_{calculated}$ The rest of the procedure is the same as that described for the modified goodness-of-fit chi-squared test, where: $$\chi^2_{calculated} = \sum \left[\frac{(|observed - expected| - 0.5)^2}{expected} \right]$$	**4. How to work out $\chi^2_{calculated}$** $$\chi^2_{calculated} = \frac{(|89 - 90.34884| - 0.5)^2}{90.34884} + \frac{(|59 - 57.65116| - 0.5)^2}{57.65116}$$ $$+ \frac{(|16 - 14.65116| - 0.5)^2}{14.65116} + \frac{(|8 - 9.34884| - 0.5)^2}{9.34884}$$ $$= 0.14672$$

BOX 7.6 Continued

5. How to find $\chi^2_{critical}$

See Appendix d, Table D1. As before, you now need to calculate the degrees of freedom (v) before looking up the critical value. In this case, v is the (number of rows − 1) × (number of columns − 1). The critical value can be found in a chi-squared table at $p = 0.05$ and the appropriate degrees of freedom.

5. How to find $\chi^2_{critical}$

In our snails example, there are two rows (hedgerow and woodland). Do not include your 'expected' rows as these are part of your calculation. There are two columns (*C. nemoralis* and *C. hortensis*). Therefore, $v = (2 − 1) \times (2 − 1) = 1 \times 1 = 1$. The critical value to test for an association between the snail species and habitat is found in the chi-squared table where $v = 1$, $p = 0.05$, and $\chi^2_{critical} = 3.84$.

6. The rule

If $\chi^2_{calculated}$ is greater than $\chi^2_{critical}$, you may reject the null hypothesis.

6. The rule

$\chi^2_{calculated}$ (0.15) is less than $\chi^2_{critical}$ (3.84) at $p = 0.05$ and therefore we do not reject the null hypothesis.

7. What does this mean in real terms?

7. What does this mean in real terms?

There is no significant association ($\chi^2_{calculated} = 0.15$, $p = 0.05$) between the distribution of snail species and the two habitats.

Further examples relating to the topic including how to use statistical software are included in the Online Resource Centre.

online resource centre

Q5 In a small review of resources, two groups of students were asked to select one answer in response to the question, 'Have you found this book helpful?' They could select 'Yes' or 'No' (Table 7.13). The investigators wished to test whether there is an association between the course that the students are taking and their answer. What are the expected values and what is $\chi^2_{calculated}$?

Table 7.13. Closed answers given by Microbiology and Forensic Science students in their response to the question, 'Have you found this book helpful?'

	Answers to the question, 'Have you found this book helpful?'		
	Yes	No	Total
Number of Microbiology students	15	5	20
Number of Forensic Science students	10	10	20
Total observed	25	15	40

7.5 **G-tests**

Key points As an alternative to chi-squared tests, G-tests are often quicker to calculate, especially with large data sets. The criteria for the use of G-tests are the same as those for the chi-squared tests.

An alternative test to the chi-squared tests described above is the G-test. The advantages of this statistic are that, in larger data sets, it is easier to work out using a calculator and mathematicians believe that extensions of this test have advantages over the chi-squared test (Sokal & Rohlf, 1994). $G_{calculated}$ tends to be slightly larger than $\chi^2_{calculated}$, so if the calculated value is close to the critical value, the G-test will tend to reject the H_0 more often.

There are several versions of the G-test and we include two of these: a goodness-of-fit test and a test for association. The criteria that need to be satisfied are the same as those for the equivalent version of the chi-squared test. These G-tests can therefore be used to examine frequency distributions of nominal, ordinal, and interval data that are organized into discrete categories.

When observed numbers are low (between 1 and 4), G rejects H_0 too often and is likely to generate Type I errors. Therefore, a correction (Williams, 1976) similar in principle to Yates' correction can be used. When the total sample size is very small (less than 25), see Sokal & Rohlf (1994).

As the chi-squared tests are still the most frequently used tests, we have only included two applications of the G-test. However, G can be used for tests for heterogeneity and 2 × 2 tests for association. If you wish to use these types of tests, you should refer to other sources, such as Sokal & Rohlf (1994).

7.5.1 *G goodness-of-fit test*

G-tests are similar to chi-squared tests in many respects; therefore, only the points at which they differ are shown in Box 7.7. Refer back to Box 7.1 to see where the similarities and differences lie. The data being evaluated are found in Table 7.1. The criteria for using this test are the same as those given in 7.1.1. Unlike chi-squared tests, it is common practice to apply the Williams' correction routinely, although it has little effect when sample sizes are large.

In this example, $G_{calculated}$ (173.13) is greater than the $\chi^2_{calculated}$ (155.47) (Box 7.1) showing that this method for calculating a goodness of fit produces a higher test statistic, but as the critical values of $\chi^2_{critical}$ and $G_{critical}$ are much smaller than either calculated values, the outcome for both tests is the same.

BOX 7.7 How to calculate a *G* goodness-of-fit test

GENERAL DETAILS	EXAMPLE 7.1
	This calculation is given in full in the Online Resource Centre. All values have been rounded to five decimal places.
1. Hypotheses to be tested H_0: There is no difference between the expected and observed values. H_1: There is a difference between the expected and observed values.	1. Hypotheses to be tested See Box 7.1.
2. Have all the criteria for using this test been met?	2. Have all the criteria for using this test been met? Yes (7.1.1).
3. How to work out expected values Expected values are calculated using the numerical formula you are expecting your data to conform to.	3. How to work out expected values See Box 7.1.
4. How to work out $G_{calculated}$ i. For each pair of observed (O) and expected (E) values, calculate O ln (O/E). The term 'ln' is the natural logarithm. Calculators usually have a function key for converting values to this scale. ii. Add these values together $\sum[O\times \ln (O/E)]$ and multiply by 2 = G.	4. How to work out $G_{calculated}$ i. First calculate O/E, then take the natural log (ln). Multiply the natural log value by O. 131/57 = 2.29825 ln 2.29825 = 0.83215 131 × 0.83215 = 109.01113 38/57 = 0.66666 ln 0.66666 = −0.40547 38 × (−0.40547) = −15.40786 2/57 = 0.03509 ln 0.03509 = −3.34990 2 × (−3.34990) = −6.69981 ii. G =2 × [109.01113 + (−15.40786) + (−6.69981)] =2 × 86.90346 = 173.80692
5. Williams' correction The Williams' correction requires two steps. First calculate *W* and then use this term to revise the *G* value. We call this corrected *G* value $G_{calculated}$. $W = 1 + (a^2 - 1)/6nv$ $G_{calculated} = \dfrac{G}{W}$ where *a* is the number of categories, *n* is the total number of observations in the sample, and *v* the degrees of freedom = *a* − 1.	5. Williams' correction For this example, there are three categories so *a* = 3. The total number of observations in the sample is 171 and the degrees of freedom are *a* − 1 = 3 − 1 = 2. Using these values we can now work out *W*. $W = 1 + (3^2 - 1)/(6 \times 171 \times 2) = 1 + 8/2052$ $= 1 + 0.00390 = 1.00390$ $G_{calculated} = \dfrac{173.80692}{1.00390} = 173.13171$
6. How to find $G_{critical}$ See Appendix d, Table D1. To find the critical value of G at *p* = 0.05, you need to know the degrees of freedom (*v*). In this goodness-of-fit test, the degrees of freedom are the number of categories (*a*) − 1. Use a chi-squared table to find the critical value.	6. How to find $G_{critical}$ When *p* = 0.05 and *v* = 2, then $G_{critical}$ is 5.99.

BOX 7.7 Continued

7. The rule	7. The rule
If $G_{calculated}$ is greater than $G_{critical}$, you may reject the null hypothesis.	$G_{calculated}$ (173.13) is greater than $G_{critical}$ (5.99) at $p = 0.05$ and therefore we reject the null hypothesis.
8. What does this mean in real terms?	8. What does this mean in real terms? See Box 7.1.

Further examples relating to the topic including how to use statistical software are included in the Online Resource Centre.

online resource centre

7.5.2 An $r \times c$ G-test for association

Here we are using the data from Example 7.3 where the association between habitat and distribution of snail shell patterns was examined (Box 7.8). The observed data are found in Table 7.6. The criteria for using this test are given in 7.3.1. The process of calculating the $G_{calculated}$ statistic differs critically from the chi-squared. Again, the Williams' correction has been used. Where steps are similar to those in Box 7.4, this is indicated.

BOX 7.8 How to calculate an $r \times c$ G-test for association

GENERAL DETAILS	EXAMPLE 7.3
	This calculation is given in full in the Online Resource Centre. All values have been rounded to five decimal places.
1. Hypotheses to be tested H_0: There is no association between the two variables. H_1: There is an association between the two variables.	1. Hypotheses to be tested See Box 7.4.
2. Have all the criteria for using this test been met?	2. Have all the criteria for using this test been met? Yes (7.3.1).
3. How to work out expected values Not needed.	3. How to work out expected values Not needed.

BOX 7.8 Continued

4. How to work out $G_{calculated}$

i. First calculate the totals for each row and column and the grand total

ii. Calculate $\Sigma(O \times \ln O)$ for all categories and sum. (The term ln refers to the natural logarithm. This function button can be found on most statistical calculators)

iii. Calculate the same for the grand total. i.e. $N \times \ln N$

iv. For each row total and column total, calculate the same and add together.

v. Add the value from (ii) to the value from (iii) and take away the value from (iv).

vi. $G = 2 \times$ value from (v).

4. How to work out $G_{calculated}$

i. See Table 7.6.

ii. First take the ln of an observed value. Then multiply this by the same observed value. e.g. $10 \times \ln 10 = 10 \times 2.30258 = 23.02585$ Repeat this for each observed value and add all these values together = 278.50015

iii. The grand total is 105. So this step is: $105 \times \ln 105 = 105 \times 4.65396 = 488.66584$

iv. The first row total is 50, so for this row: $50 \times \ln 50 = 195.60115$ The second row total is 55, so for this row: $55 \times \ln 55 = 220.40333$. Continue for the four column totals (27, 27, 24, and 27) and add all the resulting values together. The total for rows and columns in this example = 759.24055.

v. In this example, this step will be: $278.50015 + 488.66584 - 759.24055 = 7.92544$

vi. $G = 2 \times 7.92544 = 15.85088$

5. Williams' correction

i. First work out 1/each row total and add these values together. Multiply by the grand total. Subtract 1.

ii. Work out 1/each column total and add these together. Multiply by the grand total. Subtract 1.

iii. Calculate (i) $\times$ (ii).

iv. Next calculate $6n - (\text{rows} - 1)(\text{columns} - 1)$.

v. $W = 1 + \dfrac{\text{(iii)}}{\text{(iv)}}$ $G_{calculated} = \dfrac{G}{W}$

5. Williams' correction

i. $1/50 + 1/55 = 0.03818$
$0.03818 \times 105 = 4.00909$
$4.00909 - 1 = 3.00909$

ii. $1/27 + 1/27 + 1/24 + 1/27 = 0.15277$
$0.15277 \times 105 = 16.04167$
$16.04167 - 1 = 15.04167$

iii. (i) $\times$ (ii) $= 3.00909 \times 15.04167 = 45.26173$

iv. $6 \times 105 \times (2 - 1) \times (4 - 1) = 1890$

v. $W = 1 + \dfrac{45.26173}{1890} = 1.02395$

$G_{calculated} = \dfrac{15.85088}{1.02395} = 15.48016$

6. How to find $G_{critical}$

See Appendix d, Table D1. v is the (number of rows – 1)(number of columns – 1). Look up $G_{critical}$ in the chi-squared table at $p = 0.05$.

6. How to find $G_{critical}$

When $v = 3$ and $p = 0.05$, $G_{critical} = 7.81$.

7. The rule

If $G_{calculated}$ is greater than $G_{critical}$, you may reject the null hypothesis.

7. The rule

$G_{calculated}$ (15.48) is greater than $G_{critical}$ (7.81) at $p = 0.05$ and therefore we reject the null hypothesis.

BOX 7.8 Continued	
8. What does this mean in real terms?	8. What does this mean in real terms? See Box 7.4.

Further examples relating to the topic including how to use statistical software are included in the Online Resource Centre.

online resource centre

If you compare the results from using the chi-squared test for association (Box 7.4) and the G-test for association (Box 7.8), it is again clear that the two calculated values differ. For the same data set, the $\chi^2_{calculated}$ (15.19) is less than $G_{calculated}$ (15.48). However, as the critical value is much smaller (7.81), the decision that is made is the same in both cases.

7.6 **The problem with small numbers**

Key points One criterion for using a chi-squared or G-test is that the 'expected' values are greater than 5. When this criterion is not met, you may be able to combine categories so that you may still analyse the data using one of these tests.

One of the criteria for using the tests described in this chapter is a requirement for all expected values to be greater than 5. There is some debate about this requirement and it is generally accepted that, as long as no more than 20% of all expected numbers are less than 5 and none is less than 1, then you may proceed with these tests.

Clearly, therefore, you need to bear this in mind when designing your investigations if you are planning to use any of the tests described in this chapter. Ideally, you need to carry out a preliminary investigation to get a sense of the relative numbers you would expect in each category. If you look at Example 7.5, you can see that not many *Cepaea hortensis* are sampled from the woodland. The sampling strategy used must therefore ensure that adequate numbers of *Cepaea hortensis* are included in the sample for analysis. Another example that we frequently come across is when students wish to use a questionnaire in their honours-year research project. For example, a student planned to ask 21 people a closed question and this closed question had seven possible answers. If these were chosen at random, you would *expect* only three respondents

Table 7.14. Maturity of *Quercus petraea* at the Wyre Forest and Mortimer Forest: the problem with small numbers in a chi-squared or *G*-test

Number of trees	Maturity of *Quercus petraea*				
	Seedling	Sapling	Immature tree	Mature tree	Total
Wyre Forest observed	51	21	29	4	105
Wyre Forest expected	51.3	23.3	26.4	3.8	
Mortimer Forest observed	15	9	5	1	30
Mortimer Forest expected	14.6	6.6	7.5	1.1	
Total	66	30	34	5	135

Table 7.15. Revised contingency table with combined classes with data from a survey of the maturity of *Quercus petraea* at the Wyre Forest and Mortimer Forest

Number of trees	Maturity of *Quercus petraea*			
	Seedling	Sapling	Mature tree and immature tree	Total
Wyre Forest observed	51	21	33	105
Wyre Forest expected	51.3	23.3	30.3	
Mortimer Forest observed	15	9	6	30
Mortimer Forest expected	14.6	6.6	8.6	
Total	66	30	39	135

to pick each answer. A sample size of 21 is therefore not adequate. This can be clearly seen in Q2/A2 earlier in this chapter.

Even with the best planning, you may still find yourself with expected values less than 5. In these circumstances, you may, if it is sensible, combine categories. For example, in Table 7.14, the results from a survey of oak trees (*Quercus petraea*) at the Wyre Forest and Mortimer Forest, Shropshire, are shown. Initially, the investigator placed the trees in one of four categories: seedling, sapling, immature tree, and mature tree.

You can see that two of the expected values are less than 5. This is more than 20% of the expected values. In addition, one of the values is about 1. To enable you to analyse these data with a chi-squared or *G*-test, it is therefore necessary to combine some of these categories and for these data it would appear both biologically acceptable and mathematically useful to combine the data from the immature and mature trees. The revised contingency table and new expected values are shown in Table 7.15.

The expected values are now all greater than 5. This does mean, however, that you can only comment on the immature and mature trees

in the two woodlands as you will have no statistics in which these two groups are examined separately.

The data in Table 7.14 are organized into nominal categories, and hypotheses relating to these data can be tested by a test for association. If the original classes result in expected values less than 5, then it is possible to combine them, as we have done, and then calculate revised expected values from the data in these new classes. This approach can be used for all cases where the expected values are less than 5 with one exception.

The exception arises when you wish to carry out a goodness-of-fit test and you have an *a priori* expectation, but have interval data, so the mid-point is used when calculating the expected values from the predicted ratio (7.1.3iii). The relative relationship between the mid-point and predicted ratio becomes disturbed when classes are combined and incorrect expected values will be calculated. In these circumstances, you should work out the expected values for the original classes and then add the expected values together in the classes you are combining (e.g. Box 7.2).

7.7 Experimental design and chi-squared and G-tests

Key points In general, experiments that meet the criteria for testing with a chi-squared test or *G*-test are simple in design with only one or two variables, usually each with only one sample and with data that is frequencies or counts recorded in discrete categories. If there are replicates or more than one sample, then the test for heterogeneity may be used.

In this chapter, we have considered chi-squared and *G*-tests for goodness of fit, heterogeneity, and association. Chi-squared and *G* goodness-of-fit tests can be used to compare the observed values from one sample with those predicted by an *a priori* expectation. The test for heterogeneity, however, allows you to compare several samples and to determine whether they are significantly different. Both these sets of tests allow you to test the significance of one variable. Where you wish to examine the effect of more than one variable, you may use the tests for association such as chi-squared and *G*-tests for association (considered in this chapter) and regressions and correlations (Chapter 8).

The chi-squared and *G*-tests can be used when you have count data organized into two or more categories, for example the number of trees in a forest at different stages of maturity (Table 7.14). The categories can be derived directly from the variable being observed (Example 7.3) or imposed as 'classes' (Example 7.1). These categories could include a control, although the information gained about the differences between the effect of the treatment and the control would be limited.

Table 7.16. Experimental design that may be suitable for analysis by a chi-squared or *G* goodness-of-fit test if all other criteria are met (7.1.1)

	Category 1	Category 2, etc.
Sample		

Table 7.17. Experimental design that may be suitable for analysis for heterogeneity in a goodness-of-fit chi-squared test if all other criteria are met (7.2.1)

	Category 1	Category 2, etc.
Sample 1		
Sample 2 etc.		

Table 7.18. Experimental design that may be suitable for analysis by a chi-squared test or *G*-test for association if all other criteria are met (7.3.1)

Treatment variable 2	Treatment variable 1	
	Category 1	Category 2, etc.
Category 1		
Category 2, etc.		

The goodness-of-fit tests require a single sample with one treatment variable where there are two or more categories (7.1, 7.4.1, and 7.5.1). In Table 7.1, for example, there is one row of observations in three categories. This experimental design has the generic structure shown in Table 7.16.

This simple design can be extended in that there may be more than one sample as seen in 7.2 on heterogeneity in a goodness-of-fit test (Table 7.17). These samples within the chi-squared tests or *G*-tests for heterogeneity are replicates within the experimental design.

The final type of design that may be suitable for testing with a chi-squared test or *G*-test for association (7.3, 7.4.2, and 7.5.2) is one with two variables each having two or more categories (Table 7.18).

From these, you can see that the applications of chi-squared and *G*-tests are limited to only a few experimental designs and you may need to investigate other more flexible tests to accommodate your design.

Q6 Which of the following designs are suitable for analysis by a chi-squared test or *G*-test, and which of these tests would be most appropriate?

a. An investigator recorded percentage germination in three varieties of seed and wished to examine whether they were significantly different.

b. Science students on three different courses were surveyed and asked whether they agreed with the government reinstating the full student grant system. They could answer either 'Yes' or 'No'. The investigator wished to compare the three groups of students.

c. In an investigation of visual responses in insects, bowls of different colours were placed in a grassy area and the numbers of insects attracted to each bowl were recorded. In total, three different types of insect (bugs, beetles, and flies) were attracted to the bowls. The numbers of each type of insect in each bowl were recorded and the researcher wished to compare the groups of insects in their responses to the particular colours.

Summary of Chapter 7

- The two groups of statistical tests considered in this chapter are the more commonly used chi-squared test (7.1–7.4) and the *G*-test (7.5). The latter is more useful when you have a large data set.

- The chi-squared and *G*-tests may be used to test two hypotheses: 'Do the data match an expected ratio?' (7.1, 7.4.1, and 7.5.1) or 'Is there an association between two or more variables?' (7.3, 7.4.2, and 7.5.2). These tests may also be used to check whether your samples are heterogeneous (7.2).

- The chi-squared and *G*-tests may be used if you have count data and wish to examine the effect of one or two treatment variables, where the data are organized into categories (7.7). If there are only two categories for the two variables, then Yates' correction may be used for the chi-squared test. In the *G*-test, the Williams' correction is always used (7.4 and 7.5).

- Both the chi-squared and *G*-test require you to work out expected values, which are then compared with your observations. In the goodness-of-fit tests, these expected values are derived from your *a priori* expectation (7.1), and in the test for association, expected values are determined by totals from the complete data set (7.3).

- There are only three experimental designs suitable for chi-squared and *G*-tests (7.7).

- Common faults in the use of these tests are the use of percentage data and organizing the data incorrectly in the contingency table so that the wrong comparisons are made (7.7 and Q6).

- The Online Resource Centre includes interactive exercises that test your understanding of this chapter with other topics, particularly those considered in Chapters 3 and 7–11.

 online resource centre

Answers to chapter questions

A1 The *a priori* expectation is that if the colour of the pitfall trap does not affect beetle behaviour, then there should be equal numbers of beetles in all the pitfall traps. As the total number of beetles observed was 20 + 34 + 10 + 40 = 104, we would expect 104/4 = 26 beetles in each trap.

A2 This is an example we included in Chapter 3 (Table 3.6). Did your expected values agree with those given? If not why not?

A3 Yes, we do have an *a priori* expectation that the plants will flower in a three purple:one white ratio. There is one treatment variable (flower colour), one sample, and the observations only fall into two discrete categories (purple and white). The number of plants has been recorded and each plant has been recorded only once; the data are independent. The final criterion relating to expected numbers can only be confirmed when you begin the calculation. If you look at Table 7.10 on the 'expected' row, these values are greater than 5. So this criterion is also met.

A4 Yes, we do wish to test for an association between two treatment variables (species and habitat) and each have only two categories. The number of snails has been recorded and each snail has only been recorded once, so the data are independent. If you look at Table 7.12. on the 'expected' rows all of these values are greater than 5. So this criterion is met.

A5 There are two variables and each has only two categories; therefore, there is only one degree of freedom. The investigators wish to test for an association and there is no *a priori* expectation; therefore, a chi-squared test for association with Yates' correction should be used. We included this example in Chapter 3 and the expected values are given in Table 3.4.

$$\chi^2_{calculated} = \frac{(|15-12.5|-0.5)^2}{12.5} + \frac{(|5-7.5|-0.5)^2}{7.5}$$
$$+ \frac{(|10-12.5|-0.5)^2}{12.5} + \frac{(|10-7.5|-0.5)^2}{7.5}$$
$$= 1.70667$$

A6 a. The first point to note is that the data are a derived variable (percentage) and therefore should not be analysed using a chi-squared or *G*-test. If the researcher had the original data, where the number of seeds sown was recorded and of these the number germinated, then a table could be drawn up (Table 7.19).

Table 7.19. Generic outline of the experimental design from Q6a, but using count data instead of percentages

	Number of seeds that germinated	Number of seeds that did not germinate
Variety 1		
Variety 2		
Variety 3		

This design is in accord with Table 7.18., and if all other criteria are met (7.3.1) then these data may be analysed using a chi-squared or *G*-test for association.

b. The data here reflect two variables, the 'answer to the question' and the 'three courses', which might suggest that a chi-squared test or *G*-test for association is appropriate. The data could be organized in two ways (Tables 7.20. and 7.21).

Table 7.20. First arrangement of data: comparing the courses for the numbers of students giving a 'yes' or 'no' answer

Answer	Number of students on the course		
	Course 1	Course 2	Course 3
Yes			
No			

Table 7.21. Second arrangement of data: comparing the answers given to a particular question by students on particular courses

Course	Number of students giving this response	
	Yes	No
Course 1		
Course 2		
Course 3		

In the chi-squared and *G*-tests, it is the rows that are compared, the distribution of values in one row being contrasted to the distribution of values in the other row(s). For Table 7.20, this does not make sense. Row 1 cannot help but be the direct opposite to row 2 because a student could only answer yes or no. In the second version (Table 7.21), the three rows can sensibly be compared: what is the difference in relative responses to the question by the students on the three courses? Therefore, for this example, the data should be organized in the form shown in Table 7.21 and a chi-squared test or *G*-test for association may then be carried out if all other criteria are met (7.3.1).

c. There are two variables here: the colour of the bowls and the number of insects of each type (Table 7.22).

Table 7.22. Experimental design for a comparison between the numbers of different groups of insects attracted to different coloured bowls

Insect type	Number of insects attracted to each coloured bowl			
	Red	White	Green	Blue
Bugs				
Beetles				
Flies				

Here we have organized the data so that we are comparing the numbers of insects found in the bowls for each given insect type. As we explained in A6(b), the chi-squared test or *G*-test for association will test whether the relative numbers of insects within each coloured bowl is statistically similar for all groups of insects. As in all chi-squared tests or *G*-tests for association, the comparison is between the rows and you learn little about the difference between the columns. If you were interested in this, then you could use a chi-squared goodness-of-fit test for each group of insects (e.g. bugs, then beetles, and then flies as three separate calculations), or we suggest you reconsider your statistical test and have a look at the two-way non-parametric ANOVA or the Scheirer–Ray–Hare test (Chapter 10).

Hypothesis testing: associations and relationships

8

In a nutshell

In this chapter, we consider the second hypothesis that may be tested: a test for an association. An association is where one variable changes in a consistent and similar manner to another variable. There are three groups of statistical tests that we cover that can be used: chi-squared and *G*-tests for association (Chapter 7), correlation analysis, and regression analysis. We discuss this type of hypothesis and how to decide which is the correct statistical test to use, as well as giving examples of each test in use. The full calculations, further examples, and a further test (ANOVA to test for a linear regression where you have unequal replicates) is included in the Online Resource Centre.

online
resource
centre

In your research, you often investigate the effect of one variable, for example, a comparison of snail shell patterns found in coastal and woodland populations of *Cepaea nemoralis* (Example 7.3). The variable being examined here is 'habitat'. But there are some investigations where you will have more than one variable, such as when comparing leg length and arm length in humans (Example 8.4). In this chapter, therefore, we look at testing hypotheses where you wish to evaluate the possibility of an association between these two variables. This is the second of the three types of hypotheses we discussed in Chapter 6 (6.1.1). We cover four groups of tests. Chi-squared and *G*-tests for association were considered in Chapter 7, and correlations and regressions are considered in detail in this chapter. If you work through this chapter and the questions, it should take about two hours. The answers for these exercises are at the end of the chapter. All our examples are based on real undergraduate research projects. If these examples are not in your subject area you will find more in the Online Resource Centre.

online
resource
centre

Associations and relationships

It is important to be clear about what we mean by the terms 'association' and 'relationship'. In some text books, these terms are used interchangeably,

Association Where one variable is found to change in a similar manner to another variable

Relationship Where one variable is directly responsible for causing the change in another variable

but we believe it is important to make a distinction. An association is where one variable is found to change in a similar manner to another variable (e.g. Figs 8.13 and 8.15). A relationship is where one variable is *directly* responsible for causing the change in another variable.

Most samples of humans in which arm and leg length are recorded show a significant association between these two variables: if you have relatively long arms, you will also have relatively long legs (Example 8.4). But you would not argue that your arm length directly causes your leg length. This is not sensible. Instead, this association reflects the actions of other phenomena that have not been observed. Therefore, although you may have a significant association, this does not indicate a relationship. You can contrast this example with a significant association observed between the hardness of eggshells and the amount of feed supplement eaten by pullets (Example 8.1). In this case, you could argue that nutritional factors in the feed supplement directly influence eggshell hardness. This is an association for which you can argue that there is a relationship.

The correlation and regression analysis included in this chapter test for the presence of a statistically significant linear association. As such, the change in one variable (y) may well be caused by the change in the second variable (x), or the change in the second variable may be caused by the change in the first variable. These are both relationships. In addition to this, a significant association may indicate that the changes in the first and second variables are due to a change in a third, unrecorded variable, as described in our arms and legs example. Alternatively, the statistical association between two variables may be a coincidence. You will need to bear these points in mind when you are interpreting the results from any significant test for association.

How to choose the correct test

In this section, we first discuss when you might consider using a chi-squared or G-test for association or when you might use a correlation or regression. We then consider why you might choose a correlation and/or a regression to analyse your data. Finally, we identify the key points to enable you to choose between regressions included in this chapter.

Should I use a chi-squared or G-test to test for an association?

In Chapter 7 (7.3, 7.4, and 7.5.2), you were introduced to the chi-squared and G-tests for association. For example, we examined the frequency of *Cepaea nemoralis* and *Cepaea hortensis* in a hedge and a woodland (Example 7.5). This was an example of an investigation where we wished to test for an association between two variables (species of snail and location). But why did we use a chi-squared test and not a correlation?

Data can be measured on a nominal, ordinal, or interval scale. In the first variable in Example 7.5, there are two snail species. The data are placed into these two discrete categories, and the categories cannot be arranged in a specific order, i.e. this is nominal data. The same is true for the second variable (location). This also has two categories ('hedge' and 'woodland'), which cannot be placed in a particular order and are therefore nominal. Where data are measured on a nominal scale and the observations are recorded as counts or frequencies within these discrete categories, then a chi-squared or G-test for association is appropriate (Chapter 7).

If the data are measured on an ordinal scale and the observations are recorded as counts or frequencies within these discrete categories, then a chi-squared or G-test for association is the more appropriate group of tests to consider, especially if the apparent association between the two variables does not appear to be linear.

With interval data, you have two choices. If you plot the data and the apparent association does not appear to be linear, you should either consider using a non-linear regression (e.g. Sokal & Rohlf, 1994) or arrange the data into categories (classes) and record the number of observations in each category. These data may then be analysed using a chi-squared or G-test for association (7.3, 7.4.2, and 7.5.2). If, when plotted, the interval data have an apparently linear association, then a regression and/or correlation should be considered first. For example, in Fig. 8.12, there is an apparently linear association between the concentration of zinc extracted in soil samples and the distance from an electricity pylon. These variables are measured on an interval scale, not organized into categories, and therefore a correlation or regression is more appropriate.

Should I use a correlation and/or a regression analysis?

When making this decision you need to consider your reasons for choosing your analysis. These tests of association can be used for three things: modelling, prediction, and establishing whether there is a significant association. The line that may be used to describe the trend in the data is a model of this association. For example, in Fig. 8.16, the line that best fits the data from Table 8.1 is shown and indicates a positive association between the amount of feed supplement eaten by pullets and eggshell hardness. This line is described by the equation $y = -2.4 + 0.5x$ (see Box 8.6). This line can also be used to predict a y value for any given x value within the range of your data. For example, if the pullets were given 10g of feed supplement (x), then the hardness of their eggshells would be expected to be about 2.6 units. If these eggshells were hard enough for the eggs to be sold, then the feed supplement given to the pullets could be more economically controlled. For both of these applications (modelling

and prediction), you should choose a regression analysis. However, if you wish to establish whether there is a significant association between the two variables, then a correlation analysis is more likely to be appropriate, but you should read on before making your final choice.

Which regression?

The regression analyses fall into two groups. The first (Model I) may be appropriate when you, the investigator, determine one of the variables. For example, you may set up an experiment to examine the number of pollen grains falling on agar plates set at specific distances from a field of oilseed rape. In this investigation, the location of the agar plates will be under your control and is not therefore subject to sampling error (6.1.3). If neither variable is under your control, then you should consider using a Model II regression. However, there is some debate as to which is the most appropriate test to use for a given set of data. We have endeavoured to summarize this debate in the table below, but for a fuller discussion you should refer to Legendre & Legendre (1998).

Which statistical test?

The following directs you to the most likely statistical test. If you are not sure, go to the specific sections indicated and look at the examples to see whether they are similar to the work you are planning. Each test has additional criteria that need to be met. These are given in the sections indicated. To check whether your data are parametric or not, go to Box 5.2. Parametric tests are more powerful; therefore, if you have parametric data, you should always use the parametric test. If you have non-parametric data, you should consider transforming them (5.9). The 'power of a test' is explained in the glossary.

You wish to test the significance of an association between two variables. The data are counts or frequencies organized into discrete categories or classes. The categories can be derived from ordinal or nominal data or can be imposed on interval data. The distribution does not need to be linear.	Chi-squared or *G*-test for association (Chapter 7)
You wish to test the significance of an association between two variables. The distribution appears to be linear. The data are non-parametric.	Spearman's rank correlation (8.2)
You wish to test the significance of an association between two variables. The distribution appears to be linear. The data are parametric.	Pearson's product moment correlation (8.3)

You have two treatment variables, one of which is usually under the control of the investigator so that the observations are not subject to sampling error. Both variables are measured on interval scales. There is only one y value for each value of x. The distribution appears to be linear. You wish to test the significance of an association and/or you wish to model the association and/or predict y values for given x values within the range of your observations.	Simple linear regression (8.5)
You have two treatment variables, one of which is usually under the control of the investigator so that the observations are not subject to sampling error. Both variables are measured on interval scales and the y values are parametric. There is more than one y value for each value of x, with equal replication in all categories. The association appears to be linear. You wish to test the significance of an association and/or you wish to model the association and/or predict y values for given x values within the range of your observations.	One-way ANOVA to test for a linear regression: more than one y for each x (equal replicates) (8.6)
You have two treatment variables and neither is under the control of the investigator. The two variables are measured on the same interval scale, e.g. mm. You wish to model the association and/or predict y values for given x values within the range of your observations.	Principal axis regression (8.7)
You have two treatment variables and neither is under the control of the investigator. The two variables are measured on different scales, but both scales are interval, e.g. g and cm. You wish to model the association and/or predict y values for given x values within the range of your observations.	Ranged principal axis regression (8.8)
None of these tests seems to be right for your data. For example, you have three or more treatment variables.	See Grafen & Hails (2002); Legendre & Legendre (1998); Sokal & Rohlf (1994).

8.1 Correlations

Key points To test the significance of a linear association, you may use a correlation analysis. We consider two statistical tests: Spearman's rank correlation for data that is non-parametric and Pearson's correlation for data that is parametric. Calculations for a correlation generate an *r* value that indicates the nature and strength of the correlation, and can be used in a test of significance. Significance is dependent on sample size. A key underlying assumption is that the association is linear.

If you have data with two treatment variables and you plot this on a scatter plot, it may suggest that there is a linear association between the two variables. A correlation analysis examines the data mathematically to see how probable this is. The two important values for this analysis are the correlation coefficient (r) and the probability value (p). The correlation coefficient may be either positive or negative in sign. A positive sign indicates that there is a positive association between the two variables: as one increases so does the other (Fig. 8.1). A negative sign indicates a negative association, as one variable increases the other decreases (Fig. 8.2).

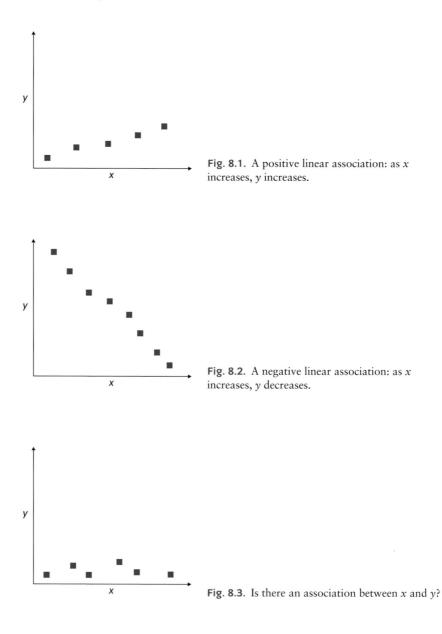

Fig. 8.1. A positive linear association: as x increases, y increases.

Fig. 8.2. A negative linear association: as x increases, y decreases.

Fig. 8.3. Is there an association between x and y?

Q1   Is there an apparent linear association between the two variables illustrated in Fig. 8.3?

A correlation coefficient will lie somewhere between −1 and +1. If you have a value greater than +1 or less than −1, then you have made a mistake somewhere in your calculation. A value of $r = 0$ means that there is zero correlation, i.e. no association between the two variables, whereas $r = +1$ or r = −1 indicates a 'perfect' association between the two variables.

A correlation coefficient in itself does not tell you whether this is a statistically significant association: a comparison between the calculated value and a critical value for a given level of probability must be made. For example, $r = 0.2$ indicates a weak association and $r = 0.8$ indicates a stronger association. A strong association is likely to be statistically significant, even with a small sample size. However, a small correlation such as $r = 0.2$ may be statistically significant, but only if the sample size is large enough. You need to bear this in mind when interpreting findings from your own or other people's research.

8.2 Spearman's rank correlation

Key points This test for the significance of a linear association can be used with ordinal or interval non-parametric data. The first key step involves ranking the observations. The criteria that need to be met by the data and the method are given.

This is a ranking test. There are many ranking tests that can be used to test hypotheses. These tests are commonly used for the analysis of non-parametric data where the data does not have to fit a particular type of distribution (5.2). These tests are collectively known as ranking tests, because the first step, which they all have in common, requires you to assign ranks to your observations. If you are not familiar with this process, you should first read Box 5.3.

8.2.1 Using this test

To use this test you:

1) Wish to test the significance of an association between the variables.
2) Need two treatment variables recorded for each item.
3) Have non-parametric data.

4) Have an apparently linear distribution where the points are reasonably scattered (see Q2).

5) Do not need to have an independent and a dependent variable.

6) Have variables that are measured on rankable scales.

7) Have seven or more pairs of observations. (Having more than 30 pairs of observations does not add to the accuracy of this test.)

8) Should have few tied values, i.e. few of the numbers in either one of the two data sets should be the same as each other. This may look like an association, but it is not (see Q1).

 Q2 Examine Figs 8.1, 8.2, and 8.4(a–d). Which of these fulfil criterion 4? For those distributions that do not meet criterion 4, what can be done about it?

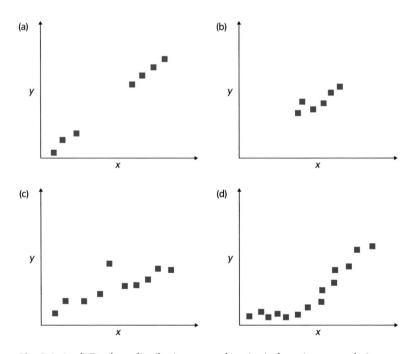

Fig. 8.4. (a–d) Do these distributions meet the criteria for using a correlation analysis?

EXAMPLE 8.1. The hardness of eggshells in pullets

A student believes that there is a relationship between the hardness of shells in eggs laid by Maren pullets and the amount of a particular layer's pellet eaten as part of their mixed diet. She selected 13 of her pullets at random and on one day recorded the amount of each layer's pellets consumed and the hardness of the eggs laid, on a scale of 0.0 (softest) to 10.0 (hardest) (Table 8.1).

Table 8.1. The hardness of eggshells produced by 13 Maren pullets and their consumption of a food supplement

Pullet	Amount of food supplement (g)	Hardness of shells
1	19.5	7.1
2	11.2	3.4
3	14.0	4.5
4	15.1	5.1
5	9.5	2.1
6	7.0	1.2
7	9.8	2.1
8	11.6	3.4
9	17.5	6.1
10	11.2	3.0
11	8.2	1.7
12	12.4	3.4
13	14.2	4.2

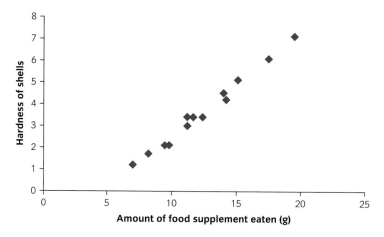

Fig. 8.5. Amount of food supplement eaten (g) and hardness of eggshells in Maren pullets.

The results in Example 8.1 meet the criteria for using Spearman's rank correlation. For each pullet (an item), there are two observations, one for each variable: 'g' and 'hardness'. Observations on these scales can be ranked. There are 13 pairs of observations. When plotted on a scatter plot, there appears to be a linear association and the points are reasonably scattered (Fig. 8.5).

8.2.2 The calculation

We show you the general calculation and a specific example (Example 8.1) for Spearman's rank correlation in Box 8.1. In addition, you will need to refer to the calculation table (Table 8.2). Full details of all steps in these calculations are given in the Online Resource Centre.

BOX 8.1 How to carry out Spearman's rank correlation

GENERAL DETAILS	EXAMPLE 8.1
	This calculation is given in full in the Online Resource Centre. All values have been rounded to 5 decimal places.
1. Hypotheses to be tested H_0: There is no correlation between variable 1 and variable 2 in the population from which the samples are taken. H_1: There is a correlation between variable 1 and variable 2 in the population from which the samples are taken.	**1. Hypotheses to be tested** H_0: There is no correlation between the hardness of shells and the amount of food supplement (g) eaten by 13 Maren pullets. H_1: There is a correlation between the hardness of shells and the amount of food supplement (g) eaten by 13 Maren pullets.

BOX 8.1 Continued	
2. Have the criteria for using this test been met?	**2. Have the criteria for using this test been met?** Yes (8.2.1).
3. How to work out $r_{s\,calculated}$ i. For this test it does not matter if you have a dependent and an independent variable (8.5); therefore, let the variable on the x-axis be variable 1 and the variable on the y-axis be variable 2. ii. Arrange the data for variable 1 in numerical order from the smallest to the largest. iii. Assign each observation a rank. Where there is a tie, assign the middle (average) rank. If you are unfamiliar with this process, see Box 5.3. iv. Repeat steps (i) and (ii) for variable 2. v. Return to your original table of data and note down the ranks for each observation. Make sure you do not reorganize the data. The two observations for each item must remain in the same row as each other. vi. Calculate the difference between the ranks for each pair of observations (d). vii. Square each d value and then sum all the d^2 to work out Σd^2. viii. Calculate r_s: $$r_s = 1 - \frac{6\Sigma d^2}{n(n^2 - 1)}$$ where n is the number of pairs of observations.	**3. How to work out $r_{s\,calculated}$** i. Let the amount of food supplement eaten by the pullets be variable 1 and shell hardness be variable 2 (Table 8.1). ii–v. If you are working this calculation out by hand, it is simplest to use a calculation table. The outcomes from steps (ii)–(vi) are shown in Table 8.2 using the data from Example 8.1. A common error is to reorganize the data in this table to reflect the ranks. This can lead to the two observations from one item no longer being paired on the same row. If you are not sure, compare Table 8.1 with Table 8.2. The observations remain in the same order in relation to each other. vi, vii. The difference between the ranks and sum of ranks has been added to the calculation table (Table 8.2). viii. In this example, $n = 13$ pairs of observations. So: $$r_s = 1 - \frac{6 \times 6.0}{13(13^2 - 1)} = 0.98352$$
4. How to find $r_{s\,critical}$ See Appendix d, Table D2. Find r_s critical in a Spearman's rank r table for a two-tailed test at $p = 0.05$ and where n = the number of pairs of observations.	**4. How to find $r_{s\,critical}$** When $p = 0.05$ and $n = 13$, then r_s critical is 0.566.
5. The rule If the absolute*, calculated value of r_s is greater than the critical r_s, we reject H_0. (*If you are not familiar with this term please refer to the glossary.)	**5. The rule** In our example, r_s was positive, so the fact that we take the absolute value will not change r_s in this example. $r_{s\,calculated}$ (0.984) is greater than $r_{s\,critical}$ (0.566) at $p = 0.05$. In fact, at $p = 0.001$, $r_{s\,critical}$ is 0.824. Therefore, we can reject the null hypothesis at this higher level of significance.
6. What does this mean in real terms?	**6. What does this mean in real terms?** There is a strong and very highly statistically significant positive correlation ($r_s = 0.984$, $p = 0.001$) between the hardness of shells of eggs and the amount of food supplement (g) eaten by Maren pullets.
Further examples relating to the topic including how to use statistical software are included in the Online Resource Centre.	

ⓦ online resource centre

Table 8.2. Calculating Spearman's rank correlation for Example 8.1: The hardness of shells and consumption of food supplement in Maren pullets

Amount of food supplement (g)	Rank	Hardness of shells	Rank	Difference (d)	Difference² (d²)
19.5	13.0	7.1	13.0	0.0	0.00
11.2	5.5	3.4	7.0	−1.5	2.25
14.0	9.0	4.5	10.0	−1.0	1.00
15.1	11.0	5.1	11.0	0.0	0.00
9.5	3.0	2.1	3.5	−0.5	0.25
7.0	1.0	1.2	1.0	0.0	0.00
9.8	4.0	2.1	3.5	0.5	0.25
11.6	7.0	3.4	7.0	0.0	0.00
17.5	12.0	6.1	12.0	0.0	0.00
11.2	5.5	3.0	5.0	0.5	0.25
8.2	2.0	1.7	2.0	0.0	0.00
12.4	8.0	3.4	7.0	1.0	1.00
14.2	10.0	4.2	9.0	1.0	1.00
$n = 13$					$\Sigma d^2 = 6.0$

8.3 Pearson's product moment correlation

Key points This test for the significance of a linear association can be used with parametric data. The key step involves the calculation of sums of squares for both variables.

This is a parametric test and, unlike Spearman's rank test, you do not rank the data; however, you do assume that your data are parametric (5.8). Like Spearman's rank correlation, the coefficient (r) may indicate the strength and sign of the association that is apparent in your data (8.1) and may be used to test the significance of the association (see Box 8.2).

8.3.1 Using this test

To use this test you:

1) Wish to test the significance of an association between the variables.
2) Have two treatment variables measured for each item.
3) Have parametric data. Both variables must be approximately normally distributed (Box 5.2 shows you how to tell whether your data are parametric).
4) Have a distribution that appears to be linear when plotted, with points that are reasonably scattered (e.g. Figs 8.1 and 8.2).

5) Have at least three pairs of observations.

6) In addition, there is no requirement for there to be an independent or a dependent variable.

Although the Maren pullet data was used in 8.2 to illustrate Spearman's rank correlation, these data also meet the more stringent criteria for using Pearson's correlation and so are used here. If you have data that meet the criteria for a parametric test, this is the one you should choose as these tests are more powerful. (If you are not familiar with this term, refer to the glossary and 5.8.1.) The data from Example 8.1 meet the criteria for Pearson's correlation. We wish to test the significance of an association between two variables (amount of food supplement eaten and hardness of shells). Both columns of data meet the criteria for normally distributed data and are therefore parametric, and as we have already seen in Fig. 8.5, the distribution is apparently linear and the points are reasonably scattered.

8.3.2 The calculation

We show you the general calculation and a specific example (Example 8.1) for Pearson's correlation in Box 8.2 and the calculation table (Table 8.3). Where steps have been abbreviated, the full calculation is included in the Online Resource Centre.

 online resource centre

BOX 8.2 How to carry out Pearson's product moment correlation

GENERAL DETAILS	EXAMPLE 8.1
	This calculation is given in full in the Online Resource Centre. All values have been rounded to 5 decimal places.
1. Hypotheses to be tested H_0: There is no correlation between variable 1 and variable 2 in the population from which the samples are taken. H_1: There is a correlation between variable 1 and variable 2 in the population from which the samples are taken.	**1. Hypotheses to be tested** H_0: There is no correlation between the hardness of shells and the amount of food supplement (g) eaten by 13 Maren pullets. H_1: There is a correlation between the hardness of shells and the amount of food supplement (g) eaten by 13 Maren pullets.
2. Have the criteria for using this test been met?	**2. Have the criteria for using this test been met?** Yes (8.3.1)
3. How to work out r i. There is no requirement for there to be a dependent or an independent variable, so you may choose which variable should be x and which y.	**3. How to work out r** i. Let the amount of food supplement eaten be the x variable and hardness of shells be the y variable.

BOX 8.2 Continued

ii. Calculate the sum of squares of x (SS (x)):

$$SS(x) = \sum x^2 - \frac{(\sum x)^2}{n}$$

where n is the number of pairs of observations. This term should be familiar to you as it has figured in the calculation for standard deviation and variances (Box 5.1).

iii. Calculate the sum of squares of y (SS(y)). For SS(y), substitute the 'y' variable data into the same formula as above, i.e.

$$SS(y) = \sum y^2 - \frac{(\sum y)^2}{n}$$

iv. Calculate the sum of products (SP(xy)). For your data, multiply each x by its related y to produce a column of values (xy). Add these together = $\sum xy$. Using this and values from (ii) and (iii), calculate SP(xy) as follows:

$$SP(xy) = \sum xy - \frac{(\sum x)(\sum y)}{n}$$

v. Now you can calculate r.

$$r = \frac{SP(xy)}{\sqrt{[SS(x)SS(y)]}}$$

Remember, r is the correlation coefficient and should fall in the range ±1. The sign indicates whether this is a positive or negative association.

---

ii. This is an important value that you will come across frequently in this book so we have shown you again in this example how to work this parameter out (Table 8.3).

$$SS(x) = 2153.88 - \frac{25985.44}{13}$$
$$= 155.0$$

iii. For details, see Table 8.3.

$$SS(y) = 208.35 - \frac{2237.29}{13}$$
$$= 36.25077$$

iv. For details, see Table 8.3.

$$SP(xy) = 661.0 - \frac{161.2 \times 47.3}{13}$$
$$= 74.48$$

v. $r = \frac{74.48}{\sqrt{(155.0 \times 36.25077)}} = 0.99361$

This indicates a strong positive correlation. But is it statistically significant?

---

4. To test the significance of the association

See Appendix d, Table D3. To find the critical value of r, you need to know the degrees of freedom (v). In this Pearson's correlation, the degrees of freedom are the number of pairs of observations (n) – 2. Use an r table of critical values to locate the value at $p = 0.05$.

4. To test the significance of the association

In this example, there are 13 pullets so $v = 13 - 2 = 11$. At $p = 0.05$, for a two-tailed test, $r_{critical} = 0.553$

---

5. The rule

If $r_{calculated}$ is greater than $r_{critical}$, you may reject the null hypothesis.

5. The rule

As $r_{calculated}$ (0.994) is greater than $r_{critical}$ (0.553), you may reject the null hypothesis. In fact, at $p = 0.01$, $r_{critical}$ is 0.684. Therefore, we can reject the null hypothesis at this higher level of significance.

---

6. What does this mean in real terms?

6. What does this mean in real terms?

There is a strong ($r = 0.99$) and highly significant positive correlation ($p < 0.01$) between the amount of food supplement (g) given to the Maren pullets and the hardness of the eggshells that they produce.

Further examples relating to the topic including how to use statistical software are included in the Online Resource Centre.

 online resource centre

Table 8.3. Calculating Pearson's product moment correlation for Example 8.1 between the hardness of shells and food consumption in Maren pullets

Amount of food supplement (g) (x)	x^2	Hardness of shells (y)	y^2	$x \times y$
7.00	49.00	1.20	1.44	8.40
8.20	67.24	1.70	2.89	13.94
9.50	90.25	2.10	4.41	19.95
9.80	96.04	2.10	4.41	20.58
11.20	125.44	3.00	9.00	33.60
11.20	125.44	3.40	11.56	38.08
11.60	134.56	3.40	11.56	39.44
12.40	153.76	3.40	11.56	42.16
14.00	196.00	4.50	20.25	63.00
14.20	201.64	4.20	17.64	59.64
15.10	228.01	5.10	26.01	77.01
17.50	306.25	6.10	37.21	106.75
19.50	380.25	7.10	50.41	138.46
$\Sigma x = 161.20$	$\Sigma(x)^2 = 2153.88$	$\Sigma y = 47.30$	$\Sigma(y)^2 = 208.35$	$\Sigma xy = 661.00$
$\Sigma(x)^2 = 25985.44$		$\Sigma(y)^2 = 2237.29$		
$\bar{x} = 12.40$		$\bar{y} = 3.63846$		
$n = 13$		$n = 13$		

8.3.3 Coefficient of determination

> **Key points** This is a useful measure that may be used with Pearson's correlation to indicate how much variation in one variable is accounted for by variation in the second variable.

$r^2 \times 100$ is a useful measure called the coefficient of determination (%). It is a measure of the proportion of variability in one treatment variable that is accounted for by variation in the other treatment variable and therefore indicates to what extent other factors are influencing x and y. The use of the coefficient of determination is dependent on whether your data satisfy the criteria for Pearson's product moment correlation and whether an r value can be calculated. From Box 8.2, we know that $r = 0.99361$. Therefore, $r^2 \times 100 = 98.7\%$. This indicates that, in Example 8.1, nearly all the variation observed in the hardness of the eggshells is due to the variation in the amount of food supplement eaten.

8.4 Regressions

Key points The regressions we consider here can be used to model a linear association for interval data. We cover four regressions. To test the significance of the regression, a different approach is required depending on the regression you are using. The criteria that need to be met by the data and the methods are given.

A regression analysis differs from a correlation analysis in that it calculates the mathematical equation that describes or models your data. This model may be used as a description of any significant association between the two variables and/or used to predict values. We describe two methods for modelling a linear regression. However, there are also ways to fit a **curvilinear** line. We have not been able to include these here, but we refer you to other texts (e.g. Legendre & Legendre, 1998; Sokal & Rohlf, 1994).

Curvilinear When data are plotted and where part or all of the distribution is found to curve rather than follow a straight line

The first of the linear regressions we consider in this chapter is the simple linear regression where the line is determined by mathematically minimizing the distance between the observations and the line (Fig. 8.6). Clearly, in this process some distances will be negative and some positive and would cancel each other out, so these values are squared. This has the effect of making all the values positive. The regression 'fits' a line that minimizes these deviations from the line. This regression is also therefore known as a least-squares regression.

The two important values that are calculated in the simple linear regression are the slope or gradient of the linear association (b) and the point at which the line cuts the y-axis (a). The general mathematical formula that describes this line is $y = a + bx$ (Fig. 8.7). This is a Model I regression. In the principal axis regression and the ranged version of the principal axis regression (Model II), the line is determined by 'drawing' an ellipse around the data. The flatter the ellipse, the more likely it is that the association will be significant (Fig. 8.8). The line that passes along the longest axis of the ellipse is called the principal axis, and this can also be found by calculating the slope b_1. In these regressions, the general mathematical

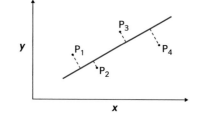

Fig. 8.6. Minimization of differences from a regression line. Differences from P_1 and P_3 to the line are positive; differences from P_2 and P_4 are negative.

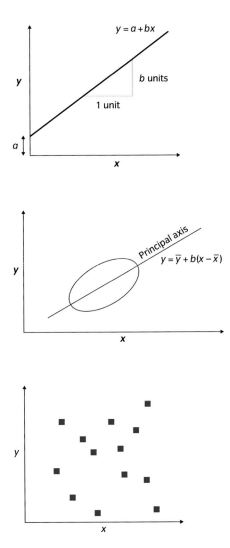

Fig. 8.7. The general line described by a Model I simple linear regression.

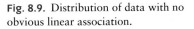

Fig. 8.8. The general line described by the Model II principal axis and ranged principal axis regressions.

Fig. 8.9. Distribution of data with no obvious linear association.

formula for the line (the principal axis) is $y = \bar{y} + b_1(x - \bar{x})$. As in correlation analyses, the slope b or b_1 can be either positive or negative. These b values are sometimes referred to as regression coefficients.

You could fit a line through any data on a scatter plot. For example, you could draw a line on Fig. 8.9, although clearly this would be meaningless. Just drawing a line is not enough to indicate an association. It is therefore necessary to show whether this line represents a statistically significant trend in the data. For a simple linear regression, a modified t-test (See Box 8.4) or ANOVA (analysis of variance) (See Box 8.6) is used. However, for the principal axis regression and ranged principal axis regression, a correlation should be carried out to test the significance of the association before making use of any regression analysis.

In the following sections, we describe the methods for drawing and testing a linear regression when you have two variables. However, regressions can be modified in a number of ways, for example if you have three or more variables. An explanation of these and other uses of regression analyses are given in, for example, Sokal & Rohlf (1994).

8.5 Model I: simple linear regression: only one *y* for each *x*

Key points With a simple linear regression, you may test the significance of an association and/or model the association and/or predict *y* values for given *x* values within the range of your observations. A simple linear regression may be used when one set of observations is not subject to sampling error and where both variables are measured on interval scales. This is a Model I regression based on the general equation $y = a + bx$.

A simple linear regression analysis is the most commonly used type of regression. In this analysis, first the coefficients *a* and *b* are calculated that allow you to mathematically describe and to draw the regression line. To test the significance of this line, a modified *t* value is calculated. One of the criteria that must be met to use a simple linear regression is that one of the variables is taken without sampling error and is therefore usually under the investigator's control. This is always denoted the *x* variable, and as regressing *y* on *x* is not the same as regressing *x* on *y*, you must take care to identify which is the variable taken without sampling error (*x*) and which is not (*y*). Although a change in *x* may not directly cause a change in *y* (8.1), *y* is often referred to as the dependent variable and *x* as the independent variable (see Glossary for a full explanation of these terms).

8.5.1 Using this test

To use this test you:

1) Wish to test for an association between two treatment variables.
2) Have an association that appears to be linear, with the points reasonably scattered (see Q2).
3) Assume that one variable (*x*) has been taken without sampling error and as such is usually under the investigator's control.
4) Can measure both variables on an interval scale.
5) Should know that each value of *y* varies normally with each value of *x* and they should have a similar variance. (This sounds complicated

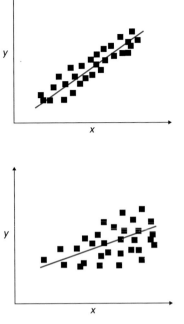

Fig. 8.10. Points evenly distributed about the line, indicating that *y* varies normally with *x*.

y

x

Fig. 8.11. Points becoming more spread out as *x* increases, indicating that *y* does not vary normally with *x*.

but you can tell whether your data probably meet this criterion if the points on your scatter plot are fairly evenly distributed along the whole line (see Fig. 8.10 compared with Fig. 8.11).)

6) Have three or more pairs of observations.

7) Should note that, if you are using the modified *t*-test to test the significance of the association, then the data should be parametric.

EXAMPLE 8.2. Heavy metal contamination of soil under electricity pylons

An undergraduate investigated heavy metal tolerance in plants growing under electricity pylons. As part of her study, she recorded the concentration of zinc in soil samples taken at regular intervals moving away from the pylons (Table 8.4). One of the objectives of her investigation was to see whether there was an association between the distance from the pylon and the concentration of zinc in the soil.

Table 8.4. Zinc concentrations in soil at specific distances from an electricity pylon

Distance from pylon (m)	Zinc concentration (µg Zn/g soil)
1.0	648
1.5	610
2.0	534
2.5	500
3.0	472

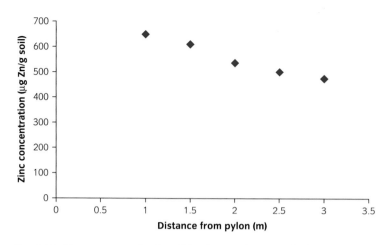

Fig. 8.12. Zinc concentration in soil (µg Zn/g soil) at specific distances from an electricity pylon.

Example 8.2 meets the criteria for using a simple linear regression analysis. For each item (sample of soil), there are two treatment variables ('distance from pylon' measured in metres and 'concentration of zinc' measured in micrograms of zinc per gram of soil). Both are interval scales of measurement. Fig. 8.12 shows that the association appears to be linear and the points are reasonably scattered. The x variable (distance from pylon) was under the investigator's control and there is no sampling error when determining the distance from the pylon. The concentrations of zinc are the y or dependent variable. There are five pairs of measurements and no replicates.

8.5.2 The calculation

online resource centre

There are two parts to this regression analysis. In the first part, you calculate the coefficients that allow you to draw the regression line through your data. Box 8.3 explains how to calculate the regression line for a simple linear regression. Full details of the calculation are included in the Online Resource Centre.

The second step of the simple linear regression is to see whether there is a statistically significant association that is described by this line. Box 8.4 shows you this part of the calculation.

BOX 8.3 How to carry out a Model I simple linear regression (one *y* value for each *x* value): drawing a regression line

GENERAL DETAILS	EXAMPLE 8.2
	This calculation is given in full in the Online Resource Centre. All values have been rounded to 5 decimal places.
1. How to work out the regression coefficient *b* i. The independent variable is that plotted on the *x*-axis. The dependent variable is plotted on the *y*-axis. ii. Calculate the following: Σx Add all *x* values together. $(\Sigma x)^2$ Square the Σx value. Σx^2 Square all the *x* values and sum. Σy Add all the *y* values together. Σxy Multiply each *x* by its corresponding *y*. Add all these *xy* values together. iii. $b = \dfrac{n\Sigma xy - \Sigma x \Sigma y}{n\Sigma x^2 - (\Sigma x)^2}$	**1. How to work out the regression coefficient *b*** i. Variable *x*, the distance from the pylon is under the investigator's control and is the independent variable. Variable *y* is the concentration of zinc in the soil and is the dependent variable. ii. Table 8.5 summarizes these calculations. iii. In this example $b = \dfrac{(5 \times 5297) - (10 \times 2764)}{(5 \times 22.5) - 100} = -92.4$
2. How to work out *a* i. Examine your data and record the means for each variable. ii. *a* can be calculated using the general formula for this straight line: $a = \bar{y} - b\bar{x}$	**2. How to work out *a*** i. The means for the data are shown in Table 8.5. ii. Therefore: $a = 552.8 - (-92.4 \times 2) = 552.8 + 184.8 = 737.6$
3. What is the regression equation? For a Model I regression, the general equation of the line is: $y = a + bx$	**3. What is the regression equation?** $y = a + bx$ $y = 737.6 + (-92.4)x$ $\quad = 737.6 - 92.4x$
4. How to draw the line The regression line indicates the apparent trend in your data. If you want to draw this line on your scatter plot you will need to calculate three pairs of data points as follows: Take any *x* value from your observed data and place this and the values for *a* and *b* into the general equation $y = a + bx$. The only unknown value will be *y*. Work out the *y* value and plot the *x* you selected and this *a* value on your graph. Repeat for at least two more points and draw a line through them within the range of your data set. This is the regression line and should be labelled with the regression equation. GO TO BOX 8.4.	**4. How to draw the line** When *x* = 1 m, $y = 737.6 - (92.4 \times 1) = 645.2$ µg Zn/g soil. When *x* = 2m, $y = 737.6 - (92.4 \times 1) = 552.8$ µg Zn/g soil. Fig. 8.13 illustrates the scatter plot with the regression line added. But you need to confirm that this is a statistically significant association: how to do this is shown in Box 8.4.

Table 8.5. Calculating a Model I simple linear regression (one *y* value for each *x* value) using the data from Example 8.2: Heavy metal contamination of soil under electricity pylons

Distance from pylon (m) (x)	Zinc concentration (μg Zn/g soil) (y)	$x \times y$
1.0	648	648
1.5	610	915
2.0	534	1068
2.5	500	1250
3.0	472	1416
$n = 5$	$n = 5$	$n = 5$
$\Sigma x = 10$	$\Sigma y = 2764$	$\Sigma xy = 5297$
$\Sigma(x)^2 = 100$	$\Sigma(y)^2 = 7639696$	
$\Sigma x^2 = 22.5$	$\Sigma y^2 = 1549944$	
$\bar{x} = 2.0$	$\bar{y} = 552.8$	

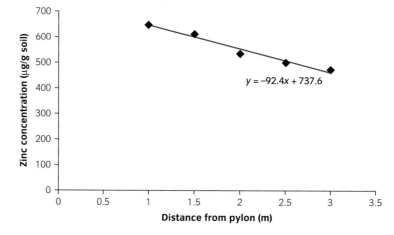

Fig. 8.13. Highly significant association ($0.01 > p > 0.001$) between the concentration of zinc in soil and the distance from the pylon so that $y = -92.4x + 737.6$.

BOX 8.4 How to carry out a Model I simple linear regression (one *y* value for each *x* value): testing the significance of the association

GENERAL DETAILS	EXAMPLE 8.2
	This calculation is given in full in the Online Resource Centre. All values have been rounded to 5 decimal places.

BOX 8.4 Continued

1. Hypotheses to be tested H_0: There is no linear association between x and y. H_1: There is a linear association between x and y, described by y = a + bx.	**1. Hypotheses to be tested** H_0: There is no linear association between the distance from the pylons (m) and concentrations of zinc in the soil (µg Zn/g soil). H_1: There is a linear association between the distance from the pylons (m) and concentrations of zinc in the soil (µg Zn/g soil).
2. Have the criteria for using this test been met?	**2. Have the criteria for using this test been met?** The criteria were considered in 8.5.1. All the criteria were met apart from the requirement for the data to be parametric. With this final criterion, it is difficult to tell as there are so few values for each variable. All that can be said is that both variables are measured on interval scales, which means they may be parametric (see Box 5.2).

3. How to work out $t_{calculated}$

i. First calculate the basic terms SS(x), SS(y) and SP(xy).

$$SS(x) = \sum x^2 - \frac{\left(\sum x^2\right)}{n}$$

$$SS(y) = \sum y^2 - \frac{\left(\sum y^2\right)}{n}$$

$$SS(xy) = \sum xy - \frac{\left(\sum x\right)\left(\sum y\right)}{n}$$

More details about how to calculate these terms are included in Box 8.2.

ii. Calculate the residual variance (s_r^2)

$$s_r^2 = \frac{1}{n-2}\left[SS(y) - \frac{(SP(xy))^2}{SS(x)}\right]$$

iii. Calculate the standard error of b (SE(b)).

$$SE(b) = \sqrt{\frac{s_r^2}{SS(x)}}$$

$$t_{calculated} = \frac{b}{SE(b)}$$

3. How to work out $t_{calculated}$

i. See Table 8.5.

$$SS(x) = 22.5 - \frac{100}{5} = 2.5$$

$$SS(y) = 1549944 - \frac{7639696}{5} = 22004.8$$

$$SP(xy) = 5297 - \frac{10 \times 2764}{5} = 231.0$$

ii. The residual variance is:

$$s_r^2 = \frac{1}{5-2}\left[22004.8 - \frac{(231.0)^2}{2.5}\right]$$

$$s_r^2 = 220.13333$$

iii. $$SE(b) = \sqrt{\frac{220.13333}{2.5}}$$

$$= 9.38367$$

$$t_{calculated} = \frac{-92.4}{9.38367} = -9.84689$$

4. How to find $t_{critical}$

See Appendix d, Table D6. Use a t table for a two-tailed test where p = 0.05. The degrees of freedom (v) are n − 2.

4. How to find $t_{critical}$

When v = 5 − 2 = 3 and p = 0.05, $t_{critical}$ = 3.182.

BOX 8.4 Continued

5. The rule	5. The rule
If the absolute value of $t_{calculated}$ is greater than the value of $t_{critical}$, then you may reject H_0.	As $t_{calculated}$ (9.847) is more than $t_{critical}$ (3.182) at $p = 0.05$, you may reject the null hypothesis. In fact, at $p = 0.01$, $t_{critical} = 5.841$, and at $p = 0.001$, $t_{critical} = 12.941$, so you may reject the null hypothesis at $p = 0.01$ but not at $p = 0.001$.
6. What does this mean in real terms?	**6. What does this mean in real terms?** There is a highly significant $(0.01 > p > 0.001)$ negative linear association between the distance from the pylons (m) and concentrations of zinc in the soil (µg Zn/g soil) described by $y = 737.6 - 92.4x$.

Further examples relating to the topic including how to use statistical software are included in the Online Resource Centre.

online resource centre

8.6 Model I: linear regression: more than one *y* for each value of *x*, with equal replicates

Key points If you have a design similar to the Model I described in 8.5 but have replication, then you may consider using an extended form of a one-way ANOVA to test the significance of the linearity in the data. In this section, we outline the method for doing this where you have equal replicates. An example for unequal replicates is given in the Online Resource Centre.

online resource centre

In some experiments, you may have more than one *y* for each value of *x*. For example, in the experiment described above (Example 8.2), the student may have taken four soil samples along transects placed at each leg of the pylon. This would provide her with four values of zinc concentration in soil (*y* values) for each distance from the pylon (*x*). This is still a Model I regression, as one of the variables (distance) is under the control of the investigator and so taken without sampling error. The test we outline in this section is for experiments like this.

EXAMPLE 8.3. The weight (g) of juvenile hamsters (4–14 weeks)

An undergraduate studying the effect of a number of commercially available hamster foods on reproduction examined, as one objective, the weight of baby hamsters produced to mothers on one particular diet. The hamsters were weighed every 2 weeks from 4 weeks to 14 weeks of age (Table 8.6).

Table 8.6. The weight (g) (*y*) of hamsters aged 4–14 weeks

	Age of hamsters (weeks) (*x*)					
	4	**6**	**8**	**10**	**12**	**14**
	18.1	27.0	26.3	35.0	35.5	41.0
	19.5	26.9	29.0	34.0	37.5	42.5
	21.0	25.4	29.7	31.1	38.5	41.5
	23.0	24.6	31.0	33.1	39.7	43.4
	23.0	22.1	31.2	35.2	39.6	42.1
$\bar{x}$	20.92	25.20	29.44	33.68	38.16	42.10
s^2	4.6570	4.0350	3.9130	2.7870	3.0180	0.8550
s	2.15800	2.00873	1.97813	1.66943	1.73724	0.92466
n	5	5	5	5	5	5
Σy	104.6	126.0	147.2	168.4	190.8	210.5
Σy^2	2206.86	3191.34	4349.22	5682.86	7293.00	8865.47

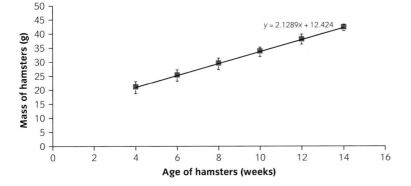

Fig 8.14. An association ($y = 12.42362 + 2.12886x$, $p < 0.001$) between the increase in weight of baby hamsters over a 10-week period.

In Example 8.3, five hamsters have been weighed at regular intervals. Therefore, for each week (*x* variable, independent variable), there are five weights (*y* variable, dependent variable). To test for an association between age and weight in these juvenile hamsters, a regression for more than one *y* for each *x* may be considered. The regression is based on a Model I linear regression and therefore the data need to be plotted to confirm an apparently linear association. Fig. 8.14 confirms this to be the case. Also, one variable has to be taken without sampling error and is usually under the control of the investigator. In this example, it is the age at which the juvenile hamsters are weighed that is under the investigator's control.

To test this regression, you use a parametric ANOVA. We cover these types of tests primarily in the next chapter (Chapter 9) and you may find it helpful to read the introduction there. Important points to note at this stage are that the ANOVA we describe is for parametric data and therefore you need to confirm that your data meet this criterion before proceeding. Secondly, the ANOVAs are usually used to test the hypothesis 'There is no difference between samples'. This is very different from the tests we have been looking at in this chapter, which examine the hypothesis 'There is no association between two variables'. This difference in hypothesis is important. Tests for association examine the data to see whether, as one variable changes, the data relating to the second variable change in a similar consistent way. (For most of the tests we describe here, this change is assumed to be linear.) However, the hypothesis that tests for 'differences between samples' is more open and is not based on the concept of any consistency in changes seen in the two variables. The ANOVA we outline here allows you to test both for a linear association and for a difference between samples. Due to the important difference in these two hypotheses, it is possible to have a statistically significant difference between samples and a statistically non-significant linear association. This can arise when the samples are different but in a non-linear way.

There are many types of ANOVA. They are known as 'one way', 'two way', or 'three way' depending on the number of variables you wish to test the effect of. In Example 8.3, there are two variables: weight of hamsters (g) and age of hamsters (weeks). In the tests for an association, we would say we are testing the significance of an *association* between these two variables. However, when testing for a difference between samples, we would say we wish to examine the *effect* of age on weight in hamsters. Therefore, we are testing an effect of one variable (age) only. This seems odd but is more convincing if you try to reverse the statement and say you wish to test the significance of the effect of weight on age. This is not sensible: the weight of the hamster cannot alter the age of the hamster. Therefore, although we have two variables, we are only testing the effect of age (one variable) and so will use a one-way ANOVA.

In our example, the five hamsters were weighed for each time point, i.e. there are equal replicates. This test can easily be extended for unequal replicates and we include an example illustrating this in the Online Resource Centre.

online resource centre

8.6.1 **Using this test**

To use this test you:

1) Wish to test for an association between two treatment variables.

2) Have an association that appears to be linear, with the points reasonably scattered (see Q2).

3) Assume that one variable (x) has been taken without sampling error and as such is usually under the investigator's control.

4) Should know that each value of y varies normally with each value of x and they should have a similar variance. (This sounds complicated but you can tell whether your data probably meet this criterion if the points on your scatter plot are fairly evenly distributed along the whole line (see Fig. 8.10. compared with Fig. 8.11) and by using an F_{max} test.)

5) Have three or more x values and both variables are measured on an interval scale.

6) Have the same number of replicates (observations) in each sample.

The data from Example 8.3 meet all the criteria for using this one-way parametric ANOVA for linear regressions with equal replicates. We do wish to test for an association between two variables (age and weight). Both variables are measured on an interval scale and age is 'under' the investigator's control. This association appears to be linear with the points reasonably scattered (Fig. 8.14). There are six x values and equal replicates. It is not possible to be confident that the y values are parametric as there are so few observations at each age point. However, the data are measured on an interval scale (Box 5.2) and we will therefore have to make an assumption that this criterion is met. To check whether the variances are similar, an F_{max} test is carried out (Box 8.5). For Example 8.3, the F_{max} test confirms that the variances are homogeneous and we may proceed with the ANOVA (Box 8.6).

BOX 8.5 How to carry out an F_{max} test to check for homogeneous variances before carrying out an ANOVA for linear regressions

GENERAL DETAILS	EXAMPLE 8.3
	This calculation is given in full in the Online Resource Centre. All values have been rounded to 5 decimal places.
1. Hypotheses to be tested H_0: There is no difference between the variances of the two samples. The variances are homogeneous. H_1: There is a difference between the variances of the two samples.	**1. Hypotheses to be tested** H_0: There is no difference between the variances of the weight of hamsters (g) at different ages. H_1: There is a difference between the variances of the weight of hamsters (g) at different ages.
2. How to work out $F_{max\ calculated}$ i. For each sample, calculate the variance (s^2) using the method described in Box 5.1.	**2. How to work out $F_{max\ calculated}$** i. See Table 8.6.

BOX 8.5 Continued

ii. $F_{\text{max calculated}}$ is the ratio between these two values found by dividing the largest variance by the smallest variance. $$F_{\text{max calculated}} = \frac{\text{largest sample variance}(s_1^2)}{\text{smallest sample variance}(s_2^2)}$$	ii. $$F_{\text{max calculated}} = \frac{4.6570}{0.8550} = 5.44678$$
3. How to find $F_{\text{max critical}}$ See Appendix d, Table D7. To look up the value for an F_{max} test that precedes an ANOVA, you will need to know the number of samples (a) and the degrees of freedom ($v = n_s - 1$). n_s is the number of observations in each sample or category.	**3. How to find $F_{\text{max critical}}$** For our example, measurements were taken at six different ages, so $a = 6$. There are five hamsters in each category so the degrees of freedom are $v = 5 - 1 = 4$. So at $p = 0.05$, $F_{\text{max critical}} = 29.5$.
4. The rule If $F_{\text{max calculated}}$ is less than $F_{\text{max critical}}$, then you may accept the null hypothesis and proceed with the ANOVA.	**4. The rule** In this example, $F_{\text{max calculated}}$ (5.45) is less than $F_{\text{max critical}}$ (29.5). Therefore, you do not reject the null hypothesis.
5. What does this mean in real terms?	**5. What does this mean in real terms?** There is no significant difference ($F_{\text{max}} = 5.45$, $p = 0.05$) between the variances of the weight of hamsters (g) at different ages. You may proceed with the ANOVA. If the null hypothesis is not accepted, then you should consider transforming your data (5.9).

8.6.2 The calculation

We have now satisfied ourselves that our data meet all the criteria for using this test. In Box 8.6, we have organized the calculation for the ANOVA to test the significance of a linear regression with equal replicates under a number of subheadings:

- A. Calculate general terms
- B. Calculate the sums of squares.
- C. Construct and complete an ANOVA calculation table.
- D. Test hypotheses.

online resource centre

At each point, we show how these steps can be applied to Example 8.3. Some of the calculation is included in Table 8.7. Where steps have been abbreviated, the full calculation is included in the Online Resource Centre.

BOX 8.6 How to carry out a one-way parametric ANOVA for a linear regression with equal replicates of *y* for each value of *x*

This calculation is given in full in the Online Resource Centre. All values have been rounded to 5 decimal places. The calculation is illustrated using data from Example 8.3.

1. **General hypotheses to be tested**
   There are three pairs of hypotheses that may be tested. These are explained at the end of the box in the context of the calculation itself.

2. **Have the criteria for using this test been met?**
   Yes. The criteria for using this test on the data from Example 8.3 have been met (8.6.1).

3. **How to work out $F_{\text{calculated}}$**

A. Calculate general terms

1. Add together all the observations in all the categories (grand total): Σy.

2. Square each observation in all the categories and add these together: Σy^2.

3. For each category add all the observations together (Σy_s). Square this value ($\Sigma y_s)^2$. Add all the squared values together $\Sigma(\Sigma y_s)^2$. Divide this number by the number of observations in any one category (n), i.e.
$$\frac{\Sigma(\Sigma y_s)^2}{n}$$

4. Take the result from step **1**, square it, and divide by N, where N is the total number of observations in all the samples combined:
$$\frac{(\text{result from step 1})^2}{N}$$

5. Add all the *x* values together (Σx) and multiply by *n*: $n(\Sigma x)$

6. Square each *x* value (x^2) and add these together ($\Sigma(x^2)$). Multiply by *n*.

7. For each *x* value, there are a number of *y* values, which have been summed (Σy). For each *x* value, multiply it by its own sum of *y*: ($x(\Sigma y)$). Add these together: $\Sigma(x(\Sigma y_s))$.

8. Square the result from step **5** and divide it by the number of categories (samples) (a) multiplied by the number of observations in each sample (n):
$$\frac{(\text{result from step 5})^2}{a \times n}.$$

9. Subtract the result from step **8** from the result from step **6**.

10. First multiply together the results from steps **5** and **1**. Divide this value by $a \times n$. Subtract the result from the result from step **7**.

In our example, the calculations are shown in the ten steps here and with reference to Table 8.6:

1. $\Sigma y = 18.1 + 27.0 + \ldots 39.6 + 42.1 = 947.50$

2. $\Sigma y^2 = (18.1)^2 + (27.0)^2 + \ldots (39.6)^2 + (42.1)^2$
   $= 2206.86 + 3191.34 + 4349.22 + 5682.86$
   $\quad + 7293.0 + 8865.47$
   $= 31588.75$

3.
$$\frac{(104.6)^2 + (126.0)^2 + (147.2)^2 + (168.4)^2 + (190.8)^2 + (210.5)^2}{5}$$
   $= (10941.16 + 15876.0 + 21667.84 +$
   $\quad 28358.56 + 36404.64 + 44310.25)/5$
   $= 157558.45/5 = 31511.69$

4. $(947.50)^2/30 = 897756.25/30 = 29925.2083$

5. $5(4 + 6 + 8 + 10 + 12 + 14) = 5 \times 54 = 270.0$

6. $((4)^2 + (6)^2 + (8)^2 + (10)^2 + (12)^2 + (14)^2) \times 5$
   $= 556.0 \times 5 = 2780.00$

BOX 8.6 Continued

7. $(4 \times 104.6) + (6 \times 126.0) + (8 \times 147.2)$
   $+ (10 \times 168.4) + (12 \times 190.8) + (14 \times 210.5)$

   $= 418.4 + 756.0 + 1177.6 + 1684.0 + 2289.6$
   $+ 294.7 = 9272.6$

8. $(270.0)^2/(6 \times 5) = 72900.0/30 = 2430.0$

9. $2780.00 - 2430.0 = 350.0$

10.

$$9272.6 - \frac{270.0 \times 947.5}{6 \times 5} = 9272.6 - (255825.0/30)$$
$$= 9272.6 - 8527.5 = 745.1$$

B. Calculate the sums of squares (SS)

11. SS_{total} = result from step **2** – result from step **4**.

12. $SS_{between^*}$ = result from step **3** – result from step **4** (*between* categories).

13. $SS_{within^{**}}$ = result from step **11** – result from step **12** (**within* categories).

14. $SS_{regression}$ = square result from step **10** and divide by result from step **9**.

15. $SS_{deviation\ from\ regression}$ = result from step **12** – result from step **14**.

In our example, the calculations are:

11. $SS_{total} = 31588.75 - 29925.2083 = 1663.5417$

12. $SS_{between} = 31511.69 - 29925.2083 = 1586.4817$

13. $SS_{within} = 1663.5417 - 1586.4817 = 77.06$

14. $SS_{regression} = (745.1)^2/350.0 = 555174.01/350.0$
    $= 1586.21146$

15. $SS_{deviation\ from\ regression} = 1586.4817 - 1586.21146$
    $= 0.27024$

C. Construct and complete an ANOVA calculation table

i. Draw an ANOVA table as illustrated here.

Source of variation	SS	v	Mean squares	F
Between categories	Step 12			
Linear regression	Step 14			
Deviation from linear regression	Step 15			
Within categories	Step 13			
Total variation	Step 11			

ii. Transfer the results from the calculations for the sums of squares into the ANOVA table in column 1 (Table 8.7).

iii. Calculate the degrees of freedom (v).

$v_{between}$ = number of categories – 1 = $a - 1$

$v_{regression} = 1$

$v_{deviations\ from\ regression} = a - 2$

v_{within} = total number of observations – number of categories = $N - a$

v_{total} = total number of observations – 1 = $N - 1$

Enter these results in column 2 (Table 8.7).

iv. Calculate the mean squares (MS) for between and within categories, regression and deviation from regression by dividing the relevant sum of squares (e.g. SS_{within}) by the associated degrees of freedom (Table 8.7).

BOX 8.6 Continued

D. Calculation of *F* values, finding critical *F* values, and testing hypotheses

There are three pairs of hypotheses we may test. The hypotheses we test are written below in relation to Example 8.3.

To test for a deviation from linearity

These hypotheses are tested in step 16.

H_0: There is no deviation from linearity between the variables 'weight of hamsters' (g) and 'age of hamsters' (weeks).

H_1: There is a deviation from linearity between the variables 'weight of hamsters' (g) and 'age of hamsters' (weeks).

Testing the significance of this null hypothesis is your first step. A significant decision at this point suggests there is significant deviation from linearity and either an association is not linear (e.g. curvilinear) or there is significant variation in the data that is masking any linearity. If you have a significant outcome in this step, you may either consider transforming your data to produce a linear distribution and/or proceed to step 18.

To test for significant linearity in the data

These hypotheses are tested in step **17**.

H_0: There is no significant linear association between the weight of hamsters (g) and their age (weeks).

H_1: There is a significant linear association between the weight of hamsters (g) and their age (weeks).

There is a debate about how the appropriate *F* value is calculated to test this hypothesis. We refer you to Sokal & Rohlf (1994) for more details. We describe the ratio that is correct most of the time and for the method we have described. A significant result here would indicate a significant regression and you may

then proceed to step 19, which shows you how to calculate the regression equation. A non-significant result here would suggest either that any association is not linear or that there is too much variation in the data, which is masking any linear trend. You should consider transforming your data and/or proceed to step 18.

To test for a difference between samples of hamsters at different ages

These hypotheses are tested in step 18.

H_0: There is no difference between the hamsters at different ages (weeks) in relation to their weights (g).

H_1: There is a difference between the hamsters at different ages (weeks) in relation to their weights (g).

This is a typical one-way ANOVA that tests the hypotheses described here. The test does not consider the linearity or any other specific type of association. The hypothesis testing instead focuses on any significant differences between samples. You may have a non-significant outcome when testing for linearity but a significant outcome here. If you wish to find out more about the nature of the difference, you may then proceed to a Tukey's test (9.6) .

16. To test for a deviation from linearity.

$F_{calculated} = MS_{deviations}$ divided by MS_{within}.

In our example, this $F_{calculated}$ value = 0.06756/3.21083 = 0.02143. The $F_{critical}$ value is found in Appendix d, Table D8, where the degrees of freedom are those for the $MS_{deviations}$ (v_1) and the MS_{within} (v_2). In our example, $v_1 = 4$ and $v_2 = 24$. When $p = 0.05$, $F_{critical} = 2.78$.

The rule for this test is that if $F_{calculated}$ is more

BOX 8.6 Continued

than $F_{critical}$, you may reject the null hypothesis. In our example, $F_{calculated}$ (0.02) is less than $F_{critical}$ (2.78) so we may not reject the null hypothesis. There is therefore no significant deviation from linearity between the variables 'weight of hamsters' (g) and 'age of hamsters' (weeks). We may now test for the significance of the regression by proceeding to the next step. If this were not the case, we would skip the next step and proceed to step **18**.

17. To test for significant linearity in the data.

$$F_{calculated} = MS_{regression} \text{ divided by } MS_{deviation}$$

In our example, this $F_{calculated}$ value = 1586.21146/0.06756 = 23478.55921. The $F_{critical}$ value is found in Appendix d, Table D8, where the degrees of freedom are those for $MS_{regression}$ (v_1) and $MS_{deviation}$ (v_2). In our example, v_1 = 1 and v_2 = 4. When p = 0.05, $F_{critical}$ = 7.71. The rule for this test is that if $F_{calculated}$ is more than $F_{critical}$, you may reject the null hypothesis. In our example, $F_{calculated}$ (23478.56) is more than $F_{critical}$ (7.71) so we may reject the null hypothesis. In fact, at p = 0.001 (Appendix d, Table D10), $F_{critical}$ = 74.13 so we may reject the null hypothesis at this higher level of significance. There is therefore a very highly significant linear association between weight (g) and age (weeks) of juvenile hamsters. We may now proceed to step **19** to calculate the equation for the significant regression.

18. To test for a difference between samples of hamsters at different ages.

If the hypothesis testing at step **16** is significant or step **17** is not significant, it is worth carrying out step **18** to test for a significant difference between the samples.

$$F_{calculated} = MS_{between} \text{ divided by } MS_{within}$$

In our example, this $F_{calculated}$ value = 317.29634/3.21083 = 98.82056. The $F_{critical}$ value is found in Appendix d, Table D8, where the degrees of freedom are those for the $MS_{between}$ (v_1) and MS_{within} (v_2). In our example, v_1 = 5 and v_2 = 24. When p = 0.05, $F_{critical}$ = 2.62. The rule for this test is that if $F_{calculated}$ is more than $F_{critical}$, you may reject the null hypothesis. In our example, $F_{calculated}$ (98.82) is more than $F_{critical}$ (2.62) so we may reject the null hypothesis. In fact, at p = 0.001 (Appendix d, Table D8), $F_{critical}$ = 5.98 so we may reject the null hypothesis at this higher level of significance. There is therefore a very highly significant difference between the weights (g) of hamster juveniles in relation to their age (weeks).

19. The regression equation.

The general equation for this regression line is $y = a + bx$. The regression coefficient is the gradient of the line (b) (Fig. 8.14) and can be calculated as b = result from step **10**/result from step 9. The point where the line crosses the y axis is a, where $a = \bar{y} - b\bar{x}$ = result from step **1**/N – [b (result from step 5/N)], where N is the total number of observations.

In this example:

$$b = \frac{745.1}{350.0} = 2.12886$$

$$a = \frac{947.50}{30} - 2.12886\left(\frac{270.0}{30}\right)$$
$$= 31.58333 - 19.15972 = 12.42362$$

The full regression equation is y = 12.42362 + 2.12886x.

Table 8.7. ANOVA table for testing significance of linear regression where there are replicates of y for each value of x, and for the testing of the hypothesis 'There is no difference between samples' for data from Example 8.3

Source of variation	SS	v	Mean squares	$F_{calculated}$
Between categories	1586.4817	6 − 1 = 5	317.29634	$MS_{between}/MS_{within}$ 317.29634/3.21083 = 98.82056
Linear regression	1586.21146	1	1586.21146	$MS_{regression}/MS_{deviation}$ 1586.21146/0.06756 = 23478.55920
Deviation from linear regression	0.27024	6 − 2 = 4	0.06756	$MS_{deviations}/MS_{within}$ 0.06756/3.21083 = 0.02143
Within categories	77.06	30 − 6 = 24	3.21083	
Total variation	1663.5417	29		

8.7 Model II: principal axis regression

Key points: If you are planning to collect data where neither of the two variables is under your control but the association appears to be linear then you may consider a Model II regression. The general equation for a Model II regression is $y = \bar{y} + b_1 (x - \bar{x})$. A Model II regression will model the line of best fit for your data but will not test the significance of association. In this section, we describe the method for a Model II regression where both variables are measured on the same scale.

Unlike the simple linear regression model, the principal axis regression model can be used when both variables are subject to sampling error. Usually this means that neither variable is under the investigator's control. However, the principal axis regression cannot be used to test the significance of an association, and in this instance, a correlation should be used. The principal axis regression and ranged principal axis regression (8.8) are part of a group of tests called principal component analyses. In these Model II regressions, the line is determined by placing an ellipse around the data and drawing a line through the longest axis, known as the principal axis. The general formula for this principal axis is $y = \bar{y} + b_1 (x - \bar{x})$. As in the simple linear regression, there is a regression coefficient (b), but to denote the difference in the method used to calculate the Model I and Model II values of b, we call the regression coefficient calculated for the principal axis regression b_1.

8.7.1 **Using this test**

To use this test you:

1) Need two treatment variables recorded for each item.

2) Have an association that appears to be linear with the points reasonably scattered.

3) Have confirmed using a correlation analysis that the association is significant.

4) Have two treatment variables that are measured on the same interval scale.

5) Have neither treatment variable under the investigator's control; therefore, both treatment variables are subject to sampling error.

6) Have three or more pairs of measurements.

EXAMPLE 8.4. The lower arm and lower leg length (cm) of a small cohort of female undergraduates

In an investigation into human morphology, the lower arm and lower leg lengths of 12 female undergraduates were recorded (Table 8.8) on one particular day.

Table 8.8. Lower arm and lower leg length (cm) in a small cohort of female undergraduates

Student	Lower arm length (cm)	Lower leg length (cm)	Student	Lower arm length (cm)	Lower leg length (cm)
1	24.0	39.0	7	26.0	41.0
2	25.0	40.0	8	25.5	40.0
3	23.0	36.0	9	24.2	36.7
4	26.0	43.0	10	25.0	38.0
5	24.0	38.0	11	24.0	41.0
6	23.0	35.0	12	24.5	40.0

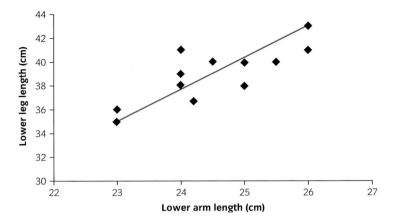

Fig. 8.15. Significant association ($r_s = 0.725$, $p = 0.05$) between the lower arm and lower leg length (cm) of female undergraduates, such that $y = -26.94 + 2.69x$.

The data from Example 8.4 meet the criteria for a principal axis regression as there are two treatment variables (arm length and leg length). The association appears to be linear and the points are reasonably scattered (Fig. 8.15). Neither variable was under the control of the investigator. There are 12 pairs of observations and both are measured on the same interval scale (cm). Spearman's rank correlation (see Online Resource Centre, Chapter 8, interactive exercise 1) confirms that this is a significant association ($r_s = 0.725$, $p = 0.05$).

online resource centre

8.7.2 The calculation

We show you the general process and a specific example of a Model II principal axis regression in Box 8.7. In addition, you will need to refer to the calculations in Table 8.9.

BOX 8.7 How to carry out a Model II principal axis regression: drawing a regression line	
GENERAL DETAILS	**EXAMPLE 8.4**
	This calculation is given in full in the Online Resource Centre. All values have been rounded to 5 decimal places.
1. Have the criteria for using this test been met?	**1. Have the criteria for using this test been met?** Yes (8.7.1).
2. How to work out b_1 i. First decide which of the variables to call x and which to call y.	**2. How to work out b_1** i. Let the measurements of arm length (cm) be x and the measurements of leg length (cm) be y (Fig. 8.15).

BOX 8.7 Continued

ii. Calculate the basic terms.

To carry out this regression analysis, you first need to calculate the following for the two samples:

Means ($\bar{x}$ and $\bar{y}$)

Sum of x values (Σx), sum of y values (Σy)

The variances (s_x^2 and s_y^2)

The number of pairs of observations n

The method for calculating these is given in Box 5.1.

iii. Calculate the variance for products xy, (s_{xy}^2), where

$$s_{xy}^2 = \frac{SP(xy)}{n-1}$$

The term $SP(xy)$ has figured already in this chapter (Box 8.2) and is

$$SP(xy) = \Sigma xy - \frac{(\Sigma x)(\Sigma y)}{n}$$

iv. The slope of the principal axis (b_1) is calculated as:

$$b_1 = \frac{s_y^2 - s_x^2 + \sqrt{\left[\left(s_y^2 - s_x^2\right)^2 + 4\left(s_{xy}^2\right)^2\right]}}{2s_{xy}^2}$$

---

ii. The general methods for calculating such terms as Σx and s_x^2 are described in Chapter 5 (Box 5.1) and therefore are not included here. The basic terms are:

$\bar{x} = 24.51666$, $\bar{y} = 38.975$

$\Sigma x = 294.2$, $\Sigma y = 467.7$

$s_x^2 = 1.03061$, $s_y^2 = 5.38932$

$n = 12$

iii. To calculate the variance for the products $xy(s_{xy}^2)$, first note or calculate each of the terms you need. Σx, Σy and n are above (ii). To calculate Σxy, see Table 8.9:

$\Sigma xy = 11487.14$

$$SP(xy) = \Sigma xy - \frac{\Sigma x \Sigma y}{n}$$

$$= 11487.14 - \frac{(294.2)(467.7)}{12}$$

$$= 20.695$$

$$s_{xy}^2 = \frac{SP(xy)}{n-1} = \frac{20.695}{11}$$

$$= 1.88136$$

iv. To calculate the slope of the principal axis b_1, first calculate the components of this equation:

$$s_y^2 - s_x^2 = 5.38932 - 1.03061$$

$$= 4.35871$$

$$\left(s_y^2 - s_x^2\right)^2 = (4.35871)^2$$

$$= 18.99837$$

$$4(s_{xy}^2)^2 = 4 \times (1.88136)^2$$

$$= 14.15812$$

$$2s_{xy}^2 = 2 \times 1.88136$$

$$= 3.76273$$

Therefore:

$$b_1 = \frac{4.35871 + \sqrt{(18.99837 + 14.15812)}}{3.76227}$$

$$= 2.68871$$

---

3. What is the regression equation?

The equation for the principal axis is:

$$y = \bar{y} + b_1(x - \bar{x})$$

Substitute the known numerical values where possible.

---

3. What is the regression equation?

Using the values you have already calculated:

$$y = \bar{y} + b_1(x - \bar{x})$$

$$= 38.975 + 2.68871(x - 24.51667)$$

$$= 38.975 + 2.68874x - 65.91822$$

$$= -26.94319 + 2.68871x$$

BOX 8.7 Continued

4. How to draw the line

In the same way as other regressions, you need to calculate the x and y values for three points using x values that lie within your data set, and your regression equation. Your line then passes through these points within the range of your data set. This is the regression line and if you add it to your figure it should be labelled with the regression equation.

4. How to draw the line

The full regression equation for this example is:
$y = -26.94319 + 2.68871x$
Therefore. If $x = 23$cm,
$y = -26.94319 + (2.68871 \times 23)$
$\quad = 34.9$cm
and if $x = 24.5$cm, then $y = 38.9$cm
and if $x = 26$cm, then $y = 43$cm.
The data from Table 8.8 and the regression line are shown in Fig. 8.15.

5. Testing the significance of the line

The principal axis regression allows you to identify a regression line that reflects the linear nature of the association between your two variables. However, unlike the simple linear regression, there is no simple associated process for testing a hypothesis relating to the significance of the association. Therefore, you should also use a correlation analysis to confirm the strength and significance of the association between the two variables.

5. Testing the significance of the line

An examination of the data indicates that they are non-parametric; therefore, a two-tailed Spearman's rank correlation was carried out. From this, it was concluded that there is a moderate ($r_s = 0.725$) statistically significant ($p = 0.05$) positive correlation between the lower arm and lower leg length in a cohort of female students, which can be described by the equation: $y = -26.94319 + 2.68871x$ (Fig. 8.15).

Further examples relating to the topic including how to use statistical software are included in the Online Resource Centre.

@ **online resource centre**

Table 8.9. Calculating a Model II principal axis regression using the data from Example 8.4: Arm and leg length (cm) in a small cohort of female students

Length of arm (cm) (x)	Length of leg (cm) (y)	$x \times y$
24.00	39.00	936.00
25.00	40.00	1000.00
23.00	36.00	828.00
26.00	43.00	1118.00
24.00	38.00	912.00
23.00	35.00	805.00
26.00	41.00	1066.00
25.50	40.00	1020.00
24.20	36.70	888.14
25.00	38.00	950.00
24.00	41.00	984.00
24.50	40.00	980.00
$n = 12$	$n = 12$	$n = 12$
$\Sigma x = 294.2$	$\Sigma y = 467.70$	$\Sigma xy = 11487.14$
$\bar{x} = 24.51667$	$\bar{y} = 38.97500$	

8.8 Model II: ranged principal axis regression

Key points If you collect data where neither of the two variables is under your control but the association appears to be linear, then you may consider a Model II regression analysis. This is the second Model II regression we consider and may be used when the two variables are measured on different scales. This regression will not test the significance of the association and a correlation is recommended.

Many investigators wish to determine whether there is an association between two variables, neither of which has been manipulated by the experimenter. In Example 8.1, neither hardness of eggshell nor the amount of food supplement eaten was under the investigator's control; therefore, a Model II regression is appropriate. Where the units of measurement for the two variables differ, as in this example, a Model II ranged principal axis regression may be used (Legendre & Legendre, 1998). Unlike the principal axis regression (8.7), the first step in the calculation is to transform the data so that each scale of measurement is revised to one that only extends from 0 to 1. These ranged data then satisfy the criteria for using a principal axis regression. The principal axis regression is carried out on this ranged data and the slope b' is then transformed back to the original units of measurement before the regression line is calculated. Like the previous principal axis regression, the line is the one that passes longitudinally through an ellipse which encompasses the data. This ranged principal axis regression cannot easily be used as the basis for a test for association so a correlation should be carried out to examine the strength and significance of any association before using this test.

8.8.1 Using this test

To use this test you:

1) Need two treatment variables recorded for each item.
2) Have an association that appears to be linear and the points reasonably scattered.
3) Have confirmed using a correlation analysis that the association is significant.
4) Have neither treatment variable under the investigator's control; therefore, both treatment variables are subject to sampling error.
5) Have three or more pairs of measurements.
6) Have two treatment variables that are measured on different interval scales.

The example we use to illustrate this Model II method of regression analysis is Example 8.1 on the association between egg shell hardness and

food supplement consumed by pullets. These data satisfy the criteria for using a ranged principal axis regression because there are two treatment variables (hardness of eggshell and amount of food supplement eaten). The association appears to be linear and the points reasonably scattered (Fig. 8.5). Neither the hardness of eggshell nor the amount of food additive eaten was manipulated by the investigator. There are 13 pairs of observations. The hardness of eggshell is an arbitrary unit of measure and the amount of food additive eaten is measured in grams. Both are interval scales. Using Pearson's product moment correlation (Box 8.2), we have confirmed that there is a strong ($r = 0.99$) and highly significant positive correlation ($p < 0.001$) between these two variables.

8.8.2 The calculation

As discussed earlier, there are more steps in this ranged principal axis regression than in the principal axis regression. However, the regression itself is the same. As in the principal axis regression it is necessary to carry out a correlation analysis to test the significance of the association. The ranged principal axis regression is shown in general and with a specific example (Example 8.1) in Box 8.8 with some calculations included in Table 8.10.

BOX 8.8 How to carry out a Model II ranged principal axis regression: drawing a regression line

GENERAL DETAILS	EXAMPLE 8.1
	This calculation is given in full in the Online Resource Centre. All values have been rounded to 5 decimal places.
1. Have the criteria for using this test been met?	**1. Have the criteria for using this test been met?** Yes (8.8.1.)
2. How to work out b′ i. First decide which of the variables to call x and which to call y.	**2. How to work out b′** i. Example 8.1 has been used to illustrate Pearson's and Spearman's rank correlations (Box 8.1 and Box 8.2). Therefore, in line with these earlier calculations, we will let the amount of food supplement eaten (g) be the x variable and hardness of shells be the y variable.
ii. Transform each x and y value into x' and y' as follows: $$x' = \frac{x - \text{lowest value of } x}{\text{highest value of } x - \text{lowest value of } x}$$ $$y' = \frac{y - \text{lowest value of } y}{\text{highest value of } y - \text{lowest value of } y}$$	ii. The transformed data are given in Table 8.10. The first original x observation was 19.5g. The lowest x value was 7.0g and the highest value was 19.5g. Therefore: $$x' = \frac{19.5 - 7.0}{19.5 - 7.0} = 1.0$$

BOX 8.8 Continued

The second original x observation was 11.20 g. Therefore:

$$x' = \frac{11.2 - 7.0}{19.5 - 7.0} = 0.336$$

The first original y observation was 7.1 units, the lowest y value was 1.2 units and the highest was 7.1 units. Therefore:

$$y' = \frac{7.1 - 1.2}{7.1 - 1.2} = 1.0$$

etc. (Table 8.10).

iii. Calculate basic terms

To carry out this regression analysis you need to first calculate the following for the two samples:
Means ($\bar{x}'$ and $\bar{y}'$)

Sum of x' values ($\sum x'$)

Sum of y' values ($\sum y'$)

The variances ($s_{x'}^2$ and $s_{y'}^2$)

Number of pairs of observations n
The method for calculating these is given in Box 5.1.

iii. Calculate basic terms

As this method has been demonstrated elsewhere (Boxes 5.1 and 8.2), we will not repeat it here. The basic terms are:

$$\bar{x}' = 0.432, \bar{y}' = 0.41320$$
$$\sum x' = 5.616, \sum y' = 5.37288$$
$$s_{x'}^2 = 0.08266, \ s_{y'}^2 = 0.08678$$
$$n = 13$$

iv. Calculate the variance ($s_{x'y'}^2$) for the products $x'y'$, where:

$$s_{x'y'}^2 = \frac{\text{SP}(x'y')}{n-1}$$

The term SP($x'y'$) has figured already in this chapter and is:

$$\text{SP}(x'y') = \sum x'y' - \frac{(\sum x')(\sum y')}{n}$$

iv. To calculate the variance ($s_{x'y'}^2$) for the products $x'y'$, first note or calculate each of the terms you need. $\sum x', \sum y'$, and n are above (iii) To calculate $\sum x'y'$, see Table 8.10.
$$\sum x'y' = 3.33098$$

$$\text{SP}(x'y') = \sum x'y' - \frac{(\sum x')(\sum y')}{n}$$

$$= 3.33098 - \frac{(5.616)(5.37288)}{13}$$

$$= 1.009896$$

$$s_{x'y'}^2 = \frac{\text{SP}(x'y')}{n-1} = \frac{1.009896}{13-1}$$

$$= 0.08416$$

v. The slope of the principal axis (b') for the ranged data is:

$$b' = \frac{s_{y'}^2 - s_{x'}^2 + \sqrt{\left(s_{y'}^2 - s_{x'}^2\right)^2 + 4\left(s_{x'y'}\right)^2}}{2s_{x'y'}^2}$$

v. To find the slope of the principal axis (b') for the ranged data.
First calculate the components of this equation.

$$s_{y'}^2 - s_{x'}^2 = 0.08678 - 0.08266 = 0.00412$$

$$\left(s_{y'}^2 - s_{x'}^2\right)^2 = (0.00412)^2$$

$$= 1.69398 \times 10^{-5}$$

$$4(s_{x'y'}^2)^2 = 4 \times (0.08416)^2$$

$$= 0.02833$$

BOX 8.8 Continued

$$\sqrt{[(s_{y'}^2 - s_{x'}^2)^2 + 4(s_{x'y'}^2)^2]}$$
$$= \sqrt{(1.69398 \times 10^{-5} + 0.02833)}$$
$$= 0.16837$$
$$2s_{x'y'}^2 = 2 \times 0.08416$$
$$= 0.16832$$

Therefore:

$$b' = \frac{0.00412 + 0.16837}{0.16832}$$
$$= 1.02475$$

vi. Transform b' to original units to obtain b_1.

$$b_1 = b' \times \frac{\text{highest } y \text{ value} - \text{lowest } y \text{ value}}{\text{highest } x \text{ value} - \text{lowest } x \text{ value}}$$

vi. Transform b' to original units to obtain b_1.

$$b_1 = 1.02475 \times \frac{7.1 - 1.2}{19.5 - 7.0}$$
$$= 0.48368$$

3. What is the regression equation?

The equation for the principal axis is:

$$y = \bar{y} + b_1(x - \bar{x})$$

You need first to calculate the means for the original x and y values, then substitute these and b_1 in the equation.

3. What is the regression equation?

Calculate the means for the original observations (Table 8.3). Then:

$$y = \bar{y} + b_1(x - \bar{x})$$
$$= 3.63846 + 0.48368(x - 12.4)$$
$$= 3.63846 + 0.48368x - 5.99763$$
$$= -2.35917 + 0.48368x$$

4. How to draw the line

In the same way as other regressions, you need to calculate the x and y values for three points using x values that lie within your data set, using your regression equation. Your line then passes through these points within the range of your data set. This is the regression line and should be labelled with the regression equation.

4. How to draw the line

The regression equation for this example is
$y = -2.35917 + 0.48368x$
Therefore, if:
$x = 8.2$g
$y = -2.35917 + (0.48368 \times 8.2)$
$= 1.6$ units
and when $x = 19.50$g, $y = 7.1$ units and
if $x = 15.10$g, $y = 4.9$ units.

5. Testing the significance of the line

The ranged principal axis regression allows you to identify a regression line that reflects the linear nature of the association between your two variables. However, unlike the simple linear regression, there is no simple associated process for testing a hypothesis relating to this association. Therefore, you should also use a correlation analysis to confirm the strength and significance of the association between the two variables.

5. Testing the significance of the line

These data were used to illustrate both Spearman's rank correlation (Box 8.1) and Pearson's correlation (Box 8.2). If we use the more powerful parametric Pearson's correlation, it can be seen that there is a strong ($r = 0.99$) and highly significant positive correlation ($p < 0.001$) between the amount of food supplement (g) given to Maren pullets and the hardness of the eggshells they produce. This linear association can be described by the regression equation: $y = -2.35917 + 0.48368x$ (Fig. 8.16).

Further examples relating to the topic including how to use statistical software are included in the Online Resource Centre.

@ **online resource centre**

Table 8.10. Calculation of ranged values for a ranged principal axis regression using data from Example 8.1: The association between the amount of food supplement eaten and eggshell hardness in Maren pullets

Pullet	Amount of food supplement (g) (x)	Ranged x' values	Hardness of shells (y)	Ranged y' values	$x' \times y'$
1	19.5	1.000	7.1	1.00000	1.0
2	11.2	0.366	3.4	0.37288	0.12529
3	14.0	0.560	4.5	0.55932	0.31322
4	15.1	0.648	5.1	0.66102	0.42834
5	9.5	0.200	2.1	0.15254	0.03051
6	7.0	0.000	1.2	0.00000	0.00000
7	9.8	0.224	2.1	0.15254	0.03417
8	11.6	0.368	3.4	0.37288	0.13722
9	17.5	0.840	6.1	0.83051	0.69763
10	11.2	0.366	3.0	0.30508	0.10251
11	8.2	0.096	1.7	0.08475	0.00814
12	12.4	0.432	3.4	0.37288	0.16108
13	14.2	0.576	4.2	0.50847	0.29288
					$\Sigma xy = 3.33098$

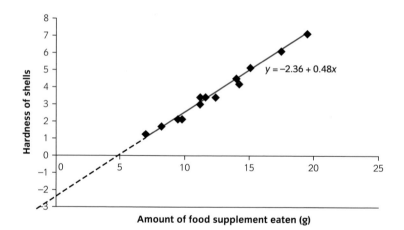

Fig. 8.16. Significant association ($r = 0.99$; $p < 0.001$) between the amount of food supplement eaten and the hardness of eggshells in Maren pullets, described by $y = -2.36 + 0.48x$. The figure also illustrates why you should not extrapolate beyond your known data (see Q3).

8.9 About regression lines

Key points We often come across problems when regression lines are reported. Here we highlight two of these relating to plotting lines and the interpretation of the information.

Reporting correlations and regressions is usually achieved using a scatter plot. For a correlation, this is sufficient if the results from the correlation, including the correlation coefficient and p value, are included in the figure legend. If a regression analysis has been used, then it is common practice to add the regression line on the scatter plot and label this with the regression equation. We consider the topic of drawing lines on figures in more detail in Chapter 11, but while we are on the subject of regressions, we would like to make two points:

- A trend may not reflect an association or a relationship. Lines (we refer to them as trend lines) are often drawn on figures, especially by the common software packages. These lines indicate the apparent trend but have no statistical meaning. There are many common errors associated with these trend lines (11.8.2). One way to overcome these errors, if you have appropriate data, is to carry out a regression analysis to model a significant association. A significant regression line is a reasonable mathematical description of the trend in your data. You should show how important the line is by adding the regression equation (e.g. Fig. 8.16 compared with Fig. 8.5). A regression line that is shown not to be significant should not be used.

- Do not extend the regression line beyond your data. A common error is to extend the significant regression line beyond the data set. Q3 illustrates why this is not appropriate.

Q3   The regression equation that we calculated for Example 8.1 in Box 8.8 between the amount of food supplement eaten by Maren pullets and the hardness of the eggshells is $y = -2.35917 + 0.48373x$ (Fig. 8.16). If the farmer wishes to economize and only gives the pullets 1g of food supplement, how hard would you predict the eggshells will be?

8.10 Experimental design, correlation, and regression

Key points Experiments that may be analysed using a test for association are often simple in design. We have only considered designs with two variables. The designs we covered in Chapter 7 were ones where counts or frequencies are recorded for discrete categories. For the correlations and regressions considered in this chapter, the two variables may be measured on the same or different scales and may have replication.

As we explained at the beginning of this chapter, regression and correlation analyses are designed to test for an association between two

Table 8.11. Experimental design for correlation and regression if all other criteria are met

Item	Observations for variable 1	Observations for variable 2
1		
2		
3		
4, etc.		

or more variables and/or model this association, thus allowing you to describe the association and make predictions. Often this will mean that each x observation has only one y value. But with replication, you may have more than one y value for each x value. This differs from the chi-squared and G-tests for association where you may also be examining two variables, but here the data recorded are the number of items in particular categories.

The experimental designs for correlation and regression are often quite simple (Table 8.11), the number of items being determined by the type of correlation or regression. For example, to use Spearman's rank correlation, you would design an experiment that had between seven and 30 pairs of observations (8.2.1). For a regression analysis, the minimum number of items is three, although you will only detect strong associations with a small sample size. To detect a weak association will require a large sample size. There is no simple way to determine sample size in this instance without carrying out a relatively comprehensive pilot study from which r may be ascertained.

When designing your experiment, one of the variables may be measured without sampling error; usually this means that it is under the investigator's control. In this case you may consider using the Model I simple linear regression.

If this is not so, then you may use a Model II regression. The scale of measurement is also important when determining which test you may use. For example, if you expect to have non-parametric data, you should plan your investigation around Spearman's rank correlation rather than Pearson's product moment correlation. Normalizing non-parametric data may be possible (5.9), although interpreting the outcome of the analysis of two treatment variables using transformed data is complex and you should refer to other texts (e.g. Sokal & Rohlf, 1984).

Correlations and regression are used to examine whether there is an association between two or more variables, only one of which may be under the control of the investigator. You may have a 'null state' within your treatment, but it cannot be called a control as the analysis makes no comparison between the various observations within one variable (i.e. it does not compare a control value with a treatment value).

Summary of Chapter 8

- In this chapter, we have considered correlation and regression analyses, which may be used to test for a linear association between two variables. The term association indicates a statistical and not necessarily a biological relationship between the two variables.

- These tests differ from the chi-squared and G-tests in that, for each item, two observations are recorded, one for each treatment variable. In the chi-squared and G-tests, the number of items in each category is determined. The chi-squared and G-tests may be used for data that are measured on a nominal scale. In the correlations and regressions, neither variable can be measured on a nominal scale (7.7 and 8.9).

- A correlation analysis will generate a coefficient that indicates the strength of association. This coefficient may then be used to test the significance of the association. The coefficient is affected by sample size; therefore, a weak correlation may not be significant if the sample size is small but may be significant if the sample size is large. We considered two correlation analyses: Spearman's rank test for non-parametric data and Pearson's product moment test for parametric data (8.1–8.3).

- Regression analyses are primarily used to determine the mathematical equation for the line that may be drawn through the data. The significance of this line as a line of best fit can be determined only

in the simple linear regression with no replication using a modified *t*-test. Where there is replication, an extended one-way ANOVA may be used. In a principal axis and ranged principal axis regression, the significance of the association can be determined using a correlation analysis (8.1 and 8.5–8.7).

- The regression analyses fall into two groups: Model I and Model II. In a Model I regression (simple linear regression), one variable must be measured without sampling error and is therefore usually under the control of the investigator. In Model II regressions (principal axis and ranged principal axis regressions), both variables may be measured with some sampling error, either using the same scale (principal axis) or different scales (ranged principal axis) (8.5–8.7).

- Correlation and regression analysis may be extended to testing or describing the association in non-linear distributions, or more than two variables. We are not able to include these tests here and refer you to other texts, such as Grafen & Hails (2002) and Sokal & Rohlf (1994).

- The Online Resource Centre includes interactive exercises that test your understanding of this chapter with other topics.

 **online resource centre**

Answers to chapter questions

A1 No. As the **x** values increase, the value of **y** stays the same. Therefore, there is no association between these two variables.

A2 Only Figs 8.1 and 8.2 meet criterion 4.

Fig. 8.1 illustrates a positive linear distribution, whereas Fig. 8.2 illustrates a negative linear distribution.

Fig. 8.4(a) has a gap in the middle of the distribution. To resolve this, you would need to repeat the experiment,

ensuring that observations are recorded for the middle part of the distribution.

In Fig. 8.4(b), the data are bunched and again you would gain a better understanding of your topic of investigation by repeating the experiment and expanding the range of measurements taken.

In Fig. 8.4(c), there is one outlying observation. You cannot ignore this datum point. If you have data like this,

you should first carry out your data analyses with this datum point included. You can then repeat the analysis with this datum point excluded. You should report on the outcomes of both analyses.

Fig. 8.4(d) shows an apparent curvilinear association. Curvilinear data can be analysed using correlation and/or regression analyses if one or both sets of observations are transformed (5.9) so that a linear distribution results, or using methods not described in this book.

A3 In our original investigation, the minimum feed supplement eaten was 7.0g, but in theory you could use the regression equation to calculate this unknown y value. In this example, if $x = 1.0$g, then $y = -2.35917 + (0.48373 \times 1.0) = -1.87544$ units. But what would an egg with a negative shell hardness look like? This illustrates the folly of working outside the boundaries of your known data set. For all we know, the pullets may die unless they eat a minimum of food supplement, or they may do brilliantly without the supplement and produce really hard eggshells. You must therefore never extrapolate outside the range of your existing observations.

Hypothesis testing: do my samples come from the same population? Parametric data

9

> ### In a nutshell
>
> Parametric tests of hypotheses are powerful and flexible but should only be used when data are normally distributed. In this chapter, we have included the tests most often used by our undergraduates to test the hypothesis 'Do the samples come from the same statistical population?' for one to three variables with or without replicates.

This chapter tells you how you may analyse parametric data to test the hypothesis that two or more samples come from the same statistical population. All these tests use the underlying mathematical relationships that come from the data having a normal distribution. If you are not sure whether your data are normally distributed, you can use the criteria in Box 5.2 to check. The normal distribution is described by the Gaussian equation, and from this, other mathematical relationships are derived that underpin the calculations in this chapter. Two of these are calculations for the sum of squares and the variance. If you are not familiar with these terms, you can refer to Box 5.1.

In many of the statistical tests described in this chapter, a variance ratio or *F*-test is calculated. For example, an *F*-test is used to check one of the criteria for using a *z*-test. Similarly, an *F*-test is used before an analysis of variance (ANOVA) to check that the data are suitable for using this test. An *F*-test is also used at the end of the ANOVA enabling you to test your hypotheses. These *F*-tests differ in the tables of critical values that are used to compare the calculated *F* values to. You must therefore always check to make sure you are using the correct table for the particular *F*-test.

Thirteen tests are covered in this chapter and worked examples are given for each. If this is the first time you have used a test, you should work through the example and check your answers before using the test on your own data. If your answer differs considerably from that given, you can check your calculation by going to the Online Resource Centre. If you work through these examples, we would expect it to take you eight hours. If you prefer to work

on parts of this chapter, we recommend you look at the *t*- and *z*-tests first, then the one- and two-way ANOVAs, and finally the three-way ANOVA and two-way nested ANOVA. To study these, we would expect you to take two hours, three hours, and three hours, respectively. The answers for these exercises are at the end of the chapter. All our examples are based on real undergraduate research projects. If these examples are not in your subject area, you will find more in the Online Resource Centre.

online resource centre

How to choose the correct test

These are largely selected on the basis of the experimental design: how many variables, how many categories in each variable, and how many replicates in each category. This table directs you to the most likely statistical test. It is assumed that you have parametric data. The terms 'matched' and 'unmatched' are explained in 9.3 and in the glossary.

If you are not clear about the criteria, refer to the specific section indicated and examine the examples to see whether they are similar to the work you are planning. Each test has additional criteria that need to be met. These are given in the sections indicated.

You wish to examine the effect of one variable with two categories or samples. The data are umatched.	t- or z-test for unmatched data (9.1 or 9.2).
You wish to examine the effect of one variable with two categories or samples. The data are matched.	t- or z-test for matched data (9.3).
You wish to examine the effect of one variable with more than two categories or samples. There are equal replicates in each category.	One-way ANOVA for equal replicates (9.5). If the outcome from this test is significant, you may follow it with a Tukey's multiple comparisons test (9.6).
You wish to examine the effect of one variable with more than two categories or samples. There are unequal replicates in each category.	One-way ANOVA for unequal replicates (9.7). If the outcome from this test is significant, you may follow it with a Tukey–Kramer multiple comparisons test (9.8).
You wish to examine the effect of one variable with more than two categories or samples. There are no replicates. There is a notable confounding variable or the data are repeated measures or matched.	Two-way ANOVA with no replicates (9.12).
You wish to examine the effect of two variables with more than two categories or samples. There are no replicates and all categories from one variable are combined with all categories from the second variable (orthogonal).	Two-way ANOVA with no replicates (9.12).

You wish to examine the effect of two variables. Each variable has at least two categories or samples and all categories from one variable are combined with all categories from the second variable (orthogonal). There are equal replicates in each category.	Two-way ANOVA with equal replicates (9.9). If the outcome from this test is significant, you may follow it with a Tukey's multiple comparisons test (9.10).
You wish to examine the effect of two variables. Each variable has at least two categories or samples and all categories from one variable are combined with all categories from the second variable (orthogonal). There is missing data or unequal replicates.	Two-way ANOVA for unequal replicates (9.11).
You wish to examine the effect of two variables. Each variable has at least two categories. One variable is randomized or nested with regard to the second variable. There are equal replicates in each category.	Two-way nested ANOVA for equal replicates (9.13). If the outcome from this test is significant, you may follow it with a Tukey's multiple comparisons test (9.13.3).
You wish to examine the effect of three variables. Each variable has at least two categories and all categories from each variable are combined with all other categories from the other variables (orthogonal). There are no replicates.	Three-way factorial ANOVA with no replicates (9.14).
You wish to examine the effect of three treatment variables. Each variable has at least two categories and all categories from each variable are combined with all other categories from the other variables (orthogonal). There are equal replicates in each category.	Three-way ANOVA with equal replicates (9.15).
None of the above.	See Chapter 10, and Sokal & Rohlf (1994) and Zar (2010).

9.1 *z*-test for unmatched data

Key points This test is appropriate when there are only two samples to be compared and there are relatively large numbers of observations. We introduce you to *F*-tests including the *F*-test that commonly precedes the *z*-test to check the criterion that the two samples have homogeneous variances.

The *z*-test relies on the known relationship between the mean and standard deviation in normally distributed data and between the mean and standard error of the mean (5.2–5.7). If samples are from the same population, then you would expect the difference between the sample means to be close

to zero. The greater the difference between the sample means, the more unlikely it is that the means are from the same population. The *z*-test estimates, in terms of standard errors, how far apart the means are. This then indicates the probability that the means are from the same population.

There are two forms of this *z*-test: one for unmatched data (9.1) and one for matched data (9.3). These two tests are quite different from each other. When samples sizes are small a *t*-test is recommended (9.2). The *t*- and *z*-tests are similar and the important differences are highlighted in the following sections.

9.1.1 Using this test

To use this test you:

1) Wish to examine the effect of one variable with two categories or samples.
2) Have unmatched parametric data (but see 3b).
3) a. Have at least 30 observations in each sample but the sample sizes need not be the same in the two samples, or
   b. The data are not normally distributed but are measured on an interval scale and there are more than 30 observations in each sample.
4) Have two samples where the variances are similar (homogeneous).

In Example 9.1, the criteria for using a *z*-test are met. There is one variable (species), two samples (mid- and lower shore), and 30 observations for each sample. Each periwinkle has been measured only once so the data are unmatched. The data are parametric. This has been checked using the criteria in Box 5.2. But are the variances similar?

One of the common major assumptions underlying many parametric tests of hypotheses is that the variances of the different samples are homogeneous (similar). A variance ratio or *F*-test may be used to examine whether this is probably true (Box 9.1).

EXAMPLE 9.1 The evolution of *Littorina littoralis* at Aberystwyth, 2002

Several years ago it was suggested that a species of periwinkle, *Littorina littoralis*, was evolving sympatrically into two species, *Littorina obtusata* and *Littorina mariae*, through niche partitioning. *Littorina obtusata* apparently grazes on the brown alga *Ascophyllum nodosum* on the mid-shore, whilst putative *Littorina mariae* feeds on the epiphytes growing on *Fucus serratus* on the lower shore. In a study of the sympatric evolution of *Littorina littoralis*, a representative sample of the two groups of individuals was collected from Aberystwyth in 2002 and their shell height (mm) recorded (Table 9.1). The investigators wished to test the hypothesis that there is no difference between shell height of the two groups of periwinkles from the mid- and lower shore.

Table 9.1. Height of shells of two putative species of periwinkles from the mid- and lower shore at Aberystwyth, 2002

<table>
<tr><th colspan="6">Shell height (mm)</th></tr>
<tr><th colspan="3">Periwinkles from the lower shore</th><th colspan="3">Periwinkles from the mid-shore</th></tr>
<tr><td>5.3</td><td>4.3</td><td>6.5</td><td>11.7</td><td>9.3</td><td>11.3</td></tr>
<tr><td>8.7</td><td>8.0</td><td>6.8</td><td>4.1</td><td>7.7</td><td>7.0</td></tr>
<tr><td>5.3</td><td>10.2</td><td>5.0</td><td>8.8</td><td>6.6</td><td>6.8</td></tr>
<tr><td>5.3</td><td>5.3</td><td>5.1</td><td>9.7</td><td>9.4</td><td>8.3</td></tr>
<tr><td>5.9</td><td>7.8</td><td>4.9</td><td>5.0</td><td>8.8</td><td>4.5</td></tr>
<tr><td>2.8</td><td>7.0</td><td>3.7</td><td>6.7</td><td>5.8</td><td>6.0</td></tr>
<tr><td>8.7</td><td>6.1</td><td>2.8</td><td>6.6</td><td>5.6</td><td>7.0</td></tr>
<tr><td>2.0</td><td>5.0</td><td>3.8</td><td>7.8</td><td>7.5</td><td>6.5</td></tr>
<tr><td>5.3</td><td>5.4</td><td>3.0</td><td>7.0</td><td>6.3</td><td>6.6</td></tr>
<tr><td>6.5</td><td>5.7</td><td>4.2</td><td>7.6</td><td>12.5</td><td>5.6</td></tr>
</table>

BOX 9.1 How to carry out an *F*-test to check for homogeneous variances before carrying out a *z*-test for unmatched data

GENERAL DETAILS	EXAMPLE 9.1
	This calculation is given in full in the Online Resource Centre. All values have been rounded to five decimal places.
1. Hypotheses to be tested H_0: There is no difference between the variances of the two samples. The variances are homogeneous. H_1: There is a difference between the variances of the two samples.	**1. Hypotheses to be tested** H_0: There is no difference between the variances of the shell height of periwinkles from the lower and mid-shore. H_1: There is a difference between the variances of the shell height of periwinkles from the lower and the mid-shore.
2. Have the criteria for using this test been met?	**2. Have the criteria for using this test been met?** Yes. The data are parametric and a variance can be calculated as described in Box 5.1.

3. How to work out $F_{calculated}$
  i. Decide which is sample 1 and which is sample 2.
	3. How to work out $F_{calculated}$
	  i. Let the periwinkles from the lower shore be sample 1 and the periwinkles from the mid-shore be sample 2.

BOX 9.1 Continued

ii. For each sample calculate the variance s_1^2 and s_2^2 using the method described in Box 5.1, where the sum of squares of x is:

$$SS(x) = (\Sigma x^2) - \frac{(\Sigma x)^2}{n}$$

The variance is:

$$s^2 = \frac{SS(x)}{n-1}$$

iii. $F_{calculated}$ is the ratio between the two variances.

$$F_{calculated} = \frac{larger\ sample\ variance}{smaller\ sample\ variance}$$

ii. For the periwinkles from the lower shore,
$\Sigma x = 166.4$, $n = 30$

$$\frac{(\Sigma x)^2}{n} = 922.96533, \quad \Sigma x^2 = 1027.08$$

$$SS(x) = 1027.08 - 922.96533$$
$$= 104.11467$$

$s_1^2 = 3.59016$
For periwinkles from the mid-shore
$\Sigma x = 224.1$, $n = 30$

$$\frac{(\Sigma x)^2}{n} = 1674.027, \quad \Sigma x^2 = 1792.85$$

$$SS(x) = 118.823,\ s_2^2 = 4.09734$$

iii. $F_{calculated} = \dfrac{4.09734}{3.59016}$

$\phantom{iii. F_{calculated}} = 1.14127$

4. How to find $F_{critical}$

You will come across three F-tests. Each uses a different table to find $F_{critical}$. You must take great care to use the correct table.
See Appendix d, Table D4. To look up the value in the F table, you will need to know the degrees of freedom (v) for each sample.
Sample 1 $v_1 = n_1 - 1$
Sample 2 $v_2 = n_2 - 1$
where n is the number of observations in that category or sample.

4. How to find $F_{critical}$

For our example, $v_1 = 30 - 1 = 29$ and $v_2 = 30 - 1 = 29$
Interpolating (6.3.6) from the F table:
$F_{critical} = 2.10$ at $p = 0.05$

5. The rule

If $F_{calculated}$ is less than $F_{critical}$, you do not reject the null hypothesis and may proceed with the z-test.

5. The rule

In this example, $F_{calculated}$ (1.14) is less than $F_{critical}$ (2.10). Therefore, you do not reject the null hypothesis.

6. What does this mean in real terms?

6. What does this mean in real terms?

There is no significant difference ($p = 0.05$) between the variances of shell height of periwinkles from the lower and mid-shores. The variances are homogeneous. You may proceed with the z-test.

If the null hypothesis is not accepted, then you should consider transforming your data (5.9) or using a non-parametric test, e.g. Mann–Whitney U test (10.1).

Further examples relating to the topic including how to use statistical software are included in the Online Resource Centre.

online resource centre

9.1.2 **The calculation**

Having satisfied ourselves that our data meet all the criteria for using this
z-test for unmatched data, we can then proceed (Box 9.2).

BOX 9.2 How to carry out a *z*-test for unmatched data

GENERAL DETAILS	EXAMPLE 9.1
	This calculation is given in full on the Online Resource Centre. All values have been rounded to five decimal places.
1. Hypotheses to be tested H_0: There is no difference between the mean values of the populations from which the samples are taken. H_1: There is a difference between the mean values of the populations from which the two samples are taken. If you use this test for data that are not normally distributed (9.1.1), then you should refer to medians not means in the hypotheses.	**1. Hypotheses to be tested** H_0: There is no difference between the mean shell height (mm) of the periwinkles from the lower and mid-shore at Aberystwyth in 2002. H_1: There is a difference between the mean shell height (mm) of the periwinkles from the lower and mid-shore at Aberystwyth in 2002.
2. Have the criteria for using this test been met?	**2. Have the criteria for using this test been met?** Yes (9.1.1).
3. How to work out $z_{\text{calculated}}$ i. Decide which is sample 1 and which is sample 2. ii. Use the variances (s^2) and *n* from Box 9.1 calculated for the *F* test. iii. Calculate the means ($\bar{x}$) for each sample. iv. The standard error of the difference of the means (SE$_D$) is calculated as: $$SE_D = \sqrt{\left(\frac{s_1^2}{n_1} + \frac{s_2^2}{n_2}\right)}$$ v. $z_{\text{calculated}} = \dfrac{\bar{x}_1 - \bar{x}_2}{SE_D}$	**3. How to work out $z_{\text{calculated}}$** i. Let the periwinkles from the lower shore be sample 1 and the periwinkles from the mid-shore be sample 2. ii. $s_1^2 = 3.59016$, $s_2^2 = 4.09734$ $n_1 = n_2 = 30$ iii. $\bar{x}_1 = 5.54667$, $\bar{x}_2 = 7.47$ iv. Standard error of the difference of the means (SE$_D$): $$SE_D = \sqrt{\left(\frac{3.59016}{30} + \frac{4.09734}{30}\right)}$$ $$= 0.50621$$ v. $z_{\text{calculated}} = \dfrac{5.54667 - 7.47}{0.50621}$ $$= -3.79947$$
4. How to find z_{critical} See Appendix d, Table D5. The *z* distribution is unusual in that it is independent of sample size. For any one *p* value, there is only one *z* value.	**4. How to find z_{critical}** For a two-tailed test at $p = 0.05$, $z_{\text{critical}} = 1.96$

BOX 9.2 Continued

5. The rule	5. The rule
If the absolute value (i.e. ignore the sign) of $z_{calculated}$ is more than or equal to $z_{critical}$, then you may reject the null hypothesis (H_0). Samples are not likely to have come from the same statistical population. (The term 'absolute' is explained in the glossary and Appendix e).	The absolute value of $z_{calculated}$ (3.8) is more than $z_{critical}$ (1.96). Therefore, you may reject the null hypothesis. In fact, at $p = 0.001$, $z_{critical} = 3.29$. You may therefore reject the null hypothesis at this higher level of significance.
6. What does this mean in real terms?	**6. What does this mean in real terms?**
	There is a very highly significant difference ($z = 3.8$, $p < 0.001$) between the mean shell heights (mm) of the periwinkles from the lower and mid-shore at Aberystwyth. *L. mariae* is the smaller of the two apparently distinct species.

Further examples relating to the topic including how to use statistical software are included in the Online Resource Centre.

online resource centre

9.2 Unequal variance *t*-test for unmatched data

Key points A statistical test in common use is Student's *t*-test. This can be used when comparing two samples with relatively small sample sizes. However, the effectiveness of Student's *t*-test is dependent on the samples having similar or homogeneous variances. The *t*-test given here can be used when the variances are unequal but can also be used when they are equal and therefore has an advantage over Student's *t*-test.

The *z*-test compares two samples of parametric data. However, when the sample sizes are relatively small, then the distribution on which the *z* statistic is based changes with the sample size. This can be taken into account by using a *t*-test. With a *t*-test, the sample sizes are used both in the estimate of the standard error and when looking up the critical value of *t* in tables. There are a number of *t*-tests. The most commonly used is Student's *t*-test after the pen name (Student) of W. Gosset who introduced this test (Student, 1908). This test requires that the two samples have similar (homogeneous) variances and we therefore carry out an *F*-test first to check that this criterion is met (Box 9.1). An alternative *t*-test (also known as the Welch test or Smith–Welch–Satterwaite test) does not require the variances to be homogeneous, although both sets of data still need to be normally distributed (Ruxton, 2006). When variances are similar, this test performs as well as the more commonly used Student's *t*-test. In this edition, we have therefore replaced the Student's *t*-test with the unequal variances *t*-test as the test of choice, but have retained Student's *t*-test in the Online Resource Centre for reference.

online resource centre

9.2.1 **Using this test**

To use this test you:

1) Wish to examine the effect of one variable with two categories or samples.

2) Have unmatched parametric data.

3) Have fewer than 30 observations in each category or sample, although the sample sizes need not be equal.

EXAMPLE 9.2 The evolution of *Littorina littoralis* at Porthcawl, 2002

The study at Aberystwyth (2002) indicated that a species of periwinkle, *Littorina littoralis*, was evolving sympatrically into two species (Example 9.1, Box 9.2). To extend this study, the researchers also sampled from Porthcawl in the same year. Here, however, they found far fewer periwinkles (Table 9.2).

Table 9.2. Height of shells in two putative species of periwinkles from the mid- and lower shore at Porthcawl, 2002

Shell height (mm)			
Periwinkles on the lower shore		Periwinkles on the mid-shore	
5.5	4.0	3.3	6.7
8.4	5.0	6.3	5.7
5.0	6.2	6.1	4.2
5.0	5.0	8.0	6.3
5.6	7.7	13.5	6.0
4.8	6.0	5.3	7.2
8.4		6.7	

The data from Example 9.2 meet the criteria for using an unequal variances *t*-test in that we wish to examine the effect of one variable (species), there are two samples (mid- and lower shore), and 13 observations in each sample. Each periwinkle has been measured only once so the data are not matched. There are few observations on which to gauge whether the data are parametric. However, the height is measured on an interval scale, and we will therefore assume that the data are parametric.

9.2.2 **The calculation**

Having satisfied ourselves that the data from Example 9.2 meet all the criteria for using this *t*-test, we can then proceed (Box 9.3). To distinguish the $t_{calculated}$ for Student's *t*-test, we use the statistic t' for the $t_{calculated}$ using this unequal variances *t*-test.

BOX 9.3 How to carry out an unequal variance *t*-test for unmatched data

GENERAL DETAILS	EXAMPLE 9.2
	This calculation is given in full in the Online Resource Centre. All values have been rounded to five decimal places.
1. Hypotheses to be tested H_0: There is no difference between the means for the two samples. H_1: There is a difference between the means for the two samples.	**1. Hypotheses to be tested** H_0: There is no difference between the mean shell height (mm) of the periwinkles from the lower and mid-shore at Porthcawl in 2002. H_1: There is a difference between the mean shell height (mm) of the periwinkles from the lower and the mid-shore at Porthcawl in 2002.
2. Have the criteria for using this test been met?	**2. Have the criteria for using this test been met?** Yes (9.2.1).
3. How to work out $t'_{calculated}$ i. Decide which is sample 1 and which is sample 2. ii. Calculate the variances (s_1^2 and s_2^2) and means ($\bar{x}_1$ and $\bar{x}_2$) and record the number of observations (n) for each sample. If you are not sure how to calculate the variances see Box 5.1. iii. Calculate the standard error of the difference between the means (SE_D). $$SE_D = \sqrt{\left(\frac{s_1^2}{n_1} + \frac{s_2^2}{n_2}\right)}$$ iv. $t'_{calculated} = \dfrac{\bar{x}_1 - \bar{x}_2}{SE_D}$	**3. How to work out $t'_{calculated}$** i. Let the periwinkles from the lower shore be sample 1 and the periwinkles from the mid-shore be sample 2. ii. $\bar{x}_1 = 5.89231$ $s_1^2 = 2.01244$ $n_1 = 13$ $\bar{x}_2 = 6.56154$ $s_2^2 = 5.82256$ $n_2 = 13$ iii. $$\begin{aligned}SE_D &= \sqrt{\left(\frac{2.01244}{13} + \frac{5.82256}{13}\right)}\\ &= \sqrt{(0.15480 + 0.44789)}\\ &= \sqrt{0.60269} = 0.77633\end{aligned}$$ iv. $\begin{aligned}t'_{calculated} &= \dfrac{5.89231 - 6.56154}{0.77633}\\ &= -0.86204\end{aligned}$
4. How to find $t_{critical}$ See Appendix d, Table D6. The critical value of t' is found using a t table and the degrees of freedom (v). $$v = \frac{\left(\dfrac{1}{n_1} + \dfrac{u}{n_2}\right)^2}{\left(\dfrac{1}{n_1^2(n_1 - 1)}\right) + \left(\dfrac{u^2}{n_2^2(n_2 - 1)}\right)}$$ This is a more complex calculation than is usual for the degrees of freedom so we have broken this down into a number of steps.	**4. How to find $t_{critical}$**

BOX 9.3 Continued

i. First calculate u

$$u = \frac{s_2^2}{s_1^2}$$

i. Calculating the degrees of freedom for Example 9.2.

$$u = \frac{s_2^2}{s_1^2} = \frac{5.82256}{2.01244} = 2.89328$$

ii. Calculate the numerator

$$\left(\frac{1}{n_1} + \frac{u}{n_2}\right)^2$$

ii.
$$\left(\frac{1}{n_1} + \frac{u}{n_2}\right)^2 = \left(\frac{1}{13} + \frac{2.89328}{13}\right)^2$$
$$= (0.07692 + 0.22256)^2$$
$$= (0.29948)^2 = 0.08969$$

iii. Calculate the first part of the denominator

$$\left(\frac{1}{n_1^2\,(n_1 - 1)}\right)$$

iii.
$$\left(\frac{1}{n_1^2\,(n_1 - 1)}\right) = \frac{1}{(13)^2\,(13-1)}$$
$$= \frac{1}{169 \times 12} = \frac{1}{2028} = 0.00049$$

iv. Calculate the second part of the denominator

$$\left(\frac{u^2}{n_2^2\,(n_2 - 1)}\right)$$

iv.
$$\left(\frac{u^2}{n_2^2\,(n_2 - 1)}\right) = \frac{(2.89328)^2}{(13)^2\,(13-1)}$$
$$= \frac{8.37109}{169 \times 12} = 0.00413$$

v. Using these terms, now calculate the degrees of freedom. $t'_{critical}$ can be found in Table D6 at this degrees of freedom and for $p = 0.05$.

v.
$$v = \frac{0.08969}{(0.00049) + (0.00413)}$$
$$v = 19.40990$$

So $t'_{critical}$ is 2.093

5. The rule

If the absolute value of $t_{calculated}$ is more than or equal to $t_{critical}$ you may reject the null hypothesis. The term 'absolute' is explained in the glossary and Appendix e.

5. The rule

When you use an absolute value, the negative sign is ignored. In this example, $t_{calculated}$ is therefore 0.862. You can therefore conclude that the absolute value of $t_{calculated}$ (|–0.862|) is less than $t_{critical}$ (2.093). Therefore, you do not reject the null hypothesis.

6. What does this mean in real terms?

6. What does this mean in real terms?

There is no significant difference ($t = 0.862$, $p = 0.05$) in the mean shell height (mm) between the periwinkles on the lower and mid-shores of Porthcawl. Therefore, these data do not support the notion that there are two distinct species.

Further examples relating to the topic including how to use statistical software are included in the Online Resource Centre.

online resource centre

Examine the results from Boxes 9.2 and 9.3. Suggest reasons why the outcomes from these two investigations differ.

9.3 *z*- and *t*-tests for matched data

Key points In some experiments, you may take 'before' and 'after' measurements from the same item. This generates 'matched' data (also called repeated measures). The two measurements from the same item are not independent of each other. This needs to be taken into account when analysing the data.

If you are designing an investigation in which you will measure an item before a particular event or treatment and then after, this will generate matched data (sometimes also known as 'repeated measures'). The two observations for each item are not independent of each other as they relate to a single item. This needs to be allowed for in the analysis of the data. Unlike unmatched *z*- and *t*-tests, the comparison of matched samples is based on the difference between each pair of measurements. If the samples are similar, then the differences between each pair of observations will be small. The mean and standard error of these differences are used to generate $z_{calculated}$.

Matched *z*- and *t*-tests are virtually identical; the main difference again arises in that in the *t*-test there is an allowance for the relatively small sample sizes being investigated. As these two tests are so similar, we have amalgamated the information in one box (Box 9.4). If you have taken more than two repeated measurements from one item you can consider the two-way ANOVA with no replicates (9.12).

9.3.1 Using this test

To use this test you:

1) Wish to examine the effect of one variable with two categories or samples.
2) Have parametric data and these data are matched; therefore the sample sizes are equal.
3) Have samples in which the variances are similar (homogeneous).
4) a. Have 30 or more pairs of observations (use the *z*-test for matched data), or

   b. Have fewer than 30 pairs of observations (use the *t*-test for matched data).

EXAMPLE 9.3. Weight loss by members of a fencing club during a 1-day competition

The members of a fencing club were aware that on days when they competed, they lost weight (Table 9.3). The weight loss was thought to be the result of dehydration. However, as not all members lost weight, the team wished to know whether the weight loss was significant (T. Richards, personal communication, 9.1.05).

Table 9.3. Weight loss during a 1-day fencing competition by members of a fencing club

Competitor	Weight before competition (kg)	Weight after competition (kg)	Difference between weights (kg)
1	60.00	59.55	0.45
2	59.15	58.70	0.45
3	60.20	59.80	0.40
4	62.40	61.90	0.50
5	57.20	57.20	0.00
6	60.35	59.85	0.50
7	59.80	59.40	0.40
8	60.10	60.10	0.00
9	60.20	59.90	0.30
10	59.90	59.90	0.00
11	60.00	60.00	0.00
12	61.20	60.75	0.45
13	58.50	58.05	0.45
$\bar{x}$	59.92308	59.62308	0.30
s^2	1.51656	1.35942	0.04583

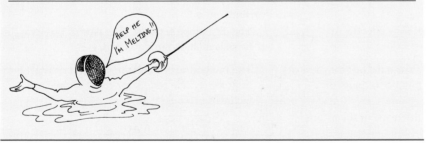

The data for Example 9.3 meet the criteria for using a *t*-test for matched data as there are fewer than 30 observations, we wish to examine the effect of one variable (weight), and there are two samples (before and after the competition). The data are matched as two observations have been recorded for each person. The test for homogeneity of variances is described in Box 9.1. For Example 9.3, $F_{\text{calculated}}$ = 1.51656/1.35942 = 1.11554, v = 13 − 1 = 12 for both samples, and at p = 0.05, F_{critical} = 3.28. As $F_{\text{calculated}}$ is less than F_{critical}, there is no significant difference between the two sample variances and we may proceed with the *t*-test.

9.3.2 **The calculation**

online
resource
centre

Having satisfied ourselves that the data for Example 9.3 meet all the criteria for using this *t*-test, we can then proceed (Box 9.4). If you are carrying out a *z*-test for matched data, the procedure is exactly the same until finding the critical values. Some of the calculations are included in Table 9.3. The full calculation is given in the Online Resource Centre.

BOX 9.4 How to carry out a *z*- and *t*-test for matched data

GENERAL DETAILS	**EXAMPLE 9.3**
	This calculation is given in full in the Online Resource Centre. All values have been rounded to five decimal places.
1. **Hypotheses to be tested** H_0: There is no difference between the means of the populations from which sample 1 and sample 2 are taken. H_1: There is a difference between the means of the populations from which sample 1 and sample 2 are taken.	1. **Hypotheses to be tested** H_0: There is no difference between the mean weights (kg) of individuals before and after a fencing competition. H_1: There is a difference between the mean weights (kg) of individuals before and after a fencing competition.
2. **Have the criteria for using this test been met?**	2. **Have the criteria for using this test been met?** Yes (9.3.1).
3. **How to work out $z_{\text{calculated}}$ or $t_{\text{calculated}}$** i. Work out the difference (D) for each pair of observations. ii. Calculate the mean ($\overline{D}$) and variance (s_D^2) for these differences, where n is the number of pairs of observations. iii. The standard error of the difference of the means $$SE_D = \sqrt{\dfrac{s_D^2}{n}}$$ iv. $$z_{\text{calculated}} \text{ or } t_{\text{calculated}} = \dfrac{\overline{D}}{SE_D}$$	3. **How to work out $t_{\text{calculated}}$*** (*In this example we are working out t; the steps would be the same here if you were working out z.) i. The difference between each pair of observations is included in Table 9.3. ii. The method for calculating a variance is given in Box 5.1. The results for this example are included in Table 9.3. iii. $$SE_D = \sqrt{\dfrac{0.04583}{13}} = 0.05938$$ iv. For this example, we are using a *t*-test, therefore $t_{\text{calculated}} = \dfrac{0.3}{0.05938} = 5.05245$
4. **How to find z_{critical} or t_{critical}** If this is a *t*-test for matched pairs, then t_{critical} is found for a two-tailed *t*-test where $v = n - 1$ at $p = 0.05$ (Appendix d, Table D6) If this is a *z*-test for matched pairs, then z_{critical} is found from the two-tailed table of *z* values at $p = 0.05$ (Appendix d, Table D5).	4. **How to find t_{critical}** In our example, $v = 13 - 1 = 12$ and for a two-tailed *t*-test at $p = 0.05$, $t_{\text{critical}} = 2.179$.

BOX 9.4 **Continued**	
5. The rule If the absolute value of $z_{\text{calculated}}$ or $t_{\text{calculated}}$ is more than or equal to z_{critical} or t_{critical}, respectively, then you may reject the null hypothesis.	**5. The rule** $t_{\text{calculated}}$ (5.053) is greater than t_{critical} (2.179), so we reject the null hypothesis. In fact, t_{critical} at $p = 0.001$ is 4.318, so we may reject the null hypothesis at this higher level of significance.
6. What does this mean in real terms?	**6. What does this mean in real terms?** There is a very highly significant difference ($t = 5.053$, $p = 0.001$) between the mean weights (kg) of individuals before and after a fencing competition.

Further examples relating to the topic including how to use statistical software are included in the Online Resource Centre.

online resource centre

9.4 **Introduction to parametric ANOVA**

Key points ANOVAs are powerful, flexible, and suitable for a range of experimental designs (9.16). We cover those used most often by our undergraduates to examine the effect of one, two, or three variables. These ANOVAs can be used to examine the effect of repeated measures in an experiment or confounding variables. Although ANOVAs can be used when there are no replicates or unequal replicates, both of these circumstances severely limit the analysis. The variables under investigation may be fixed or random. This is a critical factor in determining the ANOVA to be used.

The acronym ANOVA stands for analysis of variance. Parametric ANOVAs are flexible and powerful. There are many versions of ANOVA that allow you to handle data with two or more samples, with or without replicates, with one or more treatment variables, and with or without incomplete data sets. In this chapter, we examine the ANOVAs our undergraduates decide to use most often. There are many versions of parametric ANOVAs that we do not cover. If you think that an ANOVA is likely to be the best test for your experimental design but you cannot find the one you need in this chapter, then we recommend Sokal & Rohlf (1994) or Zar (2010).

ANOVAs are used to test the hypothesis 'There is no difference between the mean values for a given number of samples'. These samples may come from different variables or treatments or from blocks or replicates within variables. We consider experimental designs relating to ANOVAs at the end of this chapter. Although the hypotheses refer to sample means, the ANOVA 'looks' at this by comparing variation in the data. In Fig. 9.1, data from two samples are plotted. The means for both samples are similar and the variances are similar and relatively small.

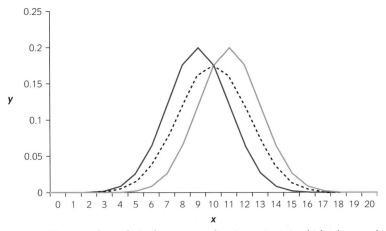

Fig. 9.1. Two samples with similar means and variances (———), which when combined (----) produce a distribution with a mean and variance similar to those of the two original samples.

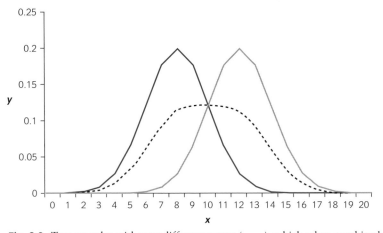

Fig. 9.2. Two samples with very different means (———), which when combined (----) produce a distribution with a variance much greater than the variance of either sample 1 or sample 2.

When the two samples are combined (Fig. 9.1), there is little change in either the mean or the total variance. Compare this with Fig. 9.2. Here the two samples have very different means, although the variances for each sample are relatively small and similar to each other. If these two samples are combined, the mean for this total data set is different and the combined variance is relatively large. In an analysis of the data from Fig. 9.1, you would probably not expect a significant difference between the two samples and this can be detected by comparing the total variance with the variance for the two samples. However, in the data from Fig. 9.2, the total variance is relatively large because the means of the two samples are so different from each other. In this case, a comparison of the two sample variances with the total variance would suggest that these two samples will be significantly different from each other. It

is this comparison of variances in the ANOVA that detects the relative differences in the means of the samples.

When asking the question 'Do samples come from the same or different populations?', it is possible to test both specific and general hypotheses (6.1). A general hypothesis asks the question 'Is there a difference?' A specific hypothesis asks a question about the nature of that difference: for example, 'Is the mean of sample 1 significantly larger than the mean of sample 2, which is larger than the mean of sample 3?' There are a number of methods that may be used to examine specific hypotheses following an ANOVA. We have included the most commonly used test, called Tukey's test.

Calculating ANOVAs by hand is straightforward but tedious. However, there are so many ANOVAs that a common mistake when using computer software is to select the wrong one. For this reason, we include the ANOVAs used most frequently by our undergraduates. We hope that this introduction will provide you with sufficient familiarity to be confident when using statistical packages.

The ANOVAs we include show you how to analyse data from experiments with one (one-way ANOVA, 9.5) to three (three-way ANOVA, 9.15) variables. Most of these designs require equal replicates and this should be considered when planning your experiment. In the two-way ANOVAs, we have a two-way ANOVA and a nested two-way ANOVA. These reflect a Model I and mixed model experimental design.

The terms Model I, Model II, and mixed models refer to the variables, which can be fixed or random. In Example 9.4, the effectiveness of weaning plantlets of *Lobelia* 'Hannah' from tissue culture onto one of four composts is considered. The four composts are a consistent or fixed treatment or variable determined by the investigator. In Example 9.6, the experiment on weaning was extended and two varieties of *Cosmos atrosanguineus* were weaned from tissue culture on one of four composts. Here, both the varieties and the composts are consistent or fixed treatments and both are determined by the investigator. An ANOVA suitable for analysing data where the variable(s) or treatment(s) is fixed is called a Model I ANOVA. The fixed treatments can be contrasted with Example 9.6. Here, soil samples were taken from three pits dug in two forests (deciduous and coniferous). The difference in this design is that the pits are 'randomized' in the forests. Pit 1 in the deciduous forest cannot be said to be the same treatment as pit 1 in the coniferous forest. In this example, the forests are fixed treatments and the pits are randomized. The randomized element is called a Model II design. The overall experimental design is called a mixed model as it has a fixed treatment variable (Model I) and a randomized treatment variable (Model II) within it. The ANOVA that is used to test a mixed model design is called a nested ANOVA. The ANOVAs we include can generally be used to test a Model I or Model II ANOVA with the exception of the nested ANOVA, which

is specifically for mixed models. It is therefore important that you know whether your design is Model I, Model II, or a mixed model.

 Q2   Look at Example 9.10. There are three variables: depths, distances, and bearings from a smelter. Which of these are fixed and which random? What model is this experimental design?

In the ANOVAs that we include, only the nested ANOVA is model-dependent. The other ANOVAs may be used for either pure Model I or pure Model II designs. Mixed models should be analysed using nested ANOVAs.

The method for calculating ANOVAs does not lend itself to the format we have followed in the boxes elsewhere in the book. However, we include both general instructions and a specific example to help you understand the calculations. In addition, there are calculation tables and ANOVA tables that illustrate both how to carry out the analysis and how to report these results.

9.5 One-way parametric ANOVA with equal numbers of replicates

Key points   This ANOVA will examine the effect of one variable usually with more than two categories. There need to be equal numbers of replicates assigned at random to each category. Preceding the ANOVA, we consider the F_{max} test, which is used to test the criteria that the variances are homogeneous.

The first and simplest ANOVA we consider is one that allows you to examine the effect of one variable, e.g. the organic content of soil in four different woodlands. Here the treatment is 'woodlands'. When there is one variable under investigation, the ANOVA is called a one-way ANOVA. This parametric one-way ANOVA can examine Model I or Model II designs with data that have an equal number of replicates (9.5) or with a small adjustment can analyse data with unequal replicates (9.7).

9.5.1 Using this test

To use this test you:

1) Wish to examine the effect of one variable.

2) Have parametric data with the same number of observations in each category or sample.

3) Have an experimental design that means that each item is assigned at random to the categories or samples.

4) Have samples where the variation is similar (homogeneous).

EXAMPLE 9.4 The effectiveness of weaning plantlets of *Lobelia* 'Hannah' from tissue culture onto one of four composts

An undergraduate studying horticulture was interested in how plants propagated by tissue culture were then weaned onto other growing media, such as soil-based compost. She designed a series of trials that examined aspects of this weaning process. In one trial, she transferred *Lobelia* 'Hannah' plantlets at random into plugs containing one of four different composts (A–D). After 8 weeks, the plantlets were examined and their fresh weight recorded (Table 9.4).

Table 9.4. The fresh weight of *Lobelia* 'Hannah' after 8 weeks' weaning in one of four types of compost (A–D): calculations using these data and summary statistics

	Fresh weight (g) of *Lobelia* 'Hannah' weaned in one of four composts (A–D)							
	A		**B**		**C**		**D**	
	7.04	2.48	3.18	2.18	3.20	5.11	4.16	3.71
	4.86	3.03	1.52	2.12	3.93	5.32	2.94	1.60
	7.47	4.21	3.85	1.27	3.21	3.00	3.62	1.60
	4.49	6.03	1.42	2.03	3.01	2.25	2.09	2.14
	3.06	4.77	4.00	4.82	3.93	3.53	3.83	3.16
$\bar{x}$	4.744		2.639		3.649		2.885	
s^2	2.84663		1.53683		0.91999		0.92307	
Σx_s	47.44		26.39		36.49		28.85	
Σx_s^2	250.675		83.4747		141.4319		91.5399	
n_s	10		10		10		10	
Σx_T					139.17			
Σx_T^2					567.1215			
N					40			

The data from Example 9.4 meet all the criteria for using this one-way parametric ANOVA in that the data are parametric and we wish to examine the effect of one variable (compost). The plantlets were assigned at random to each treatment and there are ten observations in each sample. To check whether the variances are similar, an F_{max} test is carried out (Box 9.5). This test is very similar to the one outlined in Box 9.1; however, the table of critical values is not the same. You must therefore use the correct F_{max} table.

BOX 9.5 How to carry out an F_{max} test to check for homogeneous variances, before carrying out an ANOVA

GENERAL DETAILS	EXAMPLE9.4
	This calculation is given in full in the Online Resource Centre. All values have been rounded to five decimal places.
1. Hypotheses to be tested H_0: There is no difference between the variances of the two samples. The variances are homogeneous. H_1: There is a difference between the variances of the two samples.	**1. Hypotheses to be tested** H_0: There is no difference between the variances of the fresh weight (g) of plantlets grown in the four different composts. H_1: There is a difference between the variances of the fresh weight (g) of plantlets grown in the four different composts.
2. Have the criteria for using this test been met	**2. Have the criteria for using this test been met** Yes (9.5.1).
3. How to work out $F_{max\ calculated}$ i. For each sample, calculate the variance (s^2) using the method described in Box 5.1. ii. $F_{max\ calculated}$ is the ratio between these two values found by dividing the largest variance by the smallest variance. $$F_{max\ calculated} = \frac{\text{largest sample variance}(s_1^2)}{\text{smallest sample variance}(s_2^2)}$$	**3. How to work out $F_{max\ calculated}$** i. See Table 9.4. ii. $$F_{max\ calculated} = \frac{2.84663}{0.91999} = 3.09420$$
4. How to find $F_{max\ critical}$ See Appendix d, Table D7. To look up the value for an F-test that precedes an ANOVA, you will need to know the number of categories or samples (a) and the degrees of freedom ($v = n_s - 1$). n_s is the number of observations in each category or sample.	**4. How to find $F_{max\ critical}$** For our example, there are four samples ($a = 4$). There are ten observations in each sample so the degrees of freedom are $v = 10 - 1 = 9$ So at $p = 0.05$, $F_{max\ critical} = 6.31$.
5. The rule If $F_{max\ calculated}$ is less than $F_{max\ critical}$, then you may accept the null hypothesis and proceed with the ANOVA.	**5. The rule** In this example, $F_{max\ calculated}$ (3.09) is less than $F_{max\ critical}$ (6.31). Therefore, you do not reject the null hypothesis.
6. What does this mean in real terms?	**6. What does this mean in real terms?** There is no significant difference ($F_{max} = 3.09$, $p = 0.05$) between the variances of the fresh weight of the plantlets weaned on different composts. You may proceed with the ANOVA. If the null hypothesis is not accepted, then you should consider transforming your data (5.9) or using a non-parametric test, e.g. Kruskal–Wallis (10.3).

Further examples relating to the topic including how to use statistical software are included in the Online Resource Centre.

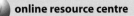 **online resource centre**

9.5.2 **The calculation**

Having satisfied ourselves that our data meet all the criteria for using this ANOVA, we can then proceed. In Box 9.6, we have organized the calculation of $F_{calculated}$ under a number of subheadings: (A) Calculate general terms; (B) Calculate the sums of squares; and (C) Construct and complete an ANOVA calculation table. At each point, we show how these steps can be applied to Example 9.4. Some of the calculations are included in Tables 9.4 and 9.5.

Table 9.5. ANOVA table for one-way parametric ANOVA of data from Example 9.4

| Source of variation | Fresh weight of *Lobelia* 'Hannah' after 8 weeks' weaning in one of four types of compost (A–D) | | | |
	SS	v	MS	F
Between composts	26.87561	3	26.87561/3 = 8.95854	5.75509
Within composts	56.03867	36	56.03867/36 = 1.55663	
Total variation	82.91428	39		

BOX 9.6. How to carry out a one-way parametric ANOVA with equal replicates

This calculation is given in full in the Online Resource Centre. All values have been rounded to five decimal places. The calculation is illustrated using data from Example 9.4.

1. **General hypotheses to be tested**

   H_0: There is no difference between the means of the populations from which the samples are taken.

   H_1: There is a difference between the means of the populations from which the samples are taken.
   In our example (Example 9.4), the hypotheses we are testing are:

   H_0: There is no difference between the mean fresh weight (g) of plantlets of *Lobelia* 'Hannah' weaned on the four composts after 8 weeks.

   H_1: There is a difference (g) between the mean fresh weight of plantlets of *Lobelia* 'Hannah' weaned on the four composts after 8 weeks.

2. **Have the criteria for using this test been met?**
   In this example, the criteria have been met (9.5.1).

3. **How to work out Fcalculated**

A. **Calculate general terms**
   1. Add together all the observations in all the categories or samples (grand total): Σx_T.

2. Square each observation in all the categories and add these together: Σx_T^2.
3. Add all the observations in a category or sample: Σx_s. This is the sample total. Do this for all samples. Square each sample total: $(\Sigma x_s)^2$. Add these together. Divide this total by the number of observations in each sample (n_s).
4. Take the result from step **1** and square it: $(\Sigma x_T)^2$. Divide by N, where N is the total number of observations in all the categories or samples combined.
   In our example, the calculations are:

   1. $\Sigma x_T = 139.17$

   2. $\Sigma(x_T^2) = (7.04)^2 + (4.86)^2 + \cdots + (2.14)^2$
      $+ (3.16)^2 = 567.1215$

   3. $\dfrac{(47.44)^2 + (26.39)^2 + (36.49)^2 + (28.85)^2}{10}$
      $= 511.8283$

   4. $(139.17)^2/40 = 484.20722$

BOX 9.6 Continued

B. Calculate the sums of squares (SS)

5. SS_{total} = result from step **2** – result from step **4**
6. $SS_{between^*}$ = result from step **3** – result from step **4** (*between samples*)
7. $SS_{within^{**}}$ = result from step **5** – result from step **6** (** *within samples*)

Optional check on calculations. *If you calculate the SS for each sample and add these together, this figure should be the same as SS_{within}. The SS for each sample is calculated as:*

$$SS_{sample} = \Sigma(x^2) - \frac{(\Sigma x)^2}{n}$$

In our example (Table 9.4) the calculations are:

5. SS_{total} = 567.1215 – 484.20722 = 82.91428
6. $SS_{between}$ = 511.8283 – 484.20722 = 26.87561
7. SS_{within} = 82.91428 – 26.87561 = 56.03867

C. Construct and complete an ANOVA calculation table

i. Draw an ANOVA table as illustrated here.

Source of variation	SS	v	MS	F
Between samples	Step 6			
Within samples	Step 7			
Total variation	Step 5			

ii. Transfer the results from the calculations for $SS_{between}$, SS_{within} and SS_{total} into the ANOVA table in column 2 (see Table 9.5).

iii. Calculate the degrees of freedom (v):
v_{total} = total number of observations – 1 = N – 1
$v_{between}$ = number of samples – 1 = a – 1
v_{within} = total number of observations – number of samples = N – a
Enter these results in column 3.

iv. Calculate the mean squares (MS) for between and within samples.
Note: Variances in this context are also known as mean squares (MS).

$$s^2 = MS = \frac{SS}{v}$$

In the 'between samples' row, take the value for SS and divide it by its value for v. In the 'within samples' row, take the value for SS and divide it by its value for v. Put the results from these calculations in column 4 (Table 9.5).

v. Finally calculate the variance ratio $F_{calculated}$ where:

$$F_{calculated} = \frac{\text{'between samples' MS}}{\text{'within samples' MS}}$$

It is usual to place the results from this calculation in column 5 in the first row (Table 9.5).

4. To find $F_{critical}$
See Appendix d, Table D7. Identify the critical F value using the F tables for an ANOVA. You may need to interpolate (6.3.6) these values. To find the critical value, you need the degrees of freedom relating to the variances 'between samples' (v_1) and 'within samples' (v_2).
In our example, 'between samples' v_1 = 3 and 'within samples' v_2 = 36.
Interpolating from the F table for $p = 0.05$, $F_{critical}$ = 2.872.

5. The rule
If $F_{calculated}$ is greater than or equal to $F_{critical}$, then you may reject the null hypothesis.
In our example, $F_{calculated}$ (5.7551) is greater than $F_{critical}$ (2.872) so you may reject the null hypothesis.
In fact, at $p = 0.01$ (Table D9), $F_{critical}$ = 4.39, so you may reject the null hypothesis at this higher level of significance.

6. What does this mean in real terms?
There is a highly significant difference ($F = 5.76$, $p = 0.01$) between the mean fresh weight (g) of *Lobelia* 'Hannah' when weaned on four different composts.

Further examples relating to the topic including how to use statistical software are included in the Online Resource Centre.

@ **online resource centre**

9.6 Tukey's test following a parametric one-way ANOVA with equal replicates

Key points An ANOVA tests the general hypothesis 'There is no difference'. However, this multiple comparisons test, which can be carried out after an ANOVA, allows a more specific comparison by examining the means from pairs of samples in all possible combinations.

The one-way ANOVA described in 9.5 has allowed us to test the general hypothesis that there was no difference between the success of weaning as indicated by mean fresh weight of *Lobelia* 'Hannah' on four composts. The analysis showed that there is a significant difference (Box 9.6). It would be helpful to be able to be more exact in our interpretation of the data. Looking at the mean values in Table 9.4, compost A appears to be most effective in this regard, but is this significant? To make a specific evaluation of this sort, you may use a multiple comparisons test. There are several of these but the one we recommend (Tukey's) is reasonably conservative, i.e. you will get fewer significant differences and are therefore able to be more discriminating in your interpretation and are less likely to make a Type I error.

The Tukey's test we describe is suitable for use following an ANOVA where the number of replicates is the same for all samples, as in our example. If you wish to use a multiple comparison test following a one-way ANOVA with unequal replicates, refer to the Tukey–Kramer test (9.8).

9.6.1 Using this test

To use this test you:

1) Wish to test specific hypotheses following a significant outcome in a parametric one-way ANOVA.
2) Should have equal numbers of observations in each category or sample.

To illustrate the Tukey's test, we will use Example 9.4 and the one-way parametric ANOVA in Box 9.6. Therefore, we are using the Tukey's test having obtained a significant difference in the ANOVA (Table 9.5) and there are ten observations in each category.

9.6.2 The calculation

In a Tukey's test, each mean value for a sample is compared with every other mean value in the investigation. This difference between the pairs of means is the calculated T value, which you then compare with a critical T value, which you also work out. We demonstrate this in Box 9.7 with supporting calculations in Table 9.6.

BOX 9.7 How to carry out Tukey's test after a significant one-way parametric ANOVA with equal replicates

GENERAL DETAILS	EXAMPLE 9.4
	This calculation is given in full in the Online Resource Centre. All values have been rounded to five decimal places.
1. Hypotheses to be tested You will be comparing two samples at a time by comparing their means. The generic hypotheses for each of these comparisons is: H_0: There is no difference between the means of the two samples being compared. H_1: There is a difference between the means of the two samples being compared.	**1. Hypotheses to be tested** In the Tukey's test, we will be comparing six pairs of mean values. Each of these comparisons is a test of hypotheses. You could write specific hypotheses for each. For example, H_0: There is no difference between the mean fresh weight for *Lobelia* 'Hannah' grown in compost A compared with compost B, etc. However, common sense suggests that where you are making many comparisons using this test, it is sensible to report the overall hypotheses only.
2. Have the criteria for using this test been met?	**2. Have the criteria for using this test been met?** Yes (9.6.1).
3. How to work out $T_{calculated}$ Calculate the difference between the means in all possible pair-wise combinations. Each of these differences is a $T_{calculated}$ value.	**3. How to work out $T_{calculated}$** See Table 9.6.
4. How to find $T_{critical}$ i. Examine the ANOVA calculation to find the following: n_s: the number of observations in a category or sample. a: number of categories or samples. MS_{within} is in the ANOVA table. For accuracy, you should use the full value and not a rounded-up value. v: degrees of freedom for MS_{within} from the ANOVA calculation table. ii. See Appendix d, Table D11. Find the q value at $p = 0.05$, in a q table for Tukey's test using a and v. iii. Calculate $T_{critical}$ where: $$T_{critical} = q \times \sqrt{\frac{s^2_{within}}{n}}$$	**4. How to find $T_{critical}$** i. From Table 9.5. $n_s = 10$ $a = 4$ $MS_{within} = 1.5566297$ $v = 36$ ii. To find q for $a = 4$ and $v = 36$, we need to interpolate from the q table: $q = 3.814$ iii. $$T_{critical} = 3.814 \times \sqrt{\frac{1.5566297}{10}}$$ $$= 1.50478 \text{ at } p = 0.05$$
5. The rule Compare $T_{critical}$ with each of the differences between the means (i.e. each $T_{calculated}$). If any $T_{calculated}$ is greater than $T_{critical}$, then you may reject the null hypothesis for these two samples.	**5. The rule** There are two $T_{calculated}$ values in Table 9.6 that are greater than $T_{critical}$ (1.505) at $p = 0.05$ and these are shown in bold in the table.

BOX 9.7 Continued	
6. What does this mean in real terms?	**6. What does this mean in real terms?** The general significant difference detected by the ANOVA (Box 9.6) is accounted for by the significant difference ($p = 0.05$) in mean fresh weight of *Lobelia* 'Hannah' growing in compost A compared with compost B and in compost A compared with compost D. From this information, the student was able to identify which components of the compost appeared to be most important in the weaning of *Lobelia* 'Hannah'.
Further examples relating to the topic including how to use statistical software are included in the Online Resource Centre.	

online resource centre

Table 9.6. The difference between each pair of means from Example 9.4: The fresh weight of *Lobelia* 'Hannah' after 8 weeks' weaning in one of four types of compost (A–D) (each difference is a $T_{calculated}$ value to be used in a Tukey's test)

Compost	A	B	C	D
	$\bar{x} = 4.744$	$\bar{x} = 2.639$	$\bar{x} = 3.649$	$\bar{x} = 2.885$
A $\bar{x} = 4.744$		4.744 – 2.639 = **2.105**	4.744 – 3.649 = 1.104	4.744 – 2.885 = **1.859**
B $\bar{x} = 2.639$			3.649 – 2.639 = 1.010	2.885 – 2.639 = 0.246
C $\bar{x} = 3.649$				3.649 – 2.885 = 0.764
D $\bar{x} = 2.885$				

Reporting your findings

In a multiple comparisons test, you are testing many hypotheses, one for each pair of means. To save you writing each hypothesis out formally at the end, an alternative method for reporting multiple comparison results is used. Firstly, rank all your mean values from the smallest to the largest. Then underscore those not significantly different from each other, or do not underscore (by the same line) any two that are significantly different.

It requires a bit of thought, but it does work in the end. In our example this would appear as:

B	D	C	A
$\bar{x} = 2.639$	$\bar{x} = 2.885$	$\bar{x} = 3.649$	$\bar{x} = 4.744$

This underscoring is commonly included in the table of results (Chapter 11).

An alternative, less cumbersome method is to use superscripts instead of a line. The same superscript is used for all means that are significantly different from each other, hence:

B[a]	D[b]	C	A[ab]
$\bar{x} = 2.639$	$\bar{x} = 2.885$	$\bar{x} = 3.649$	$\bar{x} = 4.744$

This alternative approach is particularly useful when you are comparing many means.

9.7 One-way parametric ANOVA with unequal replicates

Key points Although it is not ideal to design an experiment with unequal replicates for testing with an ANOVA, where this has occurred the amendments in this section allow such data to be analysed.

In all experiments where you are planning to analyse the data using an ANOVA, you should endeavour to obtain equal replicates. This is particularly so when testing two or more variables. In Example 9.5, we outline an investigation where an undergraduate planned to have equal replicates. However, technical problems meant that this was not achieved.

EXAMPLE 9.5 Differences in the peroxide levels of popular cooking oils

An increase in peroxide levels in cooking oils is indicative of reduced nutritional value and is associated with an increase in an unpleasant rancid taste. As part of a larger study, an undergraduate assessed the peroxide levels in a number of different oils from samples taken directly from a new bottle.

There is very little difference in the calculation of a one-way ANOVA for equal replicates (Box 9.6) and unequal replicates (Box 9.8). In this

Table 9.7. Peroxide levels (mEq of peroxide per kg sample) in a number of different cooking oils taken as fresh samples from a new bottle. All values × 10⁻²

	Corn oil	Extra virgin olive oil	Cold pressed rapeseed oil
	14	17	8
	15	13	7
	–	12	8
n_c	2	3	3
$\bar{x}$	14.5	14.0	7.66666
s^2	0.5	7.0	0.33333
Σx	29	42	23

section, we have therefore only highlighted the one step that differs. The criteria for using a one-way ANOVA with unequal replicates are virtually the same as those given in 9.5.1, apart from the requirement for equal replicates.

The data from Example 9.5 meet these criteria in that we wish to examine the effect of one variable (different oils). There are few observations so we will assume that the data are parametric as the observations are measured on an interval scale (Box 5.2). As this is a test that may be used when there are unequal replicates, it does not matter that two of the four treatments have only two rather than three replicates. To test for homogeneity of variances (Box 9.5), $F_{\text{max calculated}} = 70.33333/4.33333 = 16.23077$. To find $F_{\text{max critical}}$, calculate the degrees of freedom for the two variances used in the F ratio. (In this example, these were the variances from the 'extra virgin olive oil' and the 'cold pressed rape seed oil'.) The degrees of freedom are calculated as before as $n_s - 1$. The degrees of freedom for the cold pressed rapeseed oil are $3 - 1 = 2$, whilst the degrees of freedom for the extra virgin olive oil are also $3 - 1 = 2$. Therefore, the degrees of freedom we use to find $F_{\text{max critical}}$ are 2. With unequal sample sizes, the n_s values may not be the same for each category. For example, if we had used the variance from the corn oil in our F ratio, these degrees of freedom would be $2 - 1 = 1$. Where this is the case, you use the smallest of the degrees of freedom values. In Appendix d, Table D7, when looking up the critical values, a (number of samples) = 3, $v = 2$, and $F_{\text{max critical}} = 87.5$ at $p = 0.05$. As $F_{\text{max calculated}}$ is less than $F_{\text{max critical}}$, we may not reject the null hypothesis, the variances are homogeneous, and we may proceed with the one-way ANOVA.

BOX 9.8 How to carry out a one-way parametric ANOVA with unequal replicates

This calculation is given in full in the Online Resource Centre. All values have been rounded to five decimal places. The calculation is illustrated using data from Example 9.5.

1. **General hypotheses to be tested**

   H_0: There is no difference between the means of the populations from which the samples are taken.

   H_1: There is a difference between the means of the populations from which the samples are taken.

   In our example (Example 9.5), the hypotheses we are testing are:

   H_0: There is no difference in the mean peroxide levels (mEq of peroxide per kg sample) in different cooking oils.

   H_1: There is a difference in the mean peroxide levels (mEq of peroxide per kg sample) in different cooking oils.

2. **Have the criteria for using this test been met?**
   Yes (9.7).

3. **How to work out F$_{calculated}$**
   All these steps except step 3 are the same as those described in Box 9.6. Therefore, we will not repeat them here. The only step that is different is as follows:

   A. **Calculate general terms**

4. Add all the observations in a column: Σx_c. Square this value: $(\Sigma x_2)^2$. Divide by the number of observations in that column (n_c). Repeat for each column and add these together.
   In our example, the calculation for this step is:

5. $$\frac{(29)^2}{2} + \frac{(42)^2}{3} + \frac{(23)^2}{3} = 42.05 + 588.0$$
   $$+176.33333 = 1184.83333$$

   All other steps are exactly the same as those outlined in Box 9.6.

9.8 Tukey–Kramer test following a parametric one-way ANOVA with unequal replicates

Key points If a multiple comparisons test is wanted after an ANOVA but there are unequal replicates, then a Tukey–Kramer test rather than Tukey's test is required. This is similar in many respects to Tukey's test and also tests pairs of means, but allows for the unequal sample sizes in each category.

When there is a significant outcome from a one-way ANOVA, it is often helpful to carry out Tukey's test to examine in more detail which samples are contributing to the significant difference. However, Tukey's test requires equal replication, which is not present in Example 9.5. To overcome this inconsistent replication within the experiment, a Tukey–Kramer test may be used.

9.8.1 Using this test

To use this test you:

1) Wish to test specific hypotheses following a significant outcome in a parametric one-way ANOVA.

2) May have unequal numbers of observations in the categories or samples.

To illustrate the Tukey–Kramer test, we will use Example 9.5. The criteria for using this test have been met as we have a significant outcome from a parametric one-way ANOVA. (These calculations are shown in full in the Online Resource Centre.) There are unequal numbers of observations in the categories.

online
resource
centre

9.8.2 The calculation

In a Tukey–Kramer test, each mean value for a sample is compared with every other mean value in the investigation. This difference between the pairs of means is the calculated T value, which you then compare with a critical T value, which you also work out. This is the same approach as Tukey's test. The difference is in the method used for calculating the $T_{critical}$ value when n is not the same for the two samples. We demonstrate this in Box 9.9 with supporting calculations in Table 9.8.

BOX 9.9 How to carry out a Tukey–Kramer test after a significant one-way parametric ANOVA with unequal replicates

GENERAL DETAILS	EXAMPLE 9.5
	This calculation is given in full in the Online Resource Centre. All values have been rounded to five decimal places.
1. **Hypotheses to be tested** You will be comparing two samples at a time by comparing their means. The generic hypotheses for each of these comparisons is: H_0: There is no difference between the means of the two samples being compared. H_1: There is a difference between the means of the two samples being compared.	1. **Hypotheses to be tested** In the Tukey–Kramer test, we will be comparing three pairs of mean values. Each of these comparisons is a test of hypotheses. You could write specific hypotheses for each: for example, H_0: There is no difference between the mean peroxide content (mEq of peroxide per kg sample) of corn oil compared with virgin olive oil. However, common sense suggests that where you are making many comparisons using this test, then it is sensible to report the overall hypotheses only.
2. **Have the criteria for using this test been met?**	2. **Have the criteria for using this test been met?** Yes (9.8.1).
3. **How to work out $T_{calculated}$** Calculate the difference between the means in all possible pair-wise combinations. Each of these differences is a $T_{calculated}$ value.	3. **How to work out $T_{calculated}$** See Table 9.8.

BOX 9.9 Continued

4. **How to find $T_{critical}$**

i. Examine the ANOVA calculation to find the following:

a: number of samples.

MS_{within} from the ANOVA table. (For accuracy you should use the full value and not a rounded-up value).

v: degrees of freedom for the s^2_{within} from the ANOVA calculation table.

ii. See Appendix d, Table D11. Find the q value at $p = 0.05$, in a q table for Tukey's test using a and v.

iii. Firstly, identify all comparisons between means where the numbers of replicates are the same.

Here $T_{critical}$ is calculated as in Tukey's test where:

$$T_{critical} = q \times \sqrt{\frac{MS_{within}}{n}}$$

iv. When n is not the same for the two samples, $T_{critical}$ is:

$$T_{critical} = q \times \sqrt{\frac{MS_{within}}{2}\left(\frac{1}{n_1} + \frac{1}{n_2}\right)}$$

4. **How to find $T_{critical}$**

i. In this example, $a = 3$.

From the one-way ANOVA, $MS_{within} = 3.033334$ and its degrees of freedom are $v = 5$.

ii. When $a = 3$ and $v = 5$, $q = 4.6$ at $p = 0.05$.

iii. In our example, this is the comparison between extra virgin olive oil and cold pressed rapeseed oil where $n = 3$ for both samples.

$$T_{critical} = 4.6 \times \sqrt{\frac{3.033334}{3}}$$
$$= 4.62549 \text{ at } p = 0.05$$

iv. We have two comparisons where n is not the same for both samples. These are the comparisons between the corn oil ($n = 2$) and the rapeseed oil ($n = 3$) and the corn oil with the olive oil ($n = 3$). Here:

$$T_{critical} = 4.6 \times \sqrt{\frac{3.03334}{2}\left(\frac{1}{2} + \frac{1}{3}\right)}$$
$$= 4.6 \times \sqrt{1.51667(0.5 + 0.33333)}$$
$$= 5.17145$$

5. **The rule**

Compare the correct $T_{critical}$ with each of the differences between the means (i.e. each $T_{calculated}$). If any $T_{calculated}$ is greater than a $T_{critical}$, then you may reject the null hypothesis for this comparison of means.

5. **The rule**

There are two $T_{calculated}$ values in Table 9.8 that are greater than their $T_{critical}$ values at $p = 0.05$ and these are shown in bold in the table with an asterisk to indicate significance at this level.

If we repeat the calculations at $p = 0.01$, we find that the hypotheses may be rejected at $0.05 > p > 0.01$.

6. **What does this mean in real terms?**

Standard ways used in reporting results from the Tukey–Kramer test are outlined in 9.6.2.

6. **What does this mean in real terms?**

The general significant difference detected by the ANOVA is accounted for by the significant difference ($0.05 > p > 0.01$) in mean peroxide levels (mEq of peroxide per kg sample) in the cold pressed rapeseed oil compared with the other two oils.

Further examples relating to the topic including how to use statistical software are included in the Online Resource Centre.

online resource centre

Table 9.8. The difference between each pair of means from Example 9.5: Peroxide levels (mEq of peroxide per kg sample) in a number of different cooking oils taken as fresh samples from a new bottle. All values $\times 10^{-2}$. (Each difference is a $T_{calculated}$ value to be used in a Tukey–Kramer test.)

Cooking oils	Corn oil	Extra virgin olive oil	Cold pressed rapeseed oil
	$\bar{x} = 14.5, n = 2$	$\bar{x} = 14.0, n = 3$	$\bar{x} = 7.66667, n = 3$
Cold pressed rapeseed oil $\bar{x} = 7.66667$			
Extra virgin olive oil $\bar{x} = 14.0$			14.0 – 7.66667 = 6.33333*
Corn oil $\bar{x} = 14.5$		14.5 – 14.0 = 0.5	14.5 – 7.66667 = 6.83333*

*Significant at $p = 0.05$.

(The values in Table 9.8 are given to five decimal places as we are using this table to illustrate the calculation. If you were reporting these results, you should only include the final values and round up to a more appropriate value, in this case to one decimal place.)

9.9 Two-way parametric ANOVA with equal numbers of replicates

Key points We have included a number of different two-way ANOVAs. Your choice depends on the number of replicates and model. All require orthogonal experimental designs. This ANOVA is suitable for examining the effect of two variables with equal numbers of replicates and will test the effects of variable 1, variable 2, and an interaction between these two variables.

Parametric ANOVAs can be used to analyse data from a wide range of experimental designs. For example, if you were developing a technique for extracting DNA from bone material you might wish to compare the amount of DNA (μg) extracted across a range of temperatures and pH values (Table 9.9). This design is appropriate for the ANOVA described in this section as we wish to examine the effect of two variables (temperature and pH) and each pH treatment is combined with all temperatures. A design like this is called **orthogonal**. Other features of this design are that you would have the same number of replicates for each combination of treatments and that the bone samples would be allocated at random to a particular extraction method. The ANOVA described in this section is generally appropriate for a Model I or Model II ANOVA, but not for a mixed model ANOVA where a nested ANOVA is required (9.13).

Orthogonal A particular experimental design where every category for one treatment is found in combination with every category for every other treatment

As soon as two variables are being compared in a two-way parametric ANOVA with replicates, as illustrated here, or in a two-way non-parametric ANOVA with replicates (10.5), then three pairs of hypotheses are tested. Not surprisingly, the first of these three pairs of hypotheses refers to the difference between the samples in relation to the effect of variable 1. For example, in the design shown in Table 9.9, this would be the effect of temperature. The second pair of hypotheses refers to the effect of the second variable, e.g. the effect of pH on DNA extraction. The two-way ANOVA also tests for an interaction between these two variables and this is the third pair of hypotheses. An **interaction** between treatments can be indicated by a plot of the mean values (Fig. 9.3). In Fig. 9.3, the amount of DNA extracted as the pH increases changes in a similar way for all temperatures. If there was an interaction, then these lines would not be parallel (Fig. 9.4).

Interaction The response to a treatment shown by one group of items is reversed in a second sample in such a way that it is clear that two treatment variables are not having the same effect in two groups

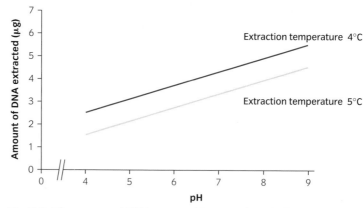

Fig. 9.3. The amount of DNA extracted increases in a similar way as pH increases, for both temperatures, indicating no interaction between pH and temperature (°C).

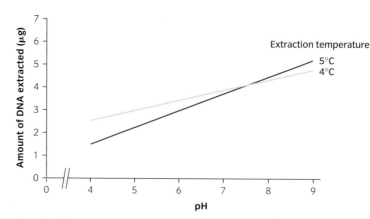

Fig. 9.4. The two temperatures do not have the same effect on the amount of DNA extracted as pH increases, indicating an interaction between temperature and pH.

Table 9.9. An orthogonal experimental design: the effect on DNA extraction of several temperatures is investigated in relation to a number of pH values (each category of one variable is combined with each category of the second variable)

pH	Temperature (°C)			
	5	10	15	20
3	✓	✓	✓	✓
7	✓	✓	✓	✓
12	✓	✓	✓	✓

EXAMPLE 9.6 The weaning of *Cosmos atrosanguineus* var. 'Pip' and var. 'Christopher' onto one of four composts following propagation by tissue culture

The undergraduate who was working on *Lobelia* 'Hannah' also examined the weaning from tissue culture of two varieties of *Cosmos atrosanguineus*, var. 'Pip' and var. 'Christopher'. As in the *Lobelia* trial (Example 9.4), she transferred plantlets at random into plugs containing one of four different composts (A–D). After 8 weeks, the plantlets were examined and the maximum height of the plants was recorded (Table 9.10).

Table 9.10. Final height of two varieties of *Cosmos atrosanguineus* (var. 'Pip' and var. 'Christopher') grown in one of four different composts (A–D) (table includes summary statistics)

	Weaning compost							
	A		B		C		D	
Final height (cm) of *C. atrosanguineus* var. 'Pip'	20.4	22.0	19.4	22.9	18.7	18.7	20.0	20.1
	20.4	21.6	28.7	25.2	18.5	22.2	25.9	18.1
	24.9	22.7	21.1	27.7	23.2	22.9	17.0	24.1
	19.6	23.5	24.5	26.3	20.6	24.1	22.7	17.5
	19.2	25.4	22.8	18.0	22.3	22.7	25.2	25.2
Summary statistics for var. 'Pip'	$\bar{x} = 21.97$		$\bar{x} = 23.66$		$\bar{x} = 21.39$		$\bar{x} = 21.58$	
	$s^2 = 4.62011$		$s^2 = 12.20266$		$s^2 = 4.39433$		$s^2 = 11.87733$	
Final height (cm) of *C. atrosanguineus* var. 'Christopher'	19.5	15.7	8.9	18.1	22.9	10.8	21.6	17.8
	21.9	21.9	8.4	15.8	6.4	17.0	15.2	15.1
	21.9	19.9	15.3	16.2	14.1	21.4	11.0	14.1
	7.9	20.7	13.1	14.8	18.0	15.6	22.9	7.0
	25.5	22.6	13.8	22.6	19.3	17.6	21.0	12.2
Summary statistics for var. 'Christopher'	$\bar{x} = 19.75$		$\bar{x} = 14.7$		$\bar{x} = 16.31$		$\bar{x} = 15.79$	
	$s^2 = 23.65167$		$s^2 = 17.1666$		$s^2 = 24.13656$		$s^2 = 25.80767$	

9.9.1 Using this test

To use this test you:

1) Wish to examine the effect of two variables with the same number of replicates (observations) in each category or sample.

2) Have parametric data.

3) Allocate all items at random to each category and the design is orthogonal.

4) Have variances that are similar (homogeneous).

The data from Example 9.6 meet the criteria for using a two-way parametric ANOVA with equal replicates, because the data is parametric, each plantlet was allocated at random to the treatments, and the design is orthogonal. You wish to examine the effect of two variables (compost and plant variety), and there are ten plants in each treatment. To check whether the variances are similar, an F_{max} test is carried out (Box 9.5). The variances for the data from Example 9.6 are included in Table 9.10. $F_{max\ calculated}$ is $25.80767/4.39433 = 5.8729$. At $p = 0.05$, $a = 8$ and $v = 10 - 1 = 9$, then $F_{max\ critical} = 8.95$ (Appendix d, Table D7). $F_{max\ calculated}$ is less than $F_{max\ critical}$, so we do not reject the null hypothesis; the variances are homogeneous and we may proceed with the ANOVA. If the variances are not homogeneous, you should consider transforming your data (5.9) or using a non-parametric two-way ANOVA (10.5).

Table 9.11. Calculation of general terms for two-way parametric ANOVA using data from Example 9.6: The weaning of *Cosmos atrosanguineus* var. 'Pip' and var. 'Christopher' onto one of four composts following propagation by tissue culture

		Composts used in weaning				Row total
		A	**B**	**C**	**D**	
Height (cm) of	Σx	219.7	236.6	213.9	215.8	886.0
C. atrosanguineus	Σx^2	4868.39	5707.78	4614.87	4763.86	19954.9
var. 'Pip'	n	10	10	10	10	40
Height (cm) of	Σx	197.5	147.0	163.1	157.9	665.5
C. atrosanguineus	Σx^2	4113.49	2315.4	2877.39	2725.51	442890.25
var. 'Christopher'	n	10	10	10	10	40
Column totals						Grand totals
	Σx	417.2	383.6	377.0	373.7	1551.5
	Σx^2	8981.88	8023.18	7492.26	7489.37	31986.69
	n	20	20	20	20	$N = 80$

9.9.2 **The calculation**

Having satisfied ourselves that our data meet all the criteria for using this two-way ANOVA, we can then proceed. In Box 9.10, we have organized the calculation of $F_{calculated}$ under a number of subheadings: (A) Calculate general terms; (B) Calculate the sums of squares; and (C) Construct and complete an ANOVA calculation table. At each point, we show how these steps can be applied to Example 9.6. Some of the calculations are included in Tables 9.11 and 9.12.

Table 9.12. ANOVA table for analysis of data from Example 9.6: The weaning of *Cosmos atrosanguineus* var. 'Pip' and var. 'Christopher' onto one of four composts following propagation by tissue culture, showing how some of the terms are calculated

Source of variation	SS	v	MS	$F_{calculated}$	$F_{critical}$	p
Between composts (columns)	59.87137	4 − 1 = 3	59.87137/3 = 19.95712	19.95712/15.48213 = 1.28904	2.7347	NS
Between varieties (rows)	607.75312	2 − 1 = 1	607.75312	607.75312/15.48213 = 39.25514	11.7740	0.001
Interaction compost × varieties	114.94938	(2 − 1)(4 − 1) = 3	114.94938/3 = 38.31646	38.31646/15.48213 = 2.47488	2.7347	NS
Within sample variation	1114.713	80 − 8 = 72	1114.713/72 = 15.48213			
Total		80 − 1 = 79				

NS, Not significant.

BOX 9.10 How to carry out a two-way parametric ANOVA with equal replicates

This calculation is given in full in the Online Resource Centre. All values have been rounded to five decimal places. The calculation is illustrated using data from Example 9.6.

1. **General hypotheses to be tested**
   As explained in 9.9, there are three pairs of general hypotheses to be tested.

   H_{0c}: There is no difference between the sample means due to variable 1 (columns).

   H_{1c}: There is a difference between the sample means due to variable 1 (columns).

   H_{0r}: There is no difference between the sample means due to variable 2 (rows).

   H_{1r}: There is a difference between the sample means due to variable 2 (rows).

H_{0i}: There is no interaction between variable A and variable B in their effects on the sample means.

H_{1i}: There is an interaction between variable A and variable B in their effects on the sample means. In our example (Example 9.6), these hypotheses will be:

H_{0c}: There is no difference between the mean height (cm) of C. atrosanguineus weaned on different composts.

H_{1c}: There is a difference between the mean height (cm) of C. atrosanguineus weaned on different composts.

H_{0r}: There is no difference between the mean heights (cm) of C. atrosanguineus var. 'Pip' compared with var. 'Christopher'.

BOX 9.10 Continued

H_{1r}: There is a difference between the mean heights (cm) of C. atrosanguineus var. 'Pip' compared with var. 'Christopher'.

H_{0I}: There is no interaction between the weaning composts and C. atrosanguineus varieties in their effects on mean height (cm).

H_{1I}: There is an interaction between the weaning composts and C. atrosanguineus varieties in their effects on mean height (cm).

2. Have the criteria for using this test been met?
In Example 9.6, the criteria have been met (9.9.1).

3. How to work out $F_{calculated}$

A. Calculate general terms

1. First add together every observation in the complete data set (grand total): Σx_T.

2. Square each and every observation and add all these squared values together: Σx_T^2

3. Add all the observations in a sample: Σx_s. This is the sample total. Do this for each sample. Square each sample total: Σx_s^2. Add these together. Divide this by the number of observations in a sample (n_s).

4. For each column add all the observations together: (Σx_c). Square the total: $(\Sigma x_c)^2$. Add these squared column values together and divide this total by the number of observations in the column (n_c).

5. For each row, add all the observations: Σx_r. Square the total: $(\Sigma x_r)^2$. Add these squared values together and divide this value by the number of observations in the row (n_r).

6. Square the result from step 1 and divide this by N, where N is the total number of observations in all the samples.
   For this example (Table 9.11), these calculations are:

1. $\Sigma x_T = 20.4 + 20.4 + 24.9 + 19.6 + \cdots + 7.0 + 21.0 + 12.2 = 1551.5$

2. $\Sigma x_T^2 = (20.4)^2 + (20.4)^2 + (24.9)^2 + (21.0)^2 + (12.2)^2 = 31986.69$

3. There are eight samples in total and ten observations in each sample. Calculate:

$$(\Sigma x_s)^2 = (19.7)^2 + (236.6)^2 + (213.9)^2 + (215.8)^2$$
$$+ (197.5)^2 + (147.0)^2 + (163.1)^2$$
$$+ (157.9)^2 = 308719.77$$

$$\Sigma \left[\frac{(\Sigma x_s)^2}{n_s} \right] = \frac{308719.77}{10} = 30871.977$$

4. There are four columns (composts) and 20 observations in each column. Calculate:

$$\frac{(417.2)^2 + (383.6)^2 + (377.0)^2 + (373.7)^2}{20}$$
$$= \frac{602985.49}{20} = 30149.2745$$

5. There are two rows (var. 'Pip' and var. 'Christopher') and 40 observations in each row. Calculate:

$$\frac{[(886.0)^2 + (665.5)^2]}{40} = 1227886.25/40$$
$$= 30697.15625$$

6. $(1551.5)^2/80 = 30089.40313$

B. Calculate the sums of squares (SS)

7. SS_{total} = result from step 2 − result from step 6

8. $SS_{samples}$ = result from step 3 − result from step 6

9. $SS_{variable\ 1\ columns}$ = result from step 4 − result from step 6

10. $SS_{variable\ 2\ rows}$ = result from step 5 − result from step 6

11. $SS_{interaction}$ = result from step 8 − result from step 9 − result from step 10

12. SS_{within} = result from 7 − result from 8

BOX 9.10 Continued

For Example 9.6, the calculations are:

7. SS_{total} = 31986.69 − 30089.40313 = 1897.28687
8. $SS_{samples}$ = 30871.977 − 30089.40313 = 782.57387
9. $SS_{variable\ 1\ columns}$ = 30149.2745 − 30089.40313 = 59.87137
10. $SS_{variable\ 2\ rows}$ = 30697.15625 − 30089.40313 = 607.75312
11. $SS_{interaction}$ = 782.57387 − 59.87137 − 607.75312 = 114.94938
12. SS_{within} = 1897.28687 − 782.57387 = 1114.713

C. Construct and complete an ANOVA calculation table

i. Draw an ANOVA table as illustrated here.

Source of variation	SS	v	MS	F
Variable 1 (columns)	Step 9			
Variable 2 (rows)	Step 10			
Interaction	Step 11			
Within sample variation	Step 12			
Total	Step 7			

ii. Transfer the results from the calculations for $SS_{variable\ 1\ columns}$, $SS_{variable\ 2\ rows}$, SS_{within}, and SS_{total} into the ANOVA table in column 2 (see Table 9.12). (You do not use SS_{sample}.)

iii. Calculate the degrees of freedom (v)
$v_{variable\ 1\ columns}$ = number of columns − 1
$v_{variable\ 2\ rows}$ = number of rows − 1
$v_{interaction}$ = (number of columns − 1) × (number of rows − 1)
v_{within} = N − (number of rows × number of columns)
Enter these results in column 3 (Table 9.12).

iv. Calculate the mean squares (MS) for the rows, columns, and interaction.
Note: Variances in this context are also known as mean squares (MS).
s^2 = MS = SS/v
For treatment 1 (columns), take the value for SS and divide it by its value for v. Repeat this for treatment 2 (rows), the interaction, and 'within

sample' variation. Put the results from these calculations in column 4 (Table 9.12).

v. Work out the F ratio. As there are three pairs of hypotheses to test, you calculate three F values.
Variable 1 (columns):

$$F_{calculated} = \frac{\text{variable 1 (columns) MS}}{\text{within samples MS}}$$

Variable 2 (rows):

$$F_{calculated} = \frac{\text{variable 2 (rows) MS}}{\text{within samples MS}}$$

Interaction: $F_{calculated} = \dfrac{\text{interaction MS}}{\text{within samples MS}}$

Place the results from these calculations in column 5 (Table 9.12).

4. To find $F_{critical}$

See Appendix d, Table D8. Again, as you are testing three pairs of hypotheses, there will be three critical F values to find. Using the F tables for an ANOVA, you may need to interpolate to find the values. The degrees of freedom to use are:
Variable 1 (columns): $v_{variable\ 1\ columns}$ and v_{within}
Variable 2 (rows): $v_{variable\ 2\ rows}$ and v_{within}
Interaction: $v_{interaction}$ and v_{within}
For Example 9.6, the degrees of freedom are:
Variable 1 (columns): 3, 72
Variable 2 (rows): 1, 72
Interaction: $v_{interaction}$ and v_{within} 3, 72
The $F_{critical}$ values are included in Table 9.12 for specific p values.

5. The rule

If $F_{calculated}$ is greater than or equal to $F_{critical}$, then you may reject the null hypothesis.
In our example, when first considering the columns (composts), $F_{calculated}$ (1.2890) is less than $F_{critical}$ (2.7347) at p = 0.05 so we may not reject the null hypothesis. There is no significant difference (NS). When comparing the two varieties of C. atrosanguineus, $F_{calculated}$ (39.255) is greater than $F_{critical}$ (11.774) at p = 0.001 so we may reject the null hypothesis.

BOX 9.10 Continued

Finally, when testing for an interaction, $F_{calculated}$ (2.6811) is less than $F_{critical}$ (2.4749) at $p = 0.05$ so we may not reject the null hypothesis. There is no significant interaction (NS).

6. **What does this mean in real terms?**
   There is no significant difference ($F = 1.3$, $p = 0.05$) between the mean height of *C. atrosanguineus* weaned on different composts and there is no significant interaction ($F = 2.5$, $p = 0.05$) between the

composts and *C. atrosanguineus* varieties in their effects on mean height (cm). However there is a very highly significant difference ($F = 39.3$, $p = 0.01$) between the mean heights of *C. atrosanguineus* var. 'Pip' compared with var. 'Christopher'.

Further examples relating to the topic including how to use statistical software are included in the Online Resource Centre.

 online resource centre

9.10 Tukey's test following a parametric two-way ANOVA with equal replicates

Key points Tukey's test, as outlined in 9.6, can be extended to test pairs of means after a two-way ANOVA in all possible combinations. This provides more detailed information about any significant difference identified in the more general ANOVA analysis.

The two-way ANOVA described in 9.9 has tested three general hypotheses of which one was significant. To test specific hypotheses, you may use a multiple comparisons test, such as Tukey's test, which allows particular samples to be compared in pairs and those contributing to the significant difference in the ANOVA can be identified. This additional testing of hypotheses allows you to make much more explicit interpretations of your data and so should always be carried out when you have a significant result in a parametric ANOVA. However, if your interaction term is significant, it is difficult to interpret the results from Tukey's test, as the element of interaction will confound any differences between pairs of means.

The method we follow has already been described in 9.6. The hypotheses are given in general terms in Box 9.7.

The calculated T values are the differences between the means for two samples. These differences for Example 9.6 are given in Table 9.13. The value for $T_{critical}$ is calculated as described in Box 9.10. For the example we are considering here, $n = 10$; $s^2_{within} = 15.482125$, $a = 8$, $v = 72$ and q when interpolated (6.3.6) from the table of q values at $p = 0.05$ is 4.346. Therefore, $T_{critical} = 4.346 \times \sqrt{(15.482125/10)} = 5.4076$.

Table 9.13. $T_{calculated}$ values for Tukey's test following a two-way ANOVA on the data from Example 9.6: The weaning of *Cosmos atrosanguineus* var. 'Pip' and var. 'Christopher' onto one of four composts following propagation by tissue culture

		Weaning composts used for C. *atrosanguineus* var. 'Pip'				Weaning composts used for C. *atrosanguineus* var. 'Christopher'			
		A	B	C	D	A	B	C	D
		$\bar{x} =$ 21.97	$\bar{x} =$ 23.66	$\bar{x} =$ 21.39	$\bar{x} =$ 21.58	$\bar{x} =$ 19.75	$\bar{x} =$ 14.70	$\bar{x} =$ 16.31	$\bar{x} =$ 15.79
Weaning composts used for C. *atrosanguineus* var. 'Pip'	A $\bar{x} = 21.97$		1.69	0.58	0.39	2.22	**7.27**	**5.66**	**6.18**
	B $\bar{x} = 23.66$			2.27	2.08	3.91	**8.96**	**7.35**	**7.87**
	C $\bar{x} = 21.39$				0.19	1.64	**6.69**	5.08	**5.60**
	D $\bar{x} = 21.58$					1.83	**6.88**	5.27	**5.79**
Weaning composts used for C. *atrosanguineus* var. 'Christopher'	A $\bar{x} = 19.75$						5.05	3.44	3.96
	B $\bar{x} = 14.70$							1.61	1.09
	C $\bar{x} = 16.31$								0.52

A significant difference between the two means is indicated where the $T_{calculated}$ value exceeds the $T_{critical}$ value. These significant differences are shown in bold in Table 9.13. It shows, for example, that in the comparison between the two species, the difference is less marked in compost A than in the other composts.

9.11 Two-way parametric ANOVA with unequal replicates

Key points There are a number of ways to approximately overcome missing data or unequal replication in an experiment with two variables. We present one method. However, this is not ideal and also undermines any use of a subsequent multiple comparisons test such as Tukey's test. It is best to avoid unequal replication where possible.

One of the criteria for using a parametric ANOVA is that replication is equal in all categories. In a one-way ANOVA, an adjustment can be made to overcome this criterion if it is not met (9.7). However, in a more complex design such as a two-way ANOVA, a simple change in the calculation cannot be used.

EXAMPLE 9.7 Changes in the peroxide levels of cooking oils in relation to storage conditions.

The undergraduate extended the small study shown in Example 9.5 by placing aliquots of each cooking oil into small sealed jars and storing the oil for 5 weeks in a number of different commonly used locations. After the 5 weeks, samples were taken from each aliquot and the peroxide levels were assessed.

Table 9.14. The effect of storage conditions on peroxide levels (mEq of peroxide per kg sample) in a number of different cooking oils. All values $\times 10^{-2}$

	Corn oil	Extra virgin olive oil	Cold pressed rape-seed oil
Control (before storing)	14.0	17.0	11.0
	16.0	13.0	7.0
	–	12.0	8.0
$\bar{x}$	15.0	14.0	8.66667
s^2	2.0	7.0	4.33333
Fridge	16.0	14.0	14.0
	16.0	13.0	7.0
	14.0	15.0	12.0
$\bar{x}$	15.3333	14.0	11.0
s^2	1.33333	1.0	13.0
Cupboard	18.0	14.0	8.0
	14.0	9.0	10.0
	14.0	8.0	9.0
$\bar{x}$	15.33333	10.33333	9.0
s^2	5.33333	10.33333	1.0
Window sill (sun)	26.0	30.0	20.0
	24.0	22.0	11.0
	22.0	20.0	–
$\bar{x}$	24.0	24.0	15.5
s^2	4.0	28.0	40.5

In Example 9.7, results are missing from corn oil control and the cold pressed rapeseed oil sample stored on the window sill. One approach we might take to balance out the number of replicates across the whole experiment would be to take values out of the data set at random to correct the imbalance. For example, we might select one of the values from the 'olive oil control' category at random, thus reducing the number of replicates to two and bringing it into line with the number of replicates in the 'corn oil control' category. This could be repeated throughout until

Table 9.15. Peroxide levels (mEq of peroxide per kg sample) in a number of different cooking oils and the effect of storage on the peroxide levels. All values × 10^{-2}. (Missing values have been replaced with the mean where necessary; mean and revised summary values are shown in bold)

	Corn oil	Extra virgin olive oil	Cold pressed rapeseed oil
Control (before storing)	14.0	17.0	11.0
	16.0	13.0	7.0
	15.0	12.0	8.0
$\bar{x}$	15.0	14.0	8.66667
s^2	**1.0**	7.0	4.33333
Fridge	16.0	14.0	14.0
	16.0	13.0	7.0
	14.0	15.0	12.0
$\bar{x}$	15.3333	14.0	11.0
s^2	1.3333	1.0	13.0
Cupboard	18.0	14.0	8.0
	14.0	9.0	10.0
	14.0	8.0	9.0
$\bar{x}$	15.3333	10.33333	9.0
s^2	5.33333	10.33333	1.0
Window sill (sun)	26.0	30.0	20.0
	24.0	22.0	11.0
	22.0	20.0	**15.5**
$\bar{x}$	24.0	24.0	15.5
s^2	4.00	28.00	**20.25**

each category only contained two replicates. Clearly this would significantly reduce the information in our data set and may introduce a bias not previously present. Alternative approaches are to estimate the missing values using an equation proposed by Shearer (1973) or by including the mean for that category. A discussion of which approach to take and when is given by Zar (2010). No solutions are ideal and should not be considered if many values are missing. For simplicity, we suggest using the mean value as an estimate of the missing value (Table 9.15).

The ANOVA then proceeds as described in Box 9.10 until you reach the calculations for the degrees of freedom. The following is an extract from Box 9.10 with the revisions required when 'missing' values' have been estimated.

---

iii. Calculate the degrees of freedom (v) for a two-way ANOVA with unequal replicates

$v_{\text{variable 1 columns}}$ = number of columns − 1

$v_{\text{variable 2 rows}}$ = number of rows − 1

$\nu_{interaction} = (\text{number of columns} - 1) \times (\text{number of rows} - 1)$
$\nu_{within} = *N - (\text{number of rows} \times \text{number of columns})$
(*Where only the actual and not the estimated values are counted within this N value: for Example 9.7, two values were estimated so N will be 34 (and not 36).)

A common practice if an ANOVA has a significant outcome is to use this and then follow it with a multiple comparisons test to obtain a more specific evaluation of the data. If you use this method for estimating missing values in a two-way ANOVA, we do not recommend using the Tukey–Kramer multiple comparisons test as MS_{within} will be affected by the inclusion of the estimated values to fill in for the missing data.

9.12 Two-way parametric ANOVA with no replicates

Key points This useful design can be used to examine the effect of two variables or the effect of a confounding variable and one treatment variable, or can be used in experiments with repeated measures and more than two samples or categories. However, unlike the two-way ANOVA with replicates, no interaction term can be tested.

There are three occasions when you may find this test relevant to your experiment. Firstly, there may be occasions when you have a similar design to that described in Example 9.6 but without replication, and where each observation is independent of all others, i.e. you have only taken one observation from any one item. To analyse the data from this type of investigation, you use a slightly different ANOVA from that outlined in 9.9. This test, however, can also be used where you know you have a notable confounding variable and you wish to be able to separate the effect of the confounding variable from that of the variable you wish to test the effect of. We consider an example like this in Box 9.11. Finally, this test may also be used if you have repeated measures. For example, you may run a number of rats through a number of different mazes and record the time taken. Clearly the results from each animal for all the mazes are likely to be influenced by that animal as well as the maze. To try to identify the effect of the individuals (the confounding variable) from the mazes (treatment variable), you could use this experimental design and this ANOVA.

Unlike the other two-way parametric ANOVAs, there will only be two and not three pairs of hypotheses to test. The lack of replication means there are no within category variances to allow us to test for an interaction between the two variables.

EXAMPLE 9.8 The effect of fertilizers on tomato production

A horticulture undergraduate wished to test a number of known tomato feeds to assess the relative effect of each on crop production. However, the only space available was a polytunnel where there was slight micro-environmental variation down the length of the tunnel. To allow some estimate of the effect of the feed as against the effect of environmental variation, the polytunnel was divided into seven lengths (blocks). Within each block, five tomato plants were arranged at random and each was given a particular feed formulation. This was repeated for each block. At the end of the trial, the fruit from each plant was harvested and the total mass recorded.

Table 9.16. The mass (kg) of tomatoes produced in cultivation in seven locations in a polytunnel and using five different tomato feeds

Polytunnel blocks	Tomato feed					n_r	Σx_r	Σx^2_r
	157a	TomPlus	General	BB	Standard			
1	3.852	3.784	3.963	4.365	3.526	5	19.490	76.34783
2	3.951	3.749	3.852	4.896	3.851	5	20.299	83.30432
3	4.083	3.681	3.795	4.865	3.365	5	19.789	79.61412
4	3.783	3.883	3.762	4.296	3.421	5	19.145	73.70028
5	3.727	3.657	3.844	4.184	3.871	5	19.283	74.53101
6	3.368	3.364	3.347	3.961	3.721	5	17.761	63.39769
7	3.279	3.217	3.236	3.678	3.278	5	16.688	55.84559
n_c	7	7	7	7	7	$N = 35$		
Σx_c	26.043	25.335	25.799	30.245	25.033			
Σx^2_c	97.41607	92.04034	95.54838	131.87094	89.86511			

9.12.1 Using this test

To use this test you:

1) Wish to examine the effect of two variables or you wish to examine the effect of one variable and one notable confounding variable.

2) Have no replicates.

3) Have parametric data.

4) Allocate one item at random to each category or, where you have repeated measures, this is treated as one of the variables (see explanation in text).

5) Have an orthogonal design.

The data from Example 9.8 meet these criteria as we wish to examine the effect of the feeds (variable) and the different polytunnel blocks (confounding variable) on the production of tomatoes. There are no replicates and the design is orthogonal. There is very little data and no replication within the experiment; therefore, the only indication that these data are

parametric is that they are measured on an interval scale (Box 5.1). We will assume that the data are parametric.

9.12.2 **The calculation**

Having satisfied ourselves that our data meet all the criteria for using this two-way ANOVA without replicates, we can then proceed. In Box 9.11, we have organized the calculation of $F_{\text{calculated}}$ under a number of sub-headings: (A) Calculate general terms; (B) Calculate the sums of squares; and (C) Construct and complete an ANOVA calculation table. At each point, we show how these steps can be applied to Example 9.8. Some of the calculations are included in Tables 9.16 and 9.17. Where steps have been abbreviated the full calculation is included in the Online Resource Centre.

online resource centre

BOX 9.11. How to carry out a two-way parametric ANOVA with no replicates

This calculation is given in full in the Online Resource Centre. All values have been rounded to five decimal places. This calculation is illustrated using the data from Example 9.8.

1. **General hypotheses to be tested**

    H_{0c}: There is no difference between the sample means due to variable 1 (columns).

    H_{1c}: There is a difference between the sample means due to variable 1 (columns).

    H_{0r}: There is no difference between the sample means due to variable 2 (rows).

    H_{1r}: There is a difference between the sample means due to variable 2 (rows).

    In Example 9.8. these hypotheses will be:

    H_{0c}: There is no difference between the mean mass of tomatoes (kg) produced in relation to the specific feed given.

    H_{1c}: There is a difference between the mean mass of tomatoes (kg) produced in relation to the specific feed given.

    H_{0r}: There is no difference between the mean mass of tomatoes (kg) produced in relation to their location in the polytunnel.

    H_{1r}: There is a difference between the mean mass of tomatoes (kg) produced in relation to their location in the polytunnel.

2. **Have the criteria for using this test been met?**

    Yes. In this example, the criteria have been met (9.12.1).

3. **How to work out F$_{\text{calculated}}$**

    A. Calculate general terms

    1. First add together every observation in the complete data set (grand total): Σx_T.
    2. Square each and every observation and add all these squared values together: Σx_T^2.
    3. Add all the observations in a column: Σx_c. Do this for each column. Square each column total: Σx_c^2.
       Add these together. Divide this by the number of observations in a column (n_c).
    4. For each row add all the observations: Σx_r. Square the total: $(\Sigma x_r)^2$. Add these squared values together and divide this value by the number of observations in the row (n_r).
    5. Square the result from step **1** and divide this by N, where N is the total number of observations in all the categories.
       For this example (Table 9.16), these calculations are:

BOX 9.11 Continued

1. $\Sigma x_T = 3.852 + 3.784 + \cdots + 3.678 + 3.278 = 132.455$

2. $\Sigma x_T^2 = (3.852)^2 + (3.784)^2 + \cdots + (3.678)^2 + (3.278)^2$
   $= 506.740853$

3. There are five columns and seven observations in each column. The column totals are given in Table 9.16.

$$\Sigma[(\Sigma x_c)^2] = (26.043)^2 + (25.335)^2 + (25.799)^2$$
$$+ (30.245)^2 + (25.033)^2 = 3527.09959$$

$$\frac{\Sigma[(\Sigma x_c)^2]}{n_c} = \frac{3527.09959}{7} = 503.87137$$

4. There are seven rows and five observations in each row. The row totals are given in Table 9.16.

$$\Sigma[(\Sigma x_r)^2] = (19.49)^2 + (20.299)^2 + (19.789)^2 + (19.145)^2$$
$$+ (19.283)^2 + (17.761)^2 + (16.688)^2 = 2515.8216$$

$$\frac{\Sigma[(\Sigma x_r)^2]}{n_r} = \frac{2515.8216}{5} = 503.16432$$

5. $(132.455)^2/35 = 501.266486$

B. Calculate the sums of squares (SS)

6. SS_{total} = result from step **2** – result from step **5**
7. $SS_{variable\ 1\ columns}$ = result from step **3** – result from step **5**
8. $SS_{variable\ 2\ rows}$ = result from step **4** – result from step **5**
9. SS_{error} = result from step **6** – result from step **7** – result from step **8**

For Example 9.8, the calculations are:

6. $SS_{total} = 506.74085 - 501.26649 = 5.47437$
7. $SS_{variable\ 1\ columns} = 503.87137 - 501.26649 = 2.60488$
8. $SS_{variable\ 2\ rows} = 503.16432 - 501.26649 = 1.89783$
9. $SS_{error} = 5.47437 - 2.60488 - 1.89783 = 0.97166$

C. Construct and complete an ANOVA calculation table

i. Draw an ANOVA table as illustrated here.

Source of variation	SS	v	MS	F
Variable 1 (columns)	Step **7**			
Variable 2 (rows)	Step **8**			
Error	Step **9**			
Total	Step **6**			

ii. Transfer the results from the calculations for $SS_{variable\ 1\ columns}$, $SS_{variable\ 2\ rows}$, SS_{error}, and SS_{total} into the ANOVA table in column 2 (see Table 9.17).

iii. Calculate the degrees of freedom (v)
$v_{variable\ 1\ columns}$ = number of columns – 1
$v_{variable\ 2\ rows}$ = number of rows – 1
v_{error} = (number of rows – 1) × (number of columns – 1)
$v_{total} = N - 1$
Enter these results in column 3 (Table 9.17).

iv. Calculate the mean squares (MS) for the rows, columns and interaction.
Note: Variances in this context are also known as mean squares (MS).
$s^2 = MS = SS/v$
For variable 1 (columns), take the value for SS and divide it by its value for v. Repeat this for variable 2 (rows) and the error SS. Put the results from these calculations in column 4 (Table 9.17).

v. Work out the F ratio. As there are two pairs of hypotheses to test, you calculate two F values.
Variable 1 (columns):

$$F_{calculated} = \frac{variable\ 1\ (columns)\ MS}{error\ MS}$$

Variable 2 (rows)

$$F_{calculated} = \frac{variable\ 2\ (rows)\ MS}{error\ MS}$$

Place the results from these calculations in column 4 (Table 9.17).

4. **To find $F_{critical}$**
See Appendix d, Table D8. Again, as you are testing two pairs of hypotheses, there will be two critical F values to find. Using the F tables for an ANOVA, you may need to interpolate to find the values. The degrees of freedom to use are:
Variable 1 (columns): $v_{variable\ 1\ columns}$ and v_{error}
Variable 2 (rows): $v_{variable\ 2\ rows}$ and v_{error}
For Example 9.8, these are:
Variable 1 (columns) 4, 24
Variable 2 (rows) 6, 24
The $F_{critical}$ values are included in Table 9.17 for specific p values.

5. **The rule**
If $F_{calculated}$ is greater than or equal to $F_{critical}$ then you may reject the null hypothesis.

BOX 9.11 Continued

In our example, when first considering the columns (tomato feeds), $F_{calculated}$ (16.08) is more than $F_{critical}$ (4.22) at $p = 0.01$ so we may reject the null hypothesis. There is a highly significant difference. When comparing the locations in the polytunnel where the plants were growing, $F_{calculated}$ (7.81) is greater than $F_{critical}$ (3.67) at $p = 0.01$ so we may reject the null hypothesis.

6. **What does this mean in real terms?**
There is a highly significant difference ($F = 16.08$, $p = 0.01$) in the mean mass (kg) of tomatoes

produced when given different tomato feeds and there is a highly significant difference ($F = 7.81$, $p = 0.01$) in the mean mass of tomatoes produced in the different locations within the polytunnel.

Further examples relating to the topic including how to use statistical software are included in the Online Resource Centre.

 online resource centre

Table 9.17. ANOVA table for analysis of data from Example 9.8: The mass (kg) of tomatoes produced when a plant is treated with a number of different feeds and grown in seven locations (blocks) in a polytunnel

Source of variation	SS	v	MS	$F_{calculated}$	$F_{critical}$	p
Between feeds (columns)	2.60488	$5 - 1 = 4$	0.65122	16.08533	4.22	0.01
Between locations (blocks) (rows)	1.89783	$7 - 1 = 6$	0.31631	7.81284	3.67	0.01
Error	0.97166	$4 \times 6 = 24$	0.04049			
Total	5.47437	$35 - 1 = 34$				

9.13 Two-way nested parametric ANOVA with equal replicates

Key points A nested ANOVA is used when there is a mixed model design. This is where the categories in one of the variables are not exactly equivalent when combined with the categories from the second variable.

A nested ANOVA is one where you do not have an orthogonal design. Instead, your second factor is randomized with respect to the first. You can easily tell whether you have a nested design, as no categories within your second treatment variable are quite the same. For example, you may wish to examine the effect of two different growth hormones on the growth of floral meristem explants from *Pharbitis nil* (morning glory). The tissue culture medium was made as one batch, split into two flasks,

and different growth hormones were added to each flask. The medium from these flasks was poured into tissue culture jars and left to cool. Four explants were grown in each jar of medium. Within each treatment, the jars were randomized. After four weeks, the number of shoots for each explant was recorded. The experimental design is:

Growth medium type 1			Growth medium type 2		
Jar 1	Jar 2	Jar 3	Jar 4	Jar 5	Jar 6
Explant 1	Explant 5	etc.			
Explant 2	etc.				
Explant 3					
Explant 4					

You can tell that this is a nested design as no jar will be exactly the same as any other jar. There will be slight differences in the depth of medium, location of the jar in the growth cabinet, etc. The four explants within each jar are subject to these unique factors and this grouping needs to be taken into account in the analysis. This grouping within the design is recognized by calling the growth media the *group* and the jars the *subgroup*.

Q3 In our comparison between shoot production by explants of *Pharbitis nil* on two different tissue culture media, how many groups (*a*) and how many subgroups (*b*) are there? How many observations are in each subgroup (n_s) and how many in each group (n_g)? How many observations are there in total (*N*)?

EXAMPLE 9.9 Hydrogen ion concentration in deciduous and coniferous forests

A soil scientist wished to compare the hydrogen ion (H^+) concentration of soils in two different forest types, one coniferous and the other deciduous. Within each forest, three different soil pits were dug and from each pit two soil samples were taken for hydrogen ion analysis. The results were recorded as mmol H^+/l per 100g soil (Table 9.18).

Table 9.18. The concentration of hydrogen ions in soil (mmol H^+/l per 100g soil) from three soil pits in two woodlands

Deciduous forest			Coniferous forest		
Pit 1	Pit 2	Pit 3	Pit 1	Pit 2	Pit 3
1.5	0.7	1.2	2.0	1.6	2.7
1.2	1.0	1.7	2.5	2.2	2.4

In this design, we have one main factor (the growth media) and one subordinate or nested factor (jars) and equal replicates in each subgroup. Clearly, there are many other possible nested designs. There could be more than one main factor or more than one nested factor or unequal replicates. In this book, we illustrate nested ANOVAs with an example with one main variable, one nested variable, and equal replicates. For information on other nested designs, we recommend Sokal & Rohlf (1994).

This experimental design is necessary because of the variable nature of soil. If only one pit was dug in each forest and multiple soil samples taken from this, there would be no assurance that the outcome of an analysis would not depend on the chance of having selected sites with soil that contained particularly high or low values of hydrogen ion concentrations. Hence, soil sampling from a number of sites within each forest is essential to ensure that any differences are due to the different type of forest and not to the location of the site. If no differences, beyond normal soil sample variability, are found between pits within the forest, any differences found can be attributed to the forest type. Hence, we would expect the difference in soil hydrogen ion concentration due to the forest type to be much greater than the differences due to location within the forests individually. Each of these aspects can be tested in a nested ANOVA.

9.13.1 Using this test

To use this test you:

1) Wish to examine the effect of two variables, one of which is subordinate or nested within the main factor.
2) Have parametric data with the same number of replicates (observations) in each category or sample.
3) Have assigned each item at random within the subordinate variable.
4) Have samples where the variation is similar (homogeneous).

In Example 9.9, it is difficult to tell whether the data are parametric or whether the variances are homogeneous as there are relatively few observations in each subgroup. The scale of measurement is continuous and could be parametric. However, as we are unable to confirm that these criteria are met, we must acknowledge this when reporting the results from the analysis. The other criteria are met in that we wish to examine the effect of two variables (forest and pits) and the pits are nested within the forests, as no one pit will be the same as any other pit. There are two replicates in each sample and these were located at random in the pits.

9.13.2 **The calculation**

Having satisfied ourselves as far as we can that our data meet all the criteria for using this nested ANOVA, we can then proceed. In Box 9.12, we have organized the calculation of $F_{calculated}$ under a number of sub-headings: (A) Calculate general terms; (B) Calculate the sums of squares; and (C) Construct and complete an ANOVA calculation table. At each point, we show how these steps can be applied to Example 9.9. Some of the calculations are included in Tables 9.19 and 9.20. Where steps have been abbreviated, the full calculation is included in the Online Resource Centre.

online resource centre

BOX 9.12 How to carry out a nested ANOVA with one main factor and one nested factor with equal replicates

This calculation is given in full in the Online Resource Centre. All values have been rounded to five decimal places. This calculation is illustrated using data from Example 9.9.

1. **General hypotheses to be tested**

    H_0: There is no difference between the means of the samples.

    H_1: There is a difference between the means of the samples.

    In our example, the two pairs of hypotheses we are testing are:

    H_0: There is no difference in soil hydrogen ion concentrations (mmol H^+/l per 100g soil) due to different forest types.

    H_1: There is a difference in soil hydrogen ion concentrations (mmol H^+/l per 100g soil) due to different forest types.

    H_0: There is no difference in soil hydrogen ion concentrations (mmol H^+/l per 100g soil) among the different soil pits within the forest types.

    H_1: There is a difference in soil hydrogen ion concentrations (mmol H^+/l per 100g soil) among the different soil pits within the forest types.

2. **Have the criteria for using this test been met?**
   Yes. In this example, the criteria have been met (9.13.1).

3. **How to work out F$_{calculated}$**

A. Calculate general terms

1. Add together all the observations in all the samples (grand total): Σx_T.
2. Square each observation in all the samples and add these together: Σx_T^2.
3. Add all the observations in a subgroup: Σx_{sg}. These are the subgroup totals. Do this for all subgroups. Square each subgroup total: $(\Sigma x_{sg})^2$. Add these together. Divide this total by the number of observations in each subgroup (n_s).
4. Add all the observations in a group: Σx_g. Square these: $(\Sigma x_g)^2$. Divide by the number of observations in that group (n_g).
5. Take the result from step 1 and square it: $(\Sigma x_T)^2$. Divide by N, where N is the total number of observations in all the samples combined.

   In our example (Table 9.19), the calculations are:
   1. $\Sigma x_T = 1.5 + 1.2 + \ldots\ldots\ldots\ldots + 2.7 + 2.4 = 20.7$
   2. $\Sigma(x_T^2) = (1.5)^2 + (1.2)^2 + \ldots\ldots\ldots\ldots + (2.4)^2 = 40.21$

   3. $\dfrac{(2.7)^2 + (1.7)^2 + (2.9)^2 + (4.5)^2 + (3.8)^2 + (5.1)^2}{2}$

      $= 79.29/2 = 39.645$

   4. $\dfrac{(7.3)^2 + (13.4)^2}{6} = 232.82/6 = 38.80833$

   5. $(20.7)^2/12 = 35.7075$

B. Calculate the sums of squares (SS)

6. SS_{total} = result from step 2 – result from step 5
7. SS_{groups} = result from step 4 – result from step 5

BOX 9.12 Continued

8. $SS_{subgroups}$ = result from step 3 – result from step 4
9. SS_{within} = result from step 2 – result from step 3
   In our example, the calculations are:
6. SS_{total} = 40.21 – 35.7075 = 4.5025
7. SS_{groups} = 38.80833 – 35.7075 = 3.10083
8. $SS_{subgroups}$ = 39.645 – 38.80833 = 0.83667
9. SS_{within} = 40.21 – 39.645 = 0.565

C. Construct and complete an ANOVA calculation table

i. Draw an ANOVA table as illustrated here.

Source of variation	SS	v	MS	F
Between groups	Step 7			
Between subgroups within groups	Step 8			
Within subgroups	Step 9			
Total variation	Step 6			

ii. Transfer the results from the calculations for SS_{groups}, $SS_{subgroups}$, and SS_{within} into the ANOVA table in column 2 (see Table 9.20).

iii. Calculate degrees of freedom (v).
Let a = the number of groups, b = the number of subgroups, N = total number of observations, and n_{sg} = number of observations in each subgroup.
$$v_{total} = N - 1$$
$$v_{groups} = a - 1$$
$$v_{subgroups} = a(b - 1)$$
$$v_{within} = ab(n_{sg} - 1)$$
Enter these results in column 3.

iv. Calculate the mean squares for the groups, subgroups and within subgroups.
Note: Variances in this context are also known as mean squares (MS).
$$s^2 = MS = \frac{SS}{v}$$

In the groups row, take the value for SS and divide it by its value for v. Repeat this for the subgroups row and the within row. Put the results from these calculations in column 4 (Table 9.20).

v. Finally, to calculate the variance ratio to test the first pair of hypotheses, you use the group and subgroup MS, where:

$$F_{calculated} = \frac{\text{group MS}}{\text{subgroup MS}}$$

It is usual to place the results from this calculation in column 5 in the first row (Table 9.20). To test the second pair of hypotheses, you use the subgroups and within subgroups MS. So:

$$F_{calculated} = \frac{\text{subgroup MS}}{\text{within subgroup MS}}$$

This value is usually placed in the second row in column 5 (Table 9.20).

4. **To find $F_{critical}$**
See Appendix d, Table D8. Identify the critical F value using the F tables for an ANOVA. You may need to interpolate (6.3.6) these values. To find the critical value you need the degrees of freedom relating to the two variances in the F-test.
In our example, to test the first pair of hypotheses, the degrees of freedom are $v_1 = 1$ and $v_2 = 4$. Therefore, at $p = 0.05$, $F_{critical} = 7.71$, and at $p = 0.01$, $F_{critical} = 21.20$.
To test the second pair of hypotheses, the degrees of freedom are $v_1 = 1$ and $v_2 = 6$. Therefore, at $p = 0.05$, $F_{critical} = 5.99$.

5. **The rule**
If $F_{calculated}$ is greater than $F_{critical}$, then you may reject the null hypothesis.
In our example, when testing the first pair of hypotheses $F_{calculated}$ (14.82) is greater than $F_{critical}$ (7.71) at $p = 0.05$, but not greater than $F_{critical}$ (21.20) at $p = 0.01$, so you may reject the null hypothesis at $0.01 < p < 0.05$.
For the second pair of hypotheses, $F_{calculated}$ (2.22) is less than $F_{critical}$ (5.99) at $p = 0.05$, so you do not reject the null hypothesis.

6. **What does this mean in real terms?**
There is a significant difference ($F = 14.82$, $0.01 < p < 0.05$) in soil hydrogen ion concentrations due to different forest types. However, there is no difference ($p = 0.05$) in soil hydrogen ion concentrations among the different soil pits within the forest types.

Further examples relating to the topic including how to use statistical software are included in the Online Resource Centre.

@ online resource centre

Table 9.19. Calculation table for the parametric nested ANOVA on data from Example 9.9: Hydrogen ion concentration in a deciduous and coniferous forest

	Deciduous forest			Coniferous forest		
	Pit 1	Pit 2	Pit 3	Pit 1	Pit 2	Pit 3
Sample 1	1.5	0.7	1.2	2.0	1.6	2.7
Sample 2	1.2	1.0	1.7	2.5	2.2	2.4
Subgroup totals (Σx_{sg})	2.7	1.7	2.9	4.5	3.8	5.1
Group totals (Σx_{g})		7.3			13.4	
Sum of all 12 observations (Σx_{T})			20.7			

Table 9.20. ANOVA table from the analysis of data from Example 9.9: Hydrogen ion concentration in a deciduous and coniferous forest, showing how some of the terms are calculated

Source of variation	SS	v	MS	F
Between forests (groups)	3.10083	$2 - 1 = 1$	3.10083	3.10083/0.20917 = 14.82469
Between pits (subgroups)	0.83667	$2(3 - 1) = 4$	0.20917	0.20917/0.09417 = 2.22123
Between replicates in pits (within subgroups)	0.565	$(2 \times 3)(2 - 1) = 6$	0.09417	
Total variation	4.5025	$12 - 1 = 11$		

Unlike the previous two-way ANOVA, there are only two pairs of hypotheses to be tested. The first relates to the main treatment, in this case forests, and the second to the nested factor, the soil pits. In this analysis, there is a need to distinguish different levels of grouping so the forests are groups and the soil pits are subgroups.

9.13.3 Tukey's test for nested parametric ANOVAs

> **Key points** In a complex experimental design such as that analysed using a nested ANOVA, it becomes increasingly difficult to subsequently use a multiple comparisons test. We indicate when this may be acceptable.

It is very difficult to use a multiple comparisons test, such as Tukey's, when one variable is nested in the other and therefore any mean values for samples will be means for subgroups within groups. One circumstance where you might use Tukey's test would be, as in our example, when the subgroups are not significantly different but the groups are. Under these circumstances, you could consider all values within a group to be replicates. If you had three or more groups, you could carry

out a one-way ANOVA where all subgroup observations are replicates in the groups. The values from this ANOVA could then be used to test specific hypotheses. If you have only two groups, there is no point in carrying out this specific test as it will be clear from the data what the relationship is. In Example 9.9, the soil hydrogen ion concentrations are significantly greater in the coniferous forest than in the deciduous forest.

9.14 Factorial three-way parametric ANOVA with no replicates

Key points A three-way ANOVA will examine the effect of three variables and the interaction between any two of these (first order) and all three (second order). If the interactions are significant, the main effects of the individual variables are not tested.

When there are more than two independent factors acting on a sample, the design is called *factorial*. It is possible to have any number of factors and to test for differences arising from these in one large experiment and calculation, but in reality the experimental design and execution may become so large that it would be unmanageable. Therefore, we will limit worked examples to a demonstration of a three-factor ANOVA (factors A, B, and C) with (9.15) and without (9.14) replicates. Here the effect of the three treatments can be tested in the usual way to determine whether there is a significant difference. We can also determine whether there is an interaction between any or all of the factors. The first three interactions, A × B, B × C, and A × C, are known as first-order interactions and the interaction A × B × C is known as a second-order interaction.

There is a difference between a factorial analysis with or without replicates. If there are no replicates and for each combination of treatments there is only one observation, then the interaction term A × B × C is used in place of the SS_{within} term in the F ratio. If replicates are present, then an SS_{within} term can be calculated and this is used in the F ratio. We illustrate both these processes in the following sections.

In the same way that a two-way ANOVA can be used to examine the effect of a confounding variable, a three-way ANOVA can also be used. However, for repeated measures you need to examine the design carefully to ensure that the repeated measures only affect one of the variables being examined, and a nested ANOVA may be more suitable.

9.14.1 **Using this test**

To use this test you:

1) Wish to examine the effect of three or more variables each with two or more categories.

2) Have parametric data.

3) Have an orthogonal experimental design.

4) Have items assigned at random to treatments, with independent measurements.

As there are three variables, the results table would be in three dimensions. As this cannot be represented on paper, a data table is constructed in which the third factor (C) is located within one of the other factors, in this case A. This does not imply that factor C is in any way dependent on factor A (Table 9.22).

The data from this example meet the criteria for using a factorial ANOVA with no replicates as we wish to examine the effect of three variables (soil depth, bearing, and distance from the smelter), each variable has three or four categories, and the design is orthogonal. Each soil sample is unique to that set of treatments, so the data are not matched. As there are only single observations for each particular set of treatments, it is difficult to tell whether the data are parametric other than that the data are measured on an interval scale (Box 5.2). We shall therefore have to assume that the data are parametric based only on this criterion.

EXAMPLE 9.10 **Lead levels in soil samples taken at various depths, distances, and bearings from a smelter**

Soil samples were taken to discover whether there was a significant difference in the extent of lead contamination due to the proximity of a smelter. Soil samples were taken at various bearings and distances to the smelter and at various depths of soil (Table 9.21). Each treatment was combined with every other treatment.

Table 9.21. Summary table of the factors (3 × 4 × 4 factorial ANOVA) in the investigation of lead levels in soil samples

Factor A: Distance (km)	Factor B: Bearing	Factor C: Soil depth (cm)
0.25	South	5
0.5	South-east	10
1.0	South-west	15
2.0		20
		30

Table 9.22. Lead concentrations in soil (µg/g) at a number of locations (bearing and distance) and soil depths in the vicinity of a smelter

Distance (km) (Factor A)	Soil depth (cm) (Factor C)	Bearing (Factor B)		
		South	South-east	South-west
0.25	5	155	96	365
	10	102	73	345
	15	77	23	248
	20	65	12	176
	30	26	8	72
0.5	5	98	54	302
	10	65	24	238
	15	45	14	154
	20	23	12	98
	30	9	10	43
1.0	5	55	32	256
	10	36	24	189
	15	22	14	112
	20	12	8	54
	30	9	5	23
2.0	5	25	18	167
	10	18	14	98
	15	15	13	43
	20	8	6	22
	30	7	4	17

9.14.2 The calculation

In this calculation, we will examine the effects of the treatments (factor A, factor B, and factor C) and the first-order interactions (A × B, A × C, and B × C). The second-order interaction (A × B × C) is used in place of the variance$_{within}$ term. Therefore, we are testing six pairs of hypotheses and carrying out an *F*-test for each. The calculation of these variance ratios has been organized into a number of steps: (A) Calculate the general terms; (B) Calculate the sums of squares; and (C) Construct and complete an ANOVA table. In Box 9.13, we illustrate this calculation using the data from Table 9.22.

BOX 9.13 How to carry out a three-way factorial parametric ANOVA without replicates

This calculation is given in full in the Online Resource Centre. All values have been rounded to five decimal places. This calculation is illustrated using data from Example 9.10.

1. **General hypotheses to be tested**
   As there are six pairs of hypotheses, we have not included general hypotheses but only those from our example.
   H_0: There is no difference between the mean lead concentration in the soil samples (μg/g) due to distance from the smelter (km) (factor A).

   H_{1A}: There is a difference between the mean lead concentration in the soil samples (μg/g) due to distance from the smelter (km) (factor A).

   H_{0B}: There is no difference between the mean lead concentration in the soil samples (μg/g) due to the bearing from the smelter (factor B).

   H_{1B}: There is a difference between the mean lead concentration in the soil samples (μg/g) due to the bearing from the smelter (factor B).

   H_{0C}: There is no difference between the mean lead concentration in the soil samples (μg/g) due to the depth at which the soil was sampled (cm) (factor C).

   H_{1C}: There is a difference between the mean lead concentration in the soil samples (μg/g) due to the depth at which the soil was sampled (cm) (factor C).

   $H_{0(A \times B)}$: There is no interaction between distance (km) (factor A) and bearing (factor B) in their effects on the mean lead concentration in the soil.

   $H_{1(A \times B)}$: There is an interaction between distance (factor A) and bearing (factor B) in their effects on the mean lead concentration (μg/g) in the soil.

   $H_{0(A \times C)}$: There is no interaction between distance (km) (factor A) and soil sample depth (cm) (factor C) in their effects on the mean lead concentration (μg/g) in the soil.

   $H_{1(A \times C)}$: There is an interaction between distance (km) (factor A) and soil sample depth (cm) (factor C) in their effects on the mean lead concentration (μg/g) in the soil.

   $H_{0(B \times C)}$: There is no interaction between bearing (factor B) and depth of sampling (cm) (factor C) in their

effects on the mean lead concentration (μg/g) in the soil.

$H_{1(B \times C)}$: There is an interaction between bearing (factor B) and depth of sampling (cm) (factor C) in their effects on the mean lead concentration (μg/g) in the soil.

2. **Have the criteria for using this test been met?**
   As far as we can tell, these criteria have been met (9.14.1).

3. **How to work out $F_{calculated}$**

A. Calculate general terms

1. Calculate the grand total (Σx_T), by adding together all the observations in the data set. Note the total number of observations (N).
   $\Sigma x_T = 155 + 96 + 365 + \ldots + 7 + 4 + 17 = 4358.0$
   $N = 60$

2. Square each and every observation and add these together: (Σx_T^2)

   $$(\Sigma x_T^2) = (155)^2 + (96)^2 + (365)^2 + \ldots + (7)^2 + (4)^2$$
   $$+ (17)^2 = 780172.0$$

3. Summarize the data from the three-way table (Table 9.22) into three two-way tables.
   B × C table
   For the B × C table (Table 9.23), total the four observations for bearing south and 5cm soil depth. This total goes into the first row, second column of the B × C table. Add the four observations for bearing south-east and 5cm soil depth. This total goes into the second row, second column of the A × C table, etc.
   The number of observations being counted = $a = 4$.

Table 9.23. B × C two-way table (bearing × soil sample depth (cm)) for lead levels in soil samples (μg/g)

Factor B: Bearing from smelter	Factor C: Depth at which soil samples taken (cm)					Row total
	5	10	15	20	30	
South	333	221	159	108	51	872
South-east	200	135	64	38	27	464
South-west	1090	870	557	350	155	3022
Column total	1623	1226	780	496	233	4358

BOX 9.13 How to carry out a three-way factorial parametric ANOVA without replicates

A × C table

For the A × C two-way table (Table 9.24), add together the three observations for 5cm soil depth and distance 0.25km. This total goes into the first row, second column of the A × C table. Add together the three observations for 5cm soil depth and distance 0.5km. This total goes in the second row, second column, etc.

The number of observations being counted = $b = 3$

Table 9.24. A × C two-way table (distance (km) × soil depth (cm)) for lead levels in soil samples (µg/g)

Factor A: Distance from smelter (km)	Factor C: Depth at which soil samples taken (cm)					Row total
	5	10	15	20	30	
0.25	616	520	348	253	106	1843
0.5	454	327	213	133	62	1189
1.0	343	249	148	74	37	851
2.0	210	130	71	36	28	475
Column total	1623	1226	780	496	233	4358

A × B table

For the A × B table (Table 9.25), first add the five values for 0.25km distance and bearing South. The total goes into the first cell (first row first column) of the two way table A × B. Add the five values for distance 0.5km, bearing South. This total goes into the second cell (second row first column) of the A × B table, etc.

The number of observations being counted = $c = 5$.

Table 9.25. A × B two-way table (distance (km) × bearing) for lead levels in soil samples (µg/g)

Factor A: Distance from smelter (km)	Factor B: Bearing from smelter			Row total
	South	South-east	South-west	
0.25	425	212	1206	1843
0.5	240	114	835	1189
1.0	134	83	634	851
2.0	73	55	347	475
Column total	872	464	3022	4358

4. Square each and every row total from the A × C two-way table (Table 9.24), add these squared values together, and divide by $(b \times c)$:

$$= \frac{(1843)^2 + (1189)^2 + (851)^2 + (475)^2}{3 \times 5}$$

$$= 384013.0667$$

5. Square each and every row total from the B × C two-way table (Table 9.23), add these squared values together, and divide by $(a \times c)$:

$$= \frac{(872)^2 + (464)^2 + (3022)^2}{4 \times 5} = 505408.2$$

6. Square each and every column total from the A × C two-way table (Table 9.24), add these squared values together, and divide by $(a \times b)$:

$$= \frac{(1623)^2 + (1226)^2 + (780)^2 + (496)^2 + (233)^2}{4 \times 3}$$

$$= 420492.5$$

7. Square each value in the A × B two-way table (Table 9.25), add these squared values together, and divide by c:

$$= \frac{(425)^2 + (212)^2 + \cdots\cdots + (55)^2 + (347)^2}{5}$$

$$= 600678.0$$

8. Square each value in the A × C two-way table (Table 9.24), add these squared values together, and divide by b:

$$= \frac{(616)^2 + (520)^2 + \cdots\cdots\cdots + (36)^2 + (28)^2}{3}$$

$$= 500890.6667$$

9. Square each value in the B × C two-way table (Table 9.23), add these squared values together, and divide by a:

$$= \frac{(333)^2 + (221)^2 + \cdots\cdots + (350)^2 + (155)^2}{4}$$

$$= 666386.0$$

10. Square the grand total (Σx_T) from step **1** and divide by the total number of observations (N):

$$= (4358)^2/60 = 316536.0667$$

B. Calculate the sums of squares (SS)

11. SS_A = result from step **4** – result from step **10**
    = 384013.0667 – 316536.0667 = 67477.0

12. SS_B = result from step **5** – result from step **10**
    = 505408.2 – 316536.0667 = 188872.1333

BOX 9.13 Continued

13. SS_C = result from step **6** – result from step **10**
    = 420492.5 – 316536.0667
    = 103956.4333

14. SS_{AB} = result from step **7** + result from step **10** – result from step **4** – result from step **5**
    = 600678.0 + 316536.0667 – 384013.0667 – 505408.2 = 27792.8

15. SS_{AC} = result from step **8** + result from step **10** – result from step **4** – result from step **6**
    = 500890.6667 + 316536.0667 – 384013.0667 – 420492.5 = 12921.1667

16. SS_{BC} = result from step **9** + result from step **10** – result from step **5** – result from step **6**
    = 666386.0 + 316536.0667 – 505408.2 – 420492.5 = 57021.3667

17. SS_{ABC} = (result from step **2** + results from steps **4** + **5** + **6**) – (results from steps **7** + **8** + **9** + **10**)
    = (780172.0 + 384013.0667 + 505408.2 + 420492.5) – (600678.0 + 500890.6667 + 666386.0 + 316536.0667)
    = 5595.0333

C. Construct and complete an ANOVA calculation table

i. Construct and complete an ANOVA table as shown in Table 9.26.
   Transfer the results from the calculations for sums of squares (SS) into the table in column 2 (Table 9.26).

ii. Calculate the degrees of freedom (v) where:
   $v_A = a - 1$ = number of samples – 1
   $v_B = b - 1$ = number of samples -1
   $v_C = c - 1$ = number of samples – 1
   $v_{A \times B} = (a - 1)(b - 1)$
   $v_{A \times C} = (a - 1)(c - 1)$
   $v_{B \times C} = (b - 1)(c - 1)$
   $v_{A \times B \times C} = (a - 1)(b - 1)(c - 1)$
   Enter these results in column 3 (Table 9.26).

iii. Calculate the mean squares (MS)
   Note: *Variances in this context are also known as mean squares (MS).*

   $$s^2 = MS = \frac{SS}{v}$$

   In the Factor A row, take the value for SS_A and divide it by the value for v_A.
   In the Factor B row, take the value for SS_B and divide it by the value for v_B, and so on. Put the results from these calculations in column 4 (Table 9.26).

If the interaction terms are not significant, then you may proceed to test the main effects. In our example, the interaction terms are all significantly different (Table 9.26).

iv. How to work out $F_{calculated}$
   As in the F_{max} test, $F_{calculated}$ is a ratio. The F value for each pair of hypotheses to be tested is calculated by the ratio of its mean square over the mean square of A × B × C (which represents the 'error' mean square in a factorial with no replicates and assuming that the third factor interaction is zero), e.g.:

   $$F_{Calculated} = \frac{MS_A}{MS_{ABC}}$$

   The results from these calculations are placed in column 5 in the first row and so on (Table 9.26).

4. To find $F_{critical}$

See Appendix d, Table D8. Identify the critical F value using the F tables for an ANOVA. You may need to interpolate (6.3.6) these values, where the degrees of freedom are those relating to the numerator (v_1) and denominator, which is MS_{ABC} (v_2). As you are testing six pairs of hypotheses, you may need to find six values for $F_{critical}$.

5. The rule

The general rule is that if $F_{calculated}$ is greater than or equal to $F_{critical}$, then you may reject the null hypothesis. In this factorial analysis, you should examine the first-order interaction terms first (i.e. A × B, A × C, and B × C). If these are significant, then it is not meaningful to test the main effects of factors A, B, and C as these elements have been shown not to be contributing separately to the overall outcome.

6. Relate your findings back to your investigations

The distance × bearing, distance × soil sample depth, and bearing × soil sample depth are all highly significant interactions ($p = 0.01$). Therefore, the effect of distance from the smelter on the lead concentration in the soil samples depends on the bearing and soil sample depth. In addition, the effect of the soil sample depth on the lead concentration depends on the bearing from the smelter.

Further examples relating to the topic including how to use statistical software are included in the Online Resource Centre.

online resource centre

Table 9.26. ANOVA table for analysis of data from Example 9.10: Lead concentrations in soil (µg/g) at a number of locations (bearing and distance) and soil depths from a smelter

Source of variation	SS	v	MS	$F_{calculated}$	$F_{critical}$ at $p = 0.01$
Factor A: Distance (km)	67477.0	3	22492.3	96.48134	
Factor B: Bearing	188872.1333	2	94436.06	405.08551	
Factor C: Soil sample depth (cm)	103956.43	4	25989.11	111.48083	
Interaction A × B: Distance (km) × bearing	27792.80	6	4632.13	19.86963	3.67
Interaction A × C: Distance (km) × soil sample depth (cm)	12921.1667	12	1076.76	4.61880	3.03
Interaction B × C: Bearing × soil sample depth	57021.37	8	7127.67	30.5743	3.36
Interaction: A × B × C	5595.03	24	233.13		
SS_{within} (Box 9.13)	Result from step **19**	$v_{within} = abc(n-1)$		$s^2_{within} = SS_{within}/v_{within}$	

9.14.3 Tukey's test and a three-way parametric ANOVA

Key points In a complex experimental design such as when three variables are being examined, it becomes increasingly difficult to subsequently use a multiple comparisons test. We indicate when this may be acceptable.

As we explained in 9.10, any significant interactions between variables will make it very difficult to interpret any multiple comparisons between pairs of means. Therefore, you should restrict your use of Tukey's test to experiments where only the main factors are significant and not the interaction terms. In these cases, you apply Tukey's test as we described in Box 9.7 and section 9.10.

9.15 Factorial three-way parametric ANOVA with replicates

Key points This is an extension of the three-way ANOVA outlined in 9.14. As in 9.14, you may examine the effect of each of the three variables but also the interaction between pairs of variables and all three variables. The use of a multiple comparisons test such as Tukey's test following this ANOVA is only appropriate in certain outcomes.

Carrying out a factorial ANOVA where you have replicates, i.e. more than one observation for any one particular combination of treatments, is very similar to the method described in 9.14. In the present section, we therefore only highlight the differences that you would need to take into account to allow for replication within the design.

9.15.1 Using this test

To use this test you need to satisfy the criteria listed in 9.14.1 with two additional criteria:

1) Have samples with homogeneous variances.
2) Have equal numbers of replicates.

With more data, you will be able to be more exact as to whether your data are parametric. To test criterion 6, you should carry out an F_{max} test (Box 9.5). If you do not have equal numbers of replicates, you should refer to Sokal & Rohlf (1994) or Zar (2010).

9.15.2 The calculation

The key difference between this test and the one outlined in 9.14 is that you will have an error term, SS_{within}. The changes required to allow for this are shown in Box 9.14.

BOX 9.14 How to carry out a three-way factorial parametric ANOVA with replicates: an extension of the method outlined in Box 9.13

1. **General hypotheses to be tested**
   Apart from the general hypotheses included in Box 9.13, you will have a further pair of hypotheses relating to the second-order interaction A × B × C.
   $H_{0(A \times B \times C)}$: There is no interaction between factor A, factor B, and factor C in their effects on the sample means.

   $H_{1(A \times B \times C)}$: There is an interaction between factor A, factor B, and factor C in their effects on the sample means.

2. **How to work out $F_{calculated}$**

   A. Calculate general terms
   18. Using the data from the original three-way table of results, sum each cell, square this value, add

all these squared values together, and divide by n (the number of observations in each cell).
For example, a small extract from a results table might look like:

Distance (km)	Soil sample depth (cm)	Bearing	
		South	South-east
0.25	5	156	96
		158	98
		155	94
	10	101	
		103	etc.
		102	

BOX 9.14 Continued

Step **18** would then be:
$((156 + 158 + 155)^2 + (96 + 98 + 94)^2 + (101 + 103 + 102)^2 + \cdots)/3$

B. Calculate the sum of squares (SS)

19. SS_{within} = result from step **2** – result from step **18**

C. Construct and complete an ANOVA calculation table

An additional row is added to the bottom of the ANOVA table (Table 9.26):

MS_{within} is now used as the denominator in working out the $F_{calculated}$ values instead of the MS_{ABC} as in Box 9.13. In addition, it is possible to test hypotheses relating to the second-order interaction A × B × C, so $F_{calculated}$ for this is MS_{ABC}/MS_{within}.
Again, it is best to consider the interactions first before the main treatment effects. If any of the interactions are significant, then it is not sensible to use Tukey's test on these data.

9.16 Experimental design and parametric statistics

Key points ANOVAs may be used to examine the effect of a number of variables each with two or more categories. The designs are usually orthogonal and should have either no replicates or equal numbers of replicates. Some designs are especially useful if you wish to compare a control with a treatment or where you have repeated measurements on the same items. In addition, particular designs can be used to examine the effect of a notable confounding variable. ANOVAs allow you to examine both the effect of variables in a biological system and also the interaction between two or more variables in their effect on a biological system. Multiple comparisons tests such as Tukey's and the Tukey–Kramer tests allow you to examine more specific hypotheses and to compare pairs of samples in all possible combinations.

There are a number of very different statistical tests covered in this chapter that reflect a number of different experimental designs.

9.16.1 One variable

In experimental designs that seek to examine the effect of one variable, we consider investigations where you will collect unmatched data or matched data. We also discuss a specific design (Latin square) for reducing the effect of non-treatment variables.

i. Unmatched

We have considered four tests that may be used to analyse parametric data with one variable. The first two (*z*- and *t*-tests) can be used when you have only two categories for your treatment variable and the one-

Table 9.27. Experimental design with one treatment variable may be tested by a *z*- or *t*-test for unmatched pairs or one-way ANOVA if all other criteria are met

Category 1	Category 2	Category 3 etc.
Observation 1	Observation 1	Observation 1, etc.
Observation 2	Observation 2	
Observation 3, etc.	Observation 3, etc.	
If two categories only, you may use a *t*- or *z*-test for unmatched data, if all other criteria are met		
If three or more categories, consider a one-way ANOVA, if all other criteria are met		

way ANOVA can be used when you have two or more categories. If you have two categories, it is common to design the investigation so that a *t*- or *z*-test may be used, because the analysis for these tests is thought to be simpler. This general design is shown in Table 9.27.

The number of observations in each category is the second key factor that determines the statistical test you are likely to use and must therefore be considered before you carry out your investigation. For example, if you decided to use a *z*-test, you will need more than 30 observations in each sample and to use a *t*-test you have under 30 observations in each sample. If this design is to be analysed by an ANOVA, the number of observations is usually determined by practical constraints. These tests may be used if you have unequal numbers of observations in your categories (samples).

If you have only two categories, one could be from a control treatment and these tests would be an effective way of testing whether your treatment makes a significant impact on your experimental system compared with the control. However, you must not use the same item for the control and the treatment: all observations must be independent measurements on different items. Each observation within a category or sample is in a sense a replicate, although they are not usually referred to as such.

ii. Matched

If you have one treatment variable where your items are examined before and after the treatment, then these are matched observations. The design in this case (Table 9.28) looks similar to those suitable for a correlation or regression analysis, but you are only examining one variable.

Again, the number of observations in a sample is critical to the statistical test you use and the statistical test you are planning to use will determine the sample size. For example, if you are planning to use a *z*-test, then you need more than 30 items in your experiment. One of the samples could be a null state and could therefore be a control. Each observation within the sample is in a sense a replicate, although rarely referred to as such.

Table 9.28. Experimental design with one treatment variable and for matched data

Item	First measure	Second measure
1	Observation 1	Observation 1
2	Observation 2, etc.	Observation 2
3		
4, etc.		

iii. Latin square

When designing experiments, you will seek to identify and control or minimize the effects of confounding variables. For example, you can minimize the effect of confounding variables by the randomization of items to treatments (2.2.4). However, there is one specific experimental design called a Latin square that is structured to minimize the effect when there are two known confounding variables that are graded across the area in which you are working. For example, you may need to organize plants in a greenhouse and know that the light levels drop off as you move away from the windows; similarly, you may have a door at one end and know that there is therefore a temperature gradient along the bench caused by a draught from the door. In these circumstances, you can organize your items or blocks in a non-random arrangement. In our example of a Latin

Gradient for non-treatment variable 1

Gradient for non-treatment variable 2

A	D	C	B
C	B	A	D
B	C	D	A
D	A	B	C

Fig. 9.5 A Latin square for two non-treatment variables and one treatment variable with four categories.

square (Fig. 9.5), you can see that there are four categories (A, B, C, and D) for a single variable. Each category is present in only one row and one column. The ANOVA that is used to analyse the results from a design of this nature partitions the variation into that due to the confounding variable 1 + variation due to confounding variable 2 + variation due to the treatment variable + sampling error. We have not included a Latin square ANOVA in this book; instead we refer you to Fowler *et al.* (1998).

9.16.2 **Two variables**

When there are two variables under investigation, you need to consider a number of factors.

i. Orthogonal and randomization

A two-way design should be orthogonal. In such a design, each category from variable 1 is tested against each category from variable 2. This structure gives you 'blocks'. For example variable 1 may be divided into five categories and variable 2 may be divided into seven categories. By combining these categories in an orthogonal design, you will have 35 blocks.

The generic design is known as the randomized block design as the items are randomized to a block and the blocks are randomized to the treatments. As we explained in 2.2.4, this randomization of items within treatments is a way of minimizing the impact of non-treatment variation and is therefore a powerful tool that can be used when designing an experiment.

Table 9.29. A randomized block design

Variable 2	Variable 1				
	1	2	3	4	5
1	B1	B2	B3	B4	B5
2	B6	B7	B8	B9	B10
3	B11	B12	B13	B14	B15
4	B16	B17	B18	B19	B20
5	B21	B22	B23	B24	B24
6	B25	B26	B27	B28	B29
7	B30	B31	B32	B33	B34

ii. Confounding variables and repeated measures

As we explained in 9.12, it is possible to use this design to investigate the effect either of a confounding variable or a variable with repeated measures. For example, the time taken by rats to run around a number of mazes would fit in Table 9.29 with the different mazes being the categories of variable 1 and the different rats each being a category in variable 2. The time taken by each rat in each maze would then be the value in the block.

Similarly, as in Example 9.8, the confounding variable of microenvironmental variation in a polytunnel can be organized as categories in variable 2 and the tomato feeds would be the categories in variable 1. The mass of tomatoes produced would be the value placed within each block.

iii. Replication

It is possible to analyse the data in a randomized block design where there are no replicates (9.12) or when there are equal numbers of replicates in each block (9.9). However, it is difficult to overcome the problem of unequal numbers of replicates. This must therefore be considered when planning an experiment.

iv. Fixed and nested

With two or more variables, you need to consider whether they are fixed or random. Table 9.29 is an example of fixed variables or a Model I design and may be contrasted with Table 9.30, which illustrates a mixed model where one variable is fixed (main effect) and the subordinate variable is randomized within the categories of the first variable (Model II). We saw this in Example 9.9 where soil samples had been taken from a number of pits dug in two forests. Clearly, pit 1 in the coniferous forest

Table 9.30. Mixed model experimental design with one fixed main variable and one random, nested variable, with equal replicates

Variable 2	Variable 1			
	Category 1		Category 2	
	Category 1	Category 2	Category 3	Category 4
	BLOCK 1	BLOCK 2	BLOCK 1	BLOCK 2
	Observation 1	Observation 1	Observation 1	Observation 1
	Observation 2	Observation 2	Observation 2	Observation 2
	Observation 3, etc.	Observation 3, etc.	Observation 3, etc.	Observation 3, etc.

is not the same as pit 1 in the deciduous forest: the pits as a treatment are randomized within the forests. This design is not orthogonal but is nested. A mixed model design should be analysed using a nested ANOVA. Again, there are many versions of mixed models. In this chapter, we have included a mixed model with one fixed (main) variable and one random variable with equal replicates (Example 9.9). For other designs, we refer you to Sokal & Rohlf (1994).

9.16.3 Three variables

The principles behind the randomized block design (two-way ANOVA) and mixed model design (two-way nested ANOVA) can, as we saw in 9.14 and 9.15, be extended to experiments involving three variables. The advantage of a comprehensive design examining many variables at one go, rather than several small experiments examining one or two variables at a time is that you reduce the likelihood of a Type I error occurring. You are also more likely to detect significant interactions between the variables. The analysis of comprehensive designs is relatively more complex and the detection of significant interactions makes it more difficult to interpret your data.

The choice of statistical test for experimental designs that examine three or more treatment variables is dependent, as in the one variable and two variable designs, on the model (pure Model I, pure Model II, or mixed model) and the numbers of replicates. Some of these alternative designs are considered in other texts, including Sokal & Rohlf (1994).

9.16.4 General and specific hypotheses

In this chapter, we introduced you to Tukey's test and the Tukey–Kramer test, which allow you to test specific hypotheses and so gain more explicit information about your experimental system. These two tests can be used after the ANOVAs included in this book, with some restrictions. For example, Tukey's can only be used if you have equal numbers of replicates. When testing experiments with two or more variables, significant interactions can make it difficult to interpret any results from Tukey's test. However, the value of these specific comparisons should not be underestimated, and where possible you should design investigations where you may be able to test general and specific hypotheses.

In all the designs we have outlined in this chapter, controls may be included. Testing specific hypotheses using Tukey's test can allow you to make more explicit comparisons between any controls and the effects of the treatments.

Summary of Chapter 9

- This chapter draws on your understanding of distributions and in particular the normal distribution (5.2.1). The normal distribution is used as the basis for parametric statistics. Data that are parametric (Box 5.2) or have been normalized following transformation (5.9) may be analysed using parametric statistics.

- The basic elements of parametric statistics are the sum of squares and the variance, which we introduced in Chapter 5 (Box 5.1).

- Parametric statistics are flexible and powerful. In 9.16, we outline the range of experimental designs that may be suitable for analysis by parametric tests, if all other criteria are met.

- We consider three groups of parametric tests in this chapter: *t*-tests for matched and unmatched data (9.1 and 9.2), *z*-tests for matched and unmatched data (9.3), and parametric ANOVAS with Tukey's test or the Tukey–Kramer test for one, two, or three variables (9.4–9.15).

- The ANOVAs followed by Tukey's test allow you to test both general and specific hypotheses. This provides you with considerably more explicit information about the effects of your treatment variables and is a design that should be used whenever possible (e.g. 9.6 and 9.10).

- In Chapter 10, we consider the non-parametric equivalents to the tests covered in this chapter. Therefore, if your data are not parametric (5.8) or cannot be normalized by transforming your data (5.9), you should refer to Chapter 10.

- The Online Resource Centre includes interactive exercises that test your understanding of this chapter with other topics, particularly those considered in Chapters 2 and 7–10.

 online resource centre

Answers to chapter questions

 A1   There really is a difference between Porthcawl and Aberystwyth in the way these periwinkles are evolving.

The sampling error in the small sample from Porthcawl has masked the effect of evolution and a Type II error has occurred.

One of the assumptions was not met. For example, the data were not parametric.

 A2   All these treatments are under the control of the investigator and fixed. Therefore, this is a Model I design.

A3 $a = 2$ (media type 1 and media type 2).

$b = 3$ (there are three jars in each group).

$n_s = 4$ (there are four explants in each jar).

$n_g = 12$ (there are three jars each with four explants in each group).

$N = 24$ (there are 24 explants in total).

Hypothesis testing: do my samples come from the same population? Non-parametric data

10

In a nutshell

Often biological data is not normally distributed or you have insufficient numbers of observations to be able to tell. In these instances, you may wish to test the hypothesis that samples come from the same statistical population using non-parametric tests. We have included the tests used most often by our undergraduates and which mirror those covered in the previous chapter for testing the same hypothesis when you have parametric data.

This chapter tells you how you may analyse non-parametric data to test the hypothesis that two or more samples come from the same statistical population. When testing this type of hypothesis, almost the first step you take is to determine whether your data are parametric or non-parametric (Box 5.2). Parametric tests are more powerful so, where possible, you should use these tests. (This concept of the power of a test is explained in the glossary.) Parametric tests suitable for testing this type of hypothesis are outlined in the previous chapter. If your data are not parametric, you may be able to transform them to 'normalize' the data and so enable you to use a parametric test (5.9). Failing that, you should consider the tests described here.

The tests described in this chapter are known as *ranking* tests. They are 'distribution free' and the data therefore do not have to be distributed normally. Look at the data below. Each sample has been organized into numerical order and you can see that there is a small overlap between the two data sets.

Sample 1	1	3	4	6	7	7	9	10	10	11						
Sample 2									10	11	11	12	13	13	13

If the two data sets are from different statistical populations, then you would expect the overlap to be small. If the two samples are from the same population, then you would expect the overlap to be greater. The ranking tests provide a measure of this overlap and an estimate of the probability that this overlap could occur by chance. If you are not familiar with the process of ranking data, you should refer to Box 5.3 where we give examples.

Seven tests are covered in this chapter. Worked examples are given for each test. If this is the first time you have used these tests, you should cover up these worked examples and use the general information and the data provided to work through these examples. Then check your answers before using the test on your own data. If your answer differs considerably from that given, you should check your calculation by going to the Online Resource Centre. We would expect it to take you about three hours to complete these worked examples and questions. The answers for these exercises are at the end of the chapter. All our examples are based on real undergraduate research projects. If these examples are not in your subject area you will find more in the Online Resource Centre.

online resource centre

How to choose the correct test

The choice of these tests is similar to choosing the parametric tests. Your selection depends on how many variables you are planning to examine, how many categories in each variable, and how many replicates in each category. It is assumed that you have non-parametric data. The terms 'matched' and 'unmatched' are explained at the beginning of section 10.2 and in the glossary.

The table below directs you to the most likely statistical test. If you are not sure, look at the specific section indicated for the test and examine the examples to see whether they are similar to the work you are planning. Each test has additional criteria that need to be met. These are given in the sections indicated.

You wish to examine the effect of one variable. You are going to compare two categories or samples. The data are unmatched.	Mann–Whitney U test (10.1).
You wish to examine the effect of one variable. You are going to compare two categories or samples. The data are unmatched. The data are measured on a continuous scale and you have more than 30 observations in each sample.	z-test for unmatched data (9.1).
You wish to examine the effect of one variable. You are going to compare two categories or samples. The data are matched. You have fewer than 30 pairs of observations.	Wilcoxon's rank paired test (10.2).
You wish to examine the effect of one variable. You are going to compare two samples. The data are matched. You have more than 30 pairs of observations.	z-test for matched data (9.3).
You wish to examine the effect of one variable. You are going to compare three or more categories or samples.	One-way ANOVA (Kruskal–Wallis test) (10.3). If the outcome from this test is significant, you may follow it with a multiple comparisons test (10.4).

You wish to examine the effect of more than one variable. You are going to compare two or more categories or samples. You have equal numbers of observations in each category.	Two-way non-parametric ANOVA (10.5). If the outcome from this test is significant, you may follow it with a multiple comparisons test (10.6).
You wish to examine the effect of more than one variable. Each variable has two or more categories.	Scheirer–Ray–Hare test (10.7)
You wish to examine the effect of two variables each with two or more categories. The design is orthogonal. There is only one observation in each category. These may be repeated measures.	Friedman's test; see Sokal & Rohlf (1994) and Zar (2010).

Q1 The impact of consuming 1.5 units of alcohol was tested on reaction times (seconds) in six men and ten women. What information do you know from this brief description that can help you identify which might be the most suitable statistical test to use? What more do you need to find out before you can make a definite decision about which test to use?

10.1 Mann–Whitney *U* test

Key points This ranking test allows you to compare two categories or samples each with replicates. This is the equivalent to the *t*- or *z*-test for parametric data.

We have already introduced you to the idea of ranking observations (Box 5.3) and applied this idea in Chapter 8 when using Spearman's rank test. The Mann–Whitney *U* test uses the rank value of observations and the sum of these ranks for each sample to establish whether the two samples are significantly different.

10.1.1 Using this test

To use this test you:

1) Wish to examine the effect of one variable and there are two categories or samples.
2) Have data that are non-parametric and unmatched.
3) Have data that can be ranked (5.1).
4) Have two samples that both have a similar shaped distribution. For example, if one distribution is skewed to the left and the other to the right (5.4.4), then you should not use this test. (If this does arise you could try transforming the data (5.9).)
5) Do not need equal sample sizes.
6) You should not use this test if one sample has only one observation or if both samples have fewer than five observations each.

We will illustrate the Mann–Whitney U test using an example from Chapter 9, Example 9.2: The evolution of *Littorina littoralis* at Porthcawl, where we know there is some doubt as to whether the data meet all the criteria for a parametric test. In this example, investigators used a systematic random sampling method to collect periwinkles from the mid- and lower shore and measured the shell height (mm) (Table 9.2). The investigators wished to test the hypothesis that there was a difference between these groups of periwinkles as had been seen in an earlier study carried out at Aberystwyth (Example 9.1).

This data set meets all the criteria for using a Mann–Whitney U test as there is one variable (periwinkles) and two samples (mid- and lower shore periwinkles). The data are unmatched as each periwinkle was measured only once. The data can be ranked and both samples have more than five observations. The distributions of the observations in both categories are similar. Table 9.2 is repeated below.

Table 9.2 Height of shells in two putative species of periwinkles from the mid- and lower shore at Porthcawl, 2002

Shell height (mm)			
Periwinkles on the lower shore		Periwinkles on the mid-shore	
5.5	4.0	3.3	6.7
8.4	5.0	6.3	5.7
5.0	6.2	6.1	4.2
5.0	5.0	8.0	6.3
5.6	7.7	13.5	6.0
4.8	6.0	5.3	7.2
8.4		6.7	

10.1.2 The calculation

We show you the general process and a specific example of a Mann–Whitney U test in Box 10.1. In addition, you will need to refer to the calculation table (Table 10.1).

BOX 10.1 How to carry out a Mann–Whitney *U* test	
GENERAL DETAILS	**EXAMPLE 9.2**
	This calculation is given in full in the Online Resource Centre. All values have been rounded to five decimal places.
1. Hypotheses to be tested H_0: There is no difference between the median values in sample 1 and sample 2. H_1: There is a difference between the median values in sample 1 and sample 2.	**1. Hypotheses to be tested** H_0: There is no difference between the median shell heights (mm) of the periwinkles from the mid- and lower shore at Porthcawl. H_1: There is a difference between the median shell heights (mm) of periwinkles from the mid- and lower shore at Porthcawl.

BOX 10.1 Continued

2. Have the criteria for using this test been met?	2. Have the criteria for using this test been met? Yes (10.1.1).

3. To work out $U_{calculated}$
i.   Decide which of the samples is sample 1 and which sample 2.
ii.   Combine all observations (sample 1 and sample 2) and arrange in numerical order.
iii. Assign ranks to these values. Where there are tied values, take the middle (average) rank for these values (Box 5.3).
iv. Add up the ranks ($\Sigma r_1 = R_1$) for sample 1. Repeat for sample 2 ($\Sigma r_2 = R_2$).
v.   Record n_1, the number of observations in sample 1, and n_2, the number of observations in sample 2.
vi. You now calculate two U terms, U_1 and U_2. Watch the formulae closely. A common error is to use R_1 when calculating U_1.
 	
$$U_1 = n_1n_2 + \frac{n_2(n_2+1)}{2} - R_2$$
$$U_2 = n_1n_2 + \frac{n_1(n_1+1)}{2} - R_1$$
 	
vii. Examine the two U values. The smallest U value is $U_{calculated}$.
viii. You can check your maths at this point since $U_1 + U_2 = n_1 \times n_2$.
	3. To work out $U_{calculated}$
	i. Let the lower shore periwinkles be sample 1 and the mid-shore periwinkles be sample 2.
	ii. See Table 10.1.
	iii. See Table 10.1.
	
	iv. See Table 10.1.
	     $R_1 = 153.5$ and $R_2 = 197.5$
	
	v. In this example:
	     $n_1 = 13$, $n_2 = 13$
	vi. Taking each calculation in stages:
	     $n_1 n_2 = 13 \times 13 = 169$
	     $n_2(n_2+1) = 13(13+1) = 13 \times 14 = 182$
	     Therefore:
	$$U_1 = 169 + \frac{182}{2} - 197.5 = 62.5$$
	$$U_2 = 169 + \frac{182}{2} - 153.5 = 106.5$$
	
	vii. Therefore $U_{calculated} = 62.5$
	
	viii. $(62.5 + 106.5) = (13 \times 13)$ correct.

4. To find $U_{critical}$ See Appendix d, Table D12. To find $U_{critical}$ using the table of critical values, you need to know n_1 and n_2. Where the n_1 row intersects the n_2 column for $p = 0.05$, this is the critical value. This is a two-tailed test.	**4. To find $U_{critical}$** $n_1 = n_2 = 13$ and at $p = 0.05$, $U_{critical} = 45.0$
5. The rule If the calculated value of U is less than the critical value of U, then you may reject the null hypothesis (H$_0$).	**5. The rule** In our example, $U_{calculated}$ (62.5) is greater than $U_{critical}$ (45.0) and therefore you do not reject the null hypothesis.
6. What does this mean in real terms?	**6. What does this mean in real terms?** There is no significant difference ($U = 62.5$, $p = 0.05$) between the median shell height (mm) of the periwinkles from the mid- and lower shore at Porthcawl. The results from this investigation do not support the notion that sympatric speciation is occurring. This concurs with the analysis of the data using a *t*-test (Box 9.3).

Further examples relating to the topic including how to use statistical software are included in the Online Resource Centre.

@ **online resource centre**

Table 10.1. Calculation of ranks for the height of shells in two groups of periwinkles from the mid- and lower shore at Porthcawl, 2002

Height of shells (mm)

Periwinkles from the lower shore		Periwinkles from the mid-shore	
Observations in numerical order	Rank	Observations in numerical order	Rank
		3.3	1.0
4.0	2.0		
		4.2	3.0
4.8	4.0		
5.0	6.5		
5.0	6.5		
5.0	6.5		
5.0	6.5		
		5.3	9.0
5.5	10.0		
5.6	11.0		
		5.7	12.0
		6.0	13.5
6.0	13.5		
		6.1	15.0
6.2	16.0		
		6.3	17.5
		6.3	17.5
		6.7	19.5
		6.7	19.5
		7.2	21.0
7.7	22.0		
		8.0	23.0
8.4	24.5		
8.4	24.5		
		13.5	26.0
	$R_1 = 153.5$		$R_2 = 197.5$

10.2 Wilcoxon's matched pairs test

Key points This ranking test allows you to compare two categories or samples where the data are matched. This is the equivalent to the *t*- or *z*-test for matched parametric data.

Occasionally you may design an investigation in which you observe an item before and after a particular event or treatment. This type of data is called **matched** and has to be handled in a different way to that described in 10.1, because the 'before' measure is not independent of the

Matched data When two or more observations are recorded for one variable for each item

EXAMPLE 10.1 Enjoyment of consuming chocolate at two times in the day

A small randomized study was carried out where undergraduates were asked to eat a particular chocolate bar at 7a.m. and at 6p.m. On each occasion, the students were asked to rate their enjoyment on a scale of 1 (low) to 10 (high). The first three columns of Table 10.2 show the results.

Table 10.2. Enjoyment rating for consumption of chocolate bars at two times in the day with calculations for the Wilcoxon's matched pairs test

Student	Enjoyment rating at 7a.m.	Enjoyment rating at 6p.m.	Difference (d)	Absolute rank for d	Signed rank for d
1	2	2	0		
2	2	5	3	3	3
3	6	2	−4	5	−5
4	2	5	3	3	3
5	4	2	−2	1	−1
6	1	4	3	3	3
7	2	8	6	6	6

'after' measure. In Example 9.2, none of the periwinkles was measured more than once. However, in Example 10.1, each item is observed twice (before and after the treatment). In Wilcoxon's matched pairs test, this lack of independence is recognized and allowed for.

Wilcoxon's test allows you to compare two samples with matched or repeated measures. If, however, you have matched data (repeated measures) for more than two samples and on a scale other than an interval scale, then you should consider using Cochran's Q test (Sokal & Rohlf, 1994; Zar 2010).

10.2.1 **Using this test**

To use this test you:

1) Wish to examine the effect of one variable with two categories or samples.
2) Have non-parametric data.
3) Have matched data that can be ranked.
4) Have at least six pairs of observations where the difference between the observations is greater than zero.
5) Have two samples or categories of data that both have a similar-shaped distribution. For example, if one distribution is skewed to the left and the other to the right, then you should not use this test.

The data from Example 10.1 meet the criteria for using a Wilcoxon's matched pairs test in that there is one variable (time of day) and two samples (7a.m. and 6p.m.). Two measurements are made for each student. These two measurements are not independent of each other and the data are 'matched'. There are fewer than 30 pairs of observations so a matched z-test may not be used. The observations are measured on a scale of 1–10 and are therefore rankable and are non-parametric. There are six pairs of observations where the difference between the observations is greater than zero. The sample size is small so it is difficult to be certain whether criterion 5 is met; however, histograms for the samples indicate that the distributions are similar in shape.

10.2.2 **The calculation**

In this test, the calculation is dependent on the difference between each pair of observations and the sum of these differences. The general principles and a worked example are shown in Box 10.2. Some steps in the calculation are included in Table 10.2, columns 4–6.

BOX 10.2 How to carry out Wilcoxon's matched pairs test

GENERAL DETAILS	EXAMPLE 10.1
	This calculation is given in full in the Online Resource Centre. All values have been rounded to five decimal places.
1. Hypotheses to be tested H_0: There is no difference between the median of sample 1 and the median of sample 2. H_1: There is a difference between the median of sample 1 and the median of sample 2.	**1. Hypotheses to be tested** H_0: There is no difference between the median chocolate enjoyment scores at 7a.m. and at 6p.m. in a small group of undergraduates. H_1: There is a difference between the median chocolate enjoyment scores at 7a.m. and at 6p.m. in a small group of undergraduates.

BOX 10.2 Continued

2. **Have the criteria for using this test been met?**	2. **Have the criteria for using this test been met?** Yes (10.2.1).

3. **To work out $T_{calculated}$**
   i.   Decide which is sample 1 and which is sample 2.
	3. **To work out $T_{calculated}$**
	   i. Let the enjoyment scores recorded at 7a.m. be sample 1 and the enjoyment scores recorded at 6p.m. be sample 2.

ii.   Calculate the difference (d) between each pair of observations. Where the observation for sample 2 is larger than the observation for sample 1, the difference will be negative.
   iii.   Ignoring any d values where $d = 0$, rank the absolute d values.
      The term 'absolute' is explained in the glossary and in Appendix e. If you are not familiar with ranking values, see Box 5.3.
   iv.   Assign to each rank the positive or negative signs from the d value.
   v.   Sum all the ranks that are negative in sign. The absolute total is used.
   vi.   Sum all the ranks that are positive in sign.
   vii.   Examine the two values. The smaller value is the calculated value of T.
	ii. See Table 10.2, column 4.
	
	   iii. See Table 10.2, column 5. Ignore the 0 values. So $d = 2$ is the lowest value and has the rank 1; there are three values of $d = 3$ so these share the average rank 3, etc.
	
	   iv. See Table 10.2, column 6.
	
	   v. The sum of negative ranks = 1 + 5 = 6
	
	   vi. The sum of positive ranks = 3 + 3 + 3 + 6 = 15
	   vii. $T_{calculated} = 6$

4. **To find $T_{critical}$** See Appendix d, Table D13. This is a two-tailed test. Use a T table of critical values where N is the number of pairs of observations used to provide ranks (i.e. not those where $d = 0$).	4. **To find $T_{critical}$** For this example, one difference is zero, therefore $N = 6$ and $T_{critical} = 2$ at $p = 0.05$.
5. **The rule** When $T_{calculated}$ is equal to or less than $T_{critical}$, then you may reject the null hypothesis (H_0).	5. **The rule** In this example, $T_{calculated}$ (6) is greater than $T_{critical}$ (2) and therefore you do not reject the null hypothesis.
6. **What does this mean in real terms?**	6. **What does this mean in real terms?** There is no significant difference ($T = 6$, $p = 0.05$) between the median chocolate enjoyment scores at 7a.m. compared with 6p.m. in a small group of undergraduates.

Further examples relating to the topic including how to use statistical software are included in the Online Resource Centre.

@ **online resource centre**

10.3 One-way non-parametric ANOVA (Kruskal–Wallis test)

Key points This ranking test allows you to compare more than two categories or samples where the data are unmatched. This is the equivalent to the one-way ANOVA for parametric data. This test examines the general hypothesis 'There is no difference'. If the outcome from this test is significant, you may wish to follow it with the more specific test in section 10.4.

The acronym ANOVA stands for analysis of variance. Parametric ANOVAs are more powerful and flexible (Chapter 9). However, it is common in science to find yourself dealing with data that do not fit the criteria for a parametric ANOVA and cannot be normalized by transforming them (5.9). Under these circumstances, non-parametric tests may be used. In these tests, you will be comparing the distributions of the samples by using ranks and sums of ranks. The one-way non-parametric ANOVA was first described by Kruskal and Wallis (1952) and is usually called the Kruskal–Wallis test.

The test statistic for a Kruskal–Wallis test is often called either the K or H statistic. The calculated value of K or H is determined using the following equation:

$$K_{calculated} = \left(\Sigma \left(\frac{R^2}{n_s} \right) \times \frac{12}{N(N+1)} \right) - 3(N+1)$$

R^2 is found by summing the rank values. When there is little variation in the data, there will be many values with the same rank. If n is relatively large, the result can then be a negative value for K (H). In these circumstances, there is insufficient variation in the data compared with sample size to detect any difference between categories (samples).

The critical test statistic for this test can be found in two tables: H (Appendix d, Table D14) and chi-squared (Appendix d, Table D1). H is the exact test statistic and a specific critical value can be found in relation to a (the number of categories or samples) and n (the number of observations in a category). Where there are equal replicates, n will be the same for all categories, but where there are unequal replicates, n will differ between categories and the H value will depend on all the n values. By contrast, the chi-squared critical value is found in relation to p and the degrees of freedom. The latter is not dependent on sample size. With relatively large sample sizes, the critical values of H approximate to the critical values of chi-squared. We have included a table of critical

H values (Appendix d, Table D14) for equal replicates up to 25. Therefore, if n is less than 25 and there are equal numbers of observations in each category, we recommend using the H table. However, if you have unequal numbers of observations in each category or n is greater than 25, we suggest using the chi-squared distribution to find the critical value.

EXAMPLE 10.2 The density of *Bellis perennis* (daisy plants) at four different locations on the University of Worcester campus, 2002

Students in their first year at the University of Worcester randomly sampled four locations on campus using a 0.5m² quadrat. The number of *Bellis perennis* in each of eight quadrats in the four areas was recorded (Table 10.3). Is there a significant difference between these four areas?

Table 10.3. The number of *Bellis perennis* growing at four locations on the campus of the University of Worcester

Number of *B. perennis* on:

Cricket pitch	Lawn	Quadrangle	Rugby pitch
8	15	10	15
9	13	16	10
9	15	18	15
12	18	13	12
4	11	12	8
5	12	16	5
5	13	8	10
7	13	12	10
Median = 7.5	Median = 13.0	Median = 12.5	Median = 10.0

10.3.1 **Using this test**

To use this test you:

1) Wish to examine the effect of one variable with three or more categories or samples.
2) Have data that is non-parametric but can be ranked.
3) Must, if there are only three samples, have more than five observations per sample.
4) Do not need equal sample sizes.
5) You may not use this test if you have little variation in a relatively large sample size (see above). Where some tied ranks are present, then a correction should be calculated (Box 10.3).

 Q2 Do the data from Example 10.2 meet the criteria for a non-parametric one-way ANOVA (Kruskal–Wallis test)?

10.3.2 **The calculation**

The method for testing general hypotheses (Kruskal–Wallis test) is outlined in Box 10.3 and the calculation table (Table 10.4).

BOX 10.3 **How to carry out a one-way non-parametric ANOVA (Kruskal–Wallis test)**	
GENERAL DETAILS	**EXAMPLE 10.2**
	This calculation is given in full in the Online Resource Centre. All values have been rounded to five decimal places.
1. Hypotheses to be tested H_0: There is no difference between the medians of the samples. H_1: There is a difference between the medians of the samples.	**1. Hypotheses to be tested** H_0: There is no difference between the median density of *Bellis perennis* at the four locations (cricket pitch, lawn, rugby pitch, and quadrangle). H_1: There is a difference in the median density of *Bellis perennis* at the four locations (cricket pitch, lawn, rugby pitch, and quadrangle).
2. Have the criteria for using this test been met?	**2. Have the criteria for using this test been met?** Yes (A2, at end of chapter).
3. To work out $K_{calculated}$ i. Combine all the observations and arrange them in numerical order. Assign the appropriate rank to each observation.	**3. To work out $K_{calculated}$** i–iv. See Table 10.4.

BOX 10.3 Continued

ii. Sum the ranks for each sample: $\Sigma r = R_s$

iii. Record n_s, the number of observations in each sample.

iv. Square each R value (R^2). Divide each R^2 by its n_s value: R^2/n_s

v. Add all the R^2/n_s values together: $\Sigma(R^2/n_s)$

v. $\Sigma \dfrac{R^2}{n_s} = 392 + 4278.125 + 3549.0313 + 1755.2813$

$= 9974.4376$

vi. Add all the n_s values together: $\Sigma n_s = N$

vi. $\Sigma n_s = N = 8 + 8 + 8 + 8 = 32$

vii. The test statistic K is calculated as:

$$K_{calculated} = \left(\Sigma \left(\frac{R^2}{n_s} \right) \times \frac{12}{N(N+1)} \right) - 3(N+1)$$

(If you have a negative K value at this point, see explanation in 10.3.)

vii. $K_{calculated} = \left(9974.4376 \times \dfrac{12}{32(32+1)} \right) - 3(32+1) = 14.345882$

4. When there are tied ranks

If there are no tied ranks in your data, you may proceed to the next step (5). When there are tied ranks, a correction may be applied.

4. When there are tied ranks

i. First record the numbers of each tied rank value (t). Then calculate $T^{kw} = \Sigma(t^3 - t)$. Each of these calculations is straightforward but tedious so we have calculated a number of these for you:

t	2	3	4	5
$(t^3 - t)$	6	24	60	120
t	6	7	8	9
$(t^3 - t)$	210	336	504	720

i. For our example the numbers of each tied rank (t) are:

Observations	Tied ranks	How many observations with this rank (t)
5	3	3
8	7	3
9	9.5	2
10	12.5	4
12	18	5
13	22.5	4
15	26.5	4
16	29.5	2
18	31.5	2

Then calculate $T^{kw} = \Sigma(t^3 - t)$.

$$T^{kw} = \left(3^3 - 3\right) + \left(3^3 - 3\right) + \left(2^3 - 2\right) + \left(4^3 - 4\right) + \left(5^3 - 5\right) +$$
$$\left(4^3 - 4\right) + \left(4^3 - 4\right) + \left(2^3 - 2\right) + \left(2^3 - 2\right)$$
$$= 24 + 24 + 6 + 60 + 120 + 60 + 60 + 6 + 6 = 366$$

BOX 10.3 Continued	
ii. The correction factor is $$C = 1 - \left(\frac{T^{kw}}{(N^3 - N)} \right)$$ where N is the total number of observations.	ii. The correction factor for this example is: $$C = 1 - \left(\frac{366}{(32^3 - 32)} \right) = 1 - \left(\frac{366}{32768 - 32} \right)$$ $$= 1 - \left(\frac{366}{32736} \right) = 1 - 0.01118 = 0.98882$$
iii. The corrected K value (K_C) is the ratio $K_C = K/C$ When a corrected K value has been calculated, this is then the K value used in the hypothesis testing.	iii. $K_C = 14.34588/0.98882 = 14.50809$
5. To find $K_{critical}$ $K_{critical}$ may be found in one of two tables depending on sample size and whether there are equal numbers of observations in each category. i. If n is less than 25 and there are equal numbers of observations in each category, use Appendix d, Table D14. Locate $H_{critical}$ for a, the number of categories (samples), and n, the number of observations in each category. This is a two-tailed test. ii. If you have unequal numbers of observations in each category or n is greater than 25, use Appendix d, Table D1. Using a table of critical values for chi-squared and $p = 0.05$. The degrees of freedom (v) = number of categories (samples) –1. This is a two-tailed test.	**5. To find $K_{critical}$** i. In our example, there are four categories (cricket pitch, lawn, rugby pitch, and quadrangle), so $a = 4$. There are eight observations in each category so we may use the H table of critical values. $H_{critical} = 7.534$ at $p = 0.05$ ii. If we wished to use the chi-squared table to find our critical value, then $v = 4 - 1 = 3$ and χ^2 at $p = 0.05 = 7.81$ Clearly there is little difference between these two values although the chi-squared value tends to be higher and therefore more conservative when used in the hypothesis testing (i.e. some significant differences will not be detected when using the chi-squared value compared with the H value).
6. The rule If the calculated value of K is greater than the critical value of H or χ^2, then you may reject the null hypothesis.	**6. The rule** $K_{calculated}$ (14.508) is greater than $H_{critical}$ (7.534) and we may therefore reject the null hypothesis. In fact, at $p = 0.01$, $H_{critical} = 10.42$. Therefore, we can reject the null hypothesis at $p = 0.01$.
7. What does this mean in real terms?	**7. What does this mean in real terms?** There is a highly significant difference ($K = 14.508$, $p < 0.01$) between the median densities of *Bellis perennis* at the four locations (cricket pitch, lawn, rugby pitch, and quadrangle).

Further examples relating to the topic including how to use statistical software are included in the Online Resource Centre.

online resource centre

Table 10.4. Calculation for one-way non-parametric ANOVA (Kruskal–Wallis test) for Example 10.2: The density of *Bellis perennis* growing at four sites on the campus of the University of Worcester

Cricket pitch		Lawn		Quadrangle		Rugby pitch	
Number	Rank	Number	Rank	Number	Rank	Number	Rank
8	7.0	15	26.5	10	12.5	15	26.5
9	9.5	13	22.5	16	29.5	10	12.5
9	9.5	15	26.5	18	31.5	15	26.5
12	18.0	18	31.5	13	22.5	12	18.0
4	1.0	11	15.0	12	18.0	8	7.0
5	3.0	12	18.0	16	29.5	5	3.0
5	3.0	13	22.5	8	7.0	10	12.5
7	5.0	13	22.5	12	18.0	10	12.5
R	56.0		185.0		168.5		118.5
n	8		8		8		8
R^2	3136.00		34225.00		28392.25		14042.25
R^2/n	$R^2/n = \dfrac{3136}{8}$ $= 392.00$		4278.125		3549.0313		1755.2813
Means of ranks: R/n	7.00		23.12		21.0625		14.8125

10.4 Multiple comparisons following a non-parametric one-way ANOVA

Key points This test allows you to compare the means of ranks in pairs for all possible combinations. This provides a more specific evaluation of your data than just a Kruskal–Wallis test. This is the equivalent to Tukey's test for parametric data.

The one-way non-parametric ANOVA tests the general hypothesis 'There is no difference between the categories'. If this is significant, it is useful to test more specific hypotheses. There are several ways to do this. In this section, we examine a multiple comparisons test that allows the comparison of two of the categories at a time. This test finds the mean of the rank values and compares pairs of means using a Q statistic.

The Kruskal–Wallis test followed by a multiple comparisons test is the non-parametric equivalent of the parametric one-way ANOVA and Tukey's test. There are alternative approaches to making more specific tests of hypotheses following a Kruskal–Wallis test. One proposed by Meddis (1984) (see also Barnard *et al.*, 2001) is included in the Online Resource Centre.

online
resource
centre

10.4.1 **Using this test**

To use this test you:

1) Have met the criteria for a Kruskal–Wallis test.
2) Do not need equal numbers of observations in each category (sample).

To illustrate this multiple comparisons test, we will use Example 10.2 and the Kruskal–Wallis test in Box 10.3. We know, therefore, that the criteria for using a Kruskal–Wallis test have been met by the data.

10.4.2 **The calculation**

The method for making these multiple comparisons after a Kruskal–Wallis test is outlined in Box 10.4 and the calculation table (Table 10.5).

BOX 10.4 How to carry out multiple comparisons after a one-way non-parametric ANOVA (Kruskal–Wallis test)

GENERAL DETAILS	EXAMPLE 10.2
	This calculation is given in full in the Online Resource Centre. All values have been rounded to five decimal places.
1. **Hypotheses to be tested** H_0: There is no difference between the means of the ranks of the samples. H_1: There is a difference between the means of the ranks of the samples.	1. **Hypotheses to be tested** As in Tukey's test (Chapter 9), you could write specific hypotheses for each comparison of means for example: H_0: There is no difference between the means of the ranks for the density of *Bellis perennis* on the rugby pitch compared with the quadrangle. H_1: There is a difference between the means of the ranks for the density of *Bellis perennis* on the rugby pitch compared with the quadrangle. However, common sense suggests that where you are making many comparisons using this test, then it is sensible to report the overall hypotheses only.
2. **Have the criteria for using this test been met?**	2. **Have the criteria for using this test been met?** Yes (10.3.1 and 10.4.1).
3. **To work out $Q_{calculated}$** i. Calculate the difference between each pairwise combination of means using the mean of the ranks.	3. **To work out $Q_{calculated}$** i. The mean of ranks for the cricket pitch is 7.00 (Table 10.5). The mean of ranks for the lawn is 23.12. The difference between these two values is 16.12. In total, six comparisons between means can be made. Table 10.5 gives all the differences between these pairs of means.

BOX 10.4 Continued

ii. For each pair of means, calculate the standard error of the difference of the means (SE).

a. If there are few tied ranks, the calculation is:

$$SE = \sqrt{\left(\left(\frac{N(N+1)}{12}\right) \times \left(\frac{1}{n_a} + \frac{1}{n_b}\right)\right)}$$

where N is the total number of observations, i.e. Σn and n_a is the number of observations in the first of the two categories being compared, and n_b is the number of observations in the second of the two categories being compared.

If n is the same for all categories (samples), then this value will be the same for all pairwise combinations.

b. If there are tied ranks, then the calculation for the standard error is modified.

$$SE = \sqrt{\left(\left(\left(\frac{N(N+1)}{12}\right) - \left(\frac{T^{kw}}{12(N-1)}\right)\right) \times \left(\left(\frac{1}{n_a}\right) + \left(\frac{1}{n_b}\right)\right)\right)}$$

where N is the total number of observations, i.e. Σn, n_a is the number of observations in the first of the two categories being compared, n_b is the number of observations in the second of the two categories being compared, and T^{kw} is the value calculated during the Kruskal–Wallis test (Box 10.3), reflecting the numbers of tied values in the data. Again, if n is the same for all categories (samples), then this value will be the same for all pairwise combinations.

iii. Divide the difference in the means by its SE. This is the $Q_{calculated}$ value for the comparison between that particular pair of means.

4. To find $Q_{critical}$

See Appendix d, Table D15. $Q_{critical}$ may be found using a, the number of categories or samples and p. This critical value for Q will be the same for each comparison of means.

ii. In our example there were tied ranks and from Box 10.3 $T^{kw} = 366$, $N = 8 + 8 + 8 + 8 = 32$, all n values for this example are the same as there are 8 observations in each category so we only need to calculate the standard error of the difference once and the value will be the same for all subsequent comparisons of means.

Therefore, using the adjusted equation

$$SE =$$

$$\sqrt{\left(\left(\left(\frac{32(32+1)}{12}\right) - \left(\frac{366}{12(32-1)}\right)\right) \times \left[\left(\frac{1}{8}\right) + \left(\frac{1}{8}\right)\right]\right)}$$

$$= \sqrt{\left[(88 - 0.98387) \times (0.125 + 0.125)\right]}$$

$$= \sqrt{\left[87.01613 \times 0.25\right]} = \sqrt{21.75403}$$

$$= 4.66412$$

iii. For the comparison between the means of the ranks from the cricket pitch and the lawn:

$Q_{calculated} = 16.12/4.66412 = 3.45617$

The $Q_{calculated}$ values for the other comparison are given in Table 10.5.

4. To find $Q_{critical}$

In our example, there are four categories (cricket pitch, lawn, rugby pitch, and quadrangle), so $a = 4$. Therefore:

$Q_{critical} = 2.639$ at $p = 0.05$

BOX 10.4 Continued

5. The rule

If the calculated value of Q is greater than the critical value of Q, then you may reject the null hypothesis.

5. The rule

When comparing the means of ranks for the cricket pitch and the lawn. $Q_{calculated}$ (3.456) is greater than $Q_{critical}$ (2.639) and we may therefore reject the null hypothesis. In fact, at $p = 0.01$, $Q_{critical} = 3.144$. Therefore, we can reject the null hypothesis at this higher level of significance. At $p = 0.001$, $Q_{critical} = 3.765$, so the null hypothesis cannot be rejected at this level of significance. All pairwise combinations have been tested against the $Q_{critical}$ value. The results are shown in Table 10.5.

6. What does this mean in real terms?

6. What does this mean in real terms?

There is a highly significant difference ($Q = 10.868$, $0.01 > p > 0.001$) between the means of the ranks of the densities of *Bellis perennis* on the cricket pitch and the lawn.

Having tested all the pairwise combinations of means, it can be seen that there is only one other significant difference ($0.05 > p > 0.01$) between the density of daisies on the cricket pitch and the quadrangle.

Further examples relating to the topic including how to use statistical software are included in the Online Resource Centre.

@ **online resource centre**

Table 10.5. Calculation of differences between the means of ranking values ($\bar{x}_{ranks}$) and Q values for samples from Example 10.2: The density of *Bellis perennis* at four locations at the University of Worcester

		Rugby pitch $\bar{x}_{ranks} = 14.8125$	Quadrangle $\bar{x}_{ranks} = 21.0625$	Lawn $\bar{x}_{ranks} = 23.12$	Cricket pitch $\bar{x}_{ranks} = 7.00$
Cricket pitch $\bar{x}_{ranks} = 7.00$	Difference between means				
	Q				
Lawn $\bar{x}_{ranks} = 23.12$	Difference between means				16.12
	Q				16.12/4.66412 = 3.45617**
Quadrangle $\bar{x}_{ranks} = 21.0625$	Difference between means			2.0575	14.0625
	Q			2.0575/4.66412 = 0.41133	14.0625/4.66412 = 3.01504*
Rugby pitch $\bar{x}_{ranks} = 14.8125$	Difference between means		6.25	8.3075	7.8125
	Q		1.34002	1.78115	1.67502

*$p \leq 0.05$; **$p \leq 0.01$.

The outcome from a multiple comparisons test allows you to be more specific in your interpretation of the significant results found from a Kruskal–Wallis test. In our example, it would lead us to examine the microhabitat particularly of the cricket pitch to enable us to identify factors that may explain the observations.

There are a number of methods that can be used for reporting results from statistical tests. We review these in Chapter 11 (11.8.3). Here, we have placed the significant values in Table 10.5 in bold and indicated the level of significance using the asterisk system (11.8.3).

10.5 Two-way non-parametric ANOVA

Key points This ranking test allows you to examine the effect of two variables each with two or more categories. The design needs to be orthogonal and the number of observations in each category needs to be the same. This non-parametric test is equivalent to the two-way parametric ANOVA (Chapter 9). This test examines three pairs of general hypotheses. For a more specific evaluation, it may be followed by a multiple comparisons test (10.6).

This test is an extension of the one-way ANOVA (Meddis, 1984; Barnard *et al.*, 2001), which allows you to examine the effect of two variables and their interaction. As explained in Chapter 9 in relation to the comparable two-way parametric ANOVA, when you compare two treatment variables in this way, you will test three pairs of hypotheses. The first test examines the effect of variable 1, the second the effect of variable 2, and the third the interaction between the two variables. All these tests are carried out within one calculation so you reduce the likelihood of generating a Type I error. (The term 'Type 1 error' is explained in the glossary). We illustrate the two-way non-parametric ANOVA using data from Example 10.3 (Table 10.6).

This test uses the same equation for calculating K as that used in the one-way non-parametric ANOVA. Therefore, it is also possible to obtain a negative $K_{calculated}$ value when N is large compared with the variation in the data. If this occurs, you should consider using the Scheirer–Ray–Hare test (10.7).

10.5.1 Using this test

To use this test you:

1) Wish to test the effect of two variables each with at least two categories or samples.
2) Have an orthogonal design.

3) Have non-parametric data that can be ranked.

4) Have equal numbers of observations in each sample.

5) You may not use this test if you have little variation in a relatively large sample size. (Many tied values in your data may result in a negative $K_{calculated}$ value.) Where some tied values are present, a correction may be applied.

The data from Example 10.3 meet the criteria for using this test as the unit of measurement is the 'number of daisy plants', which is a non-parametric measure that can be ranked. We wish to examine the effect of two variables (location and fertilizer treatment), each with two categories. Each sample has four replicates (observations).

10.5.2 The calculation

As in the two-way parametric ANOVA for an orthogonal design, three pairs of hypotheses are tested: the differences between the medians due

---

EXAMPLE 10.3 The effect of fertilizer treatment on the density of *Bellis perennis* (daisy) at two locations on the University of Worcester campus

At the University of Worcester, the lawn and cricket pitch were divided and half of each plot was treated with fertilizer. Students extended their original study (Example 10.2) and, using stratified random sampling with quadrats, recorded the numbers of *Bellis perennis* in these four locations (cricket pitch with fertilizer, cricket pitch without fertilizer, lawn with fertilizer, and lawn without fertilizer) (Table 10.6). The data are organized into blocks and these will be referred to as blocks A–D.

Table 10.6. The density of *Bellis perennis* (daisies) in two locations, with or without fertilizer treatments at the University of Worcester

	Number of *B. perennis* in each quadrat	
	Lawn	Cricket pitch
Fertilizer	8	15
	9 (A)	13 (B)
	9	15
	12	18
Median	9	15
No fertilizer	4	11
	5	12
	5 (C)	13 (D)
	7	13
Median	5	12

to treatment 1, the differences between the medians due to treatment 2, and the interaction between the two treatment variables. The idea of an interaction is explained in Chapter 9 (9.9). The following set of boxes and calculation tables is arranged to take you through testing one pair of hypotheses at a time:

a. Is there a difference between the columns (i.e. locations)? Go to Box 10.5(A) and Table 10.7.

b. Is there a difference between the rows (i.e. fertilizer treatment)? Go to Box 10.5(B) and Table 10.8.

c. Is there an interaction between variable 1 and variable 2, i.e. is there an interaction between the locations and the effect of the fertilizer? Go to Box 10.5(C) and Table 10.9.

Our analysis of these data shows us that only the location is a significant factor in the density of daisies in this experiment. Neither the fertilizer nor the interaction between fertilizer and location had a significant effect on the median density of daisies. However, this has only been a test of general hypotheses. To be able to draw more explicit conclusions about the trends in the data, we should consider using a multiple comparisons test (10.6).

BOX 10.5A How to carry out a two-way non-parametric ANOVA test: columns	
GENERAL DETAILS	**EXAMPLE 10.3**
	This calculation is given in full in the Online Resource Centre. All values have been rounded to five decimal places.
1. Hypotheses to be tested H_0: There is no difference between the median values for samples given treatment 1 (columns). H_1: There is a difference between the median values for samples given treatment 1 (columns).	**1. Hypotheses to be tested** H_0: There is no difference between the median density of *Bellis perennis* on the lawn and on the cricket pitch. H_1: There is a difference between the median density of *Bellis perennis* on the lawn and on the cricket pitch.
2. Have the criteria for using this test been met?	**2. Have the criteria for using this test been met?** Yes (10.5.1).
3. To work out $K_{calculated}$ i. Combine all the observations and arrange them in numerical order. Assign the appropriate rank to each observation. ii. Sum the ranks for all values in column 2 (R_{c1}) and then all values for column 3. Repeat for each column of data (R_{c2}) etc. iii. Sum the number of observations for column 2 (n).	**3. To work out $K_{calculated}$** i–iii. See Table 10.7. There are two columns, 'cricket pitch' and 'lawn', and eight observations in each column.

BOX 10.5A Continued

iv. Using the column totals.

For column 2, square the sum of ranks (R^2_{c1}) and divide by the number of observations in that column (R^2_{c1}/n).

In a similar way, calculate R^2_{c2}/n for column 3.

Continue for each column of data.

Add together all these R^2/n values: $\Sigma R^2/n$)

Add all the n_c values together (N).

v. The test statistic K is calculated as:

$$K_{\text{calculated columns}} = \left(\Sigma\left(\frac{R^2}{n_s}\right)\times\frac{12}{N(N+1)}\right)-3(N+1)$$

iv.

$$\frac{R^2_{c1}}{n_{c1}} = 175.78125$$

$$\frac{R^2_{c2}}{n_{c2}} = 1212.7813$$

$$\Sigma(R^2/n) = 175.78125 + 1212.7813 = 1388.5626$$

$$N = 8 + 8 = 16$$

v. $K_{\text{calculated columns}}$

$$= \left[\left(1388.5626\times\left(\frac{12}{16(16+1)}\right)\right)-3(16+1)\right]$$

$$= 10.26012$$

---

4. Where there are tied ranks

If there are no tied ranks in your data, you may proceed to the next step (5). When there are tied ranks, a correction may be applied.

i. First record the number of each tied rank (t). Then calculate $T^{kw} = \Sigma(t^3 - t)$.

Each of these calculations is straightforward but tedious so we have calculated a number of these for you:

t	2	3	4	5
$(t^3 - t)$	6	24	60	120
t	6	7	8	9
$(t^3 - t)$	210	336	504	720

4. Where there are tied ranks

i. For our example, the numbers of each tied rank (t) are:

Observations	Tied ranks	How many observations with this rank (t)
5	2.5	2
9	6.5	2
12	9.5	2
13	12.0	3
15	14.5	2

Then calculate $T^{kw} = \Sigma(t^3 - t)$.

$$T^{kw} = (2^3 - 2)+(2^3 - 2)+(2^3 - 2)+$$

$$(3^3 - 3)+(2^3 - 2)= 6+6+6+24+6 = 48$$

ii. The correction factor is

$$C = 1-\left(\frac{T^{kw}}{(N^3 - N)}\right)$$

where N is the total number of observations

ii. The correction factor for this example is

$$C = 1-\left(\frac{48}{(16^3 - 16)}\right)=1-\left(\frac{48}{4096 - 16}\right)$$

$$= 1-\frac{48}{4080}=1-0.01177 = 0.98824$$

BOX 10.5A Continued	
iii. The corrected K value (K_c) is the ratio $K_{C \text{ columns}} = K_{\text{columns}}/C$ (When a corrected K value has been calculated, this is then the K value used in the hypothesis testing.)	iii. $K_{C \text{ columns}} = 10.26012/0.98824 = 10.38226$
5. To find K_{critical} See Appendix d, Table D1. Using a table of critical values for χ^2 and $p = 0.05$, the degrees of freedom (v) = number of columns − 1.	**5. To find K_{critical}** As $v = 2 − 1 = 1$ $K_{\text{critical columns}} = 3.84$ at $p = 0.05$
6. The rule If the calculated value of K is greater than the critical value of K, then you may reject the null hypothesis.	**6. The rule** $K_{\text{calculated columns}}$ (10.38) is greater than K_{critical} (3.84). You may therefore reject the null hypothesis (H_0). In fact, at $p = 0.01$, $K_{\text{critical}} = 6.64$ and at $p = 0.001$, $K_{\text{critical}} = 10.83$. Therefore, we can reject the null hypothesis at $p = 0.01$, but not at $p = 0.001$.
7. What does this mean in real terms? NOW GO ON TO BOX 10.5(B).	**7. What does this mean in real terms?** There is a highly significant difference ($K = 10.38$, $0.01 > p > 0.001$) in the median density of *Bellis perennis* on the lawn and on the cricket pitch.

Table 10.7. Calculation table for two-way non-parametric ANOVA test (columns) for Example 10.3: The number of *Bellis perennis* per quadrat in two locations (cricket pitch and lawn)

	Cricket pitch		Lawn	
	Number	Rank	Number	Rank
Fertilizer	8	5.0	15	14.5
	9	6.5	13	12.0
	9	6.5	15	14.5
	12	9.5	18	16.0
No fertilizer	4	1.0	11	8.0
	5	2.5	12	9.5
	5	2.5	13	12.0
	7	4.0	13	12.0
Totals for the columns:				
n	8		8	
R		37.5		98.5
R^2		1406.25		9702.25
$\dfrac{R_c^2}{n_c}$		175.78125		1212.7813

BOX 10.5B How to carry out a two-way non-parametric ANOVA test: rows

GENERAL DETAILS	EXAMPLE 10.3
	This calculation is given in full in the Online Resource Centre. All values have been rounded to five decimal places.
1. Hypotheses to be tested H_0: There is no difference in the median values for samples in response to treatment 2 (rows). H_1: There is a difference in the median values for samples in response to treatment 2 (rows).	**1. Hypotheses to be tested** H_0: There is no difference in the median density of *Bellis perennis* between the areas treated with fertilizer and those not treated. H_1: There is a difference in the median density of *Bellis perennis* between the areas treated with fertilizer and those not treated.
2. Have the criteria for using this test been met?	**2. Have the criteria for using this test been met?** Yes (10.5.1).

3. To work out $K_{calculated}$
 i. Sum the ranks for each row (R_{r1}) and (R_{r2}), etc.
 	
 ii. Sum the number of observations for a row (n_r).
 iii. From the row totals.
   a. For row 1, square the sum of ranks (R^2_{r1}) and divide by the number of observations in that first row $\left(\dfrac{R^2_{r1}}{n_{r1}}\right)$.
     In a similar way, calculate $\dfrac{R^2_{r2}}{n_{r2}}$ for row 2.
     Continue for each row of data.
   b. Add together all these R^2/n values: $\Sigma(R^2/n)$
   c. Add all the n_r values together (N).
 	
 iv. The test statistic K is calculated as before.
 	
 $$K_{calculated\ rows} = \left(\Sigma\left(\dfrac{R^2}{n}\right) \times \dfrac{12}{N(N+1)}\right) - 3(N+1)$$
	3. To work out $K_{calculated}$
	 i, ii. See Table 10.8.
	 In our example, there are two rows, 'fertilizer' and 'no fertilizer', each with eight observations in the row.
	
	 iii. a.
	
	 $$\dfrac{R^2_{r1}}{n_{r1}} = 892.53125$$
	
	 $$\dfrac{R^2_{r2}}{n_{r2}} = 331.53125$$
	
	 b. $\Sigma\dfrac{R^2}{n} = 892.53125 + 331.53125$
	
	    $= 1224.0625$
	 c. $\Sigma n_r = N = 8 + 8 = 16$
	
	 iv. $K_{calculated\ rows}$
	
	 $$= \left(1224.0625 \times \dfrac{12}{16(16+1)}\right) - 3(16+1)$$
	
	 $= 3.00260$

4. Where there are tied ranks A correction may be used where there are tied ranks similar to that outlined in Box 10.5(A). The correction factor (C) will be the same and $K_{C\ rows} = K_{rows}/C$	**4. Where there are tied ranks** From Box 10.5(A), $C = 0.98824$. $K_{C\ rows} = 3.00260/0.98824 = 3.03833$
5. To find $K_{critical}$ See Appendix d, Table D1. Using a table of critical values for χ^2 and $p = 0.05$, the degrees of freedom (v) = number of rows −1	**5. To find $K_{critical}$** As $v = 2 - 1 = 1$, $K_{critical\ rows} = 3.84$ at $p = 0.05$.

BOX 10.5B Continued

6. The rule	6. The rule
If the calculated value of K is greater than the critical value of K, then you may reject the null hypothesis (H_0).	$K_{calculated\ rows}$ (3.04) is less than $K_{critical}$ (3.84), so you may not reject the null hypothesis (H_0).
7. What does this mean in real terms?	7. What does this mean in real terms?
	There is no significant difference ($K = 3.04$, $p = 0.05$) in the median density of *Bellis perennis* in areas treated by fertilizer and those not treated.
NOW GO ON TO BOX 10.5(C).	

Table 10.8. Calculation table for two-way non-parametric ANOVA test (rows) for Example 10.3: The number of *Bellis perennis* per quadrat with or without fertilizer treatment

	Cricket pitch		Lawn		Total for rows	
	Number	Rank	Number	Rank		
Fertilizer	8	5.0	15	14.5	n	8
	9	6.5	13	12.0	R	84.5
	9	6.5	15	14.5	R^2	7140.25
	12	9.5	18	16.0	$\dfrac{R^2_{r1}}{n_{r1}} = 892.53125$	
No fertilizer	4	1.0	11	8.0	n	8
	5	2.5	12	9.5	R	51.5
	5	2.5	13	12.0	R^2	2652.25
	7	4.0	13	12.0	$\dfrac{R^2_{r2}}{n_{r2}} = 331.53125$	

BOX 10.5C How to carry out a two-way non-parametric ANOVA test: interaction

GENERAL DETAILS	**EXAMPLE 10.3**
	This calculation is given in full in the Online Resource Centre. All values have been rounded to five decimal places.
1. **Hypotheses to be tested** H_0: There is no difference between the median values of all samples due to an interaction between treatment variables 1 and 2. H_1: There is a difference between the median values of all samples due to an interaction between treatment variables 1 and 2.	1. **Hypotheses to be tested** H_0: There is no difference between the median density of *Bellis perennis* due to an interaction between location and fertilizer treatment. H_1: There is a difference between the median density of *Bellis perennis* due to an interaction between location and fertilizer treatment.

BOX 10.5C Continued

2. Have the criteria for using this test been met?	**2. Have the criteria for using this test been met?** Yes (10.5.1).
3. To work out $K_{calculated}$ i. For each block, calculate R, R^2, n, and R^2/n. ii. Add all the R^2/n values together: $\Sigma(R^2/n)$ iii. Add all the n values from the blocks: $\Sigma n = N$ iv. $K_{calculated\ total}$ is calculated as before. $K_{calculated\ total} =$ $\left(\Sigma\left(\dfrac{R^2}{n_s}\right)\times\dfrac{12}{N(N+1)}\right)-3(N+1)$ v. Where there are tied ranks, $K_{calculated\ total}$ may be corrected using the same correction term calculated in Box 10.5(A) and where $K_{C\ total} = K_{total}/C$ vi. You now have three values for K: $K_{columns}$ (Box 10.5A), K_{rows} (Box 10.5B), and K_{total} (step iv. above). Using these values, calculate $K_{interaction}$: $K_{interaction} = K_{total} - K_{rows} - K_{columns}$	**3. To work out $K_{calculated}$** i. See Table 10.9. There are four blocks A–D, each with four observations. ii. $\Sigma(R^2/n)$ $= 189.0625 + 821.25 + 25.00 + 430.5625$ $= 1456.875$ iii. $N = 4 + 4 + 4 + 4 = 16$ iv. $K_{calculated\ total} = \left(1456.875\times\dfrac{12}{16(16+1)}\right)-3(16+1)$ $= 13.27380$ v. From Box 10.5(A) $C = 0.98824$ So $K_{C\ total} = K_{total}/K_{C\ total}$ $= 13.27380/0.98824 = 13.43176$ vi. $K_{columns} = 10.38226$ $K_{rows} = 3.03833$ $K_{total} = 13.43176$ $K_{interaction} = 13.43176 - 3.03833 - 10.38226$ $= 0.01117$
4. To find $K_{critical}$ The degrees of freedom (v) are v for K_{total} = number of blocks – 1 v for K_{rows} = number of rows – 1 v for $K_{columns}$ = number of columns – 1 v for $K_{interaction} = v_{total} - v_{rows} - v_{columns}$ See Appendix d, Table D1. Use the v for $K_{interaction}$ and p = 0.05 to locate the critical value of K in a χ^2 table.	**4. To find $K_{critical}$** v for K_{total} = 4 – 1 = 3 v for $K_{fertilizer}$ = 2 – 1 = 1 v for $K_{location}$ = 2 – 1 = 1 v for $K_{interaction}$ = 3 – 1 – 1 = 1 Therefore, the critical value for $K_{interaction}$ at p = 0.05 and v = 1 is 3.84.
5. The rule If the calculated value of K is greater than the critical value of K, then you may reject the null hypothesis (H_0).	**5. The rule** $K_{calculated\ interaction}$ (0.01) is less than $K_{critical}$ (3.84), so you may not reject the null hypothesis (H_0).
6. What does this mean in real terms?	**6. What does this mean in real terms?** There is no significant interaction ($K = 0.011$, $p = 0.05$) between the fertilizer treatment and location in the median density of *Bellis perennis* at the University of Worcester.

Further examples relating to the topic including how to use statistical software are included in the Online Resource Centre.

@ **online resource centre**

Table 10.9. Calculation table for a two-way non-parametric ANOVA test (interaction) for Example 10.3: The number of *Bellis perennis* per quadrat in two locations with or without fertilizer treatment

| | Location of *Bellis perennis* | | | |
| | Cricket pitch | | Lawn | |
	Number	Rank	Number	Rank
With fertilizer	BLOCK A		BLOCK B	
	8	5.0	15	14.5
	9	6.5	13	12.0
	9	6.5	15	14.5
	12	9.5	18	16.0
	n	4	n	4
	R	27.5	R	57.0
	R^2	756.25	R^2	3249.00
	R^2/n	189.0625	R^2/n	812.2500
No fertilizer	BLOCK C		BLOCK D	
	4	1.0	11	8.0
	5	2.5	12	9.5
	5	2.5	13	12.0
	7	4.0	13	12.0
	n	4	n	4
	R	10.0	R	41.5
	R^2	100.00	R^2	1722.25
	R^2/n	25.0000	R^2/n	430.5625

10.6 Multiple comparisons following a two-way non-parametric ANOVA

Key points This test allows you to compare the means of the ranks in all possible pairwise combinations. It allows a more specific evaluation of the data than just the two-way non-parametric ANOVA outlined in 10.5. This is the equivalent of Tukey's test for parametric data following a two-way parametric ANOVA (Chapter 9).

If a two-way non-parametric ANOVA has been carried out and there is a significant difference in at least one of the outcomes from Boxes 10.5(A–C), then you may wish to carry out a multiple comparisons test to allow you to identify more specifically where the differences lie in your data. The multiple comparisons test we outlined in 10.4 can be extended to cover the comparisons for a two-way non-parametric ANOVA. In this case, there are more pairwise comparisons that can be made as each block counts as one category to be compared with every other different category.

This is the non-parametric equivalent to the two-way parametric ANOVA followed by Tukey's test.

In Example 10.3 (Boxes 10.5A–C), we can see that in fact only the location had a significant effect on the density of *Bellis perennis*. There are only two locations in this comparison, and we can see by looking at the data that the density of *Bellis perennis* is greater on the lawn than on the cricket pitch. There is therefore little reason for carrying out a formal multiple comparisons test on this data. We have therefore included Example 10.4 to illustrate the use of a multiple comparisons test after a two-way non-parametric ANOVA.

 online resource centre

A two-way non-parametric ANOVA was carried out on the data from Example 10.4 This calculation is shown in the Online Resource Centre. The outcome of the analysis was that there was no significant difference

---

EXAMPLE 10.4 Retrieval of pollen grains from different fabrics

In an investigation into forensic techniques, an undergraduate examined the effectiveness and reliability of pollen retrieval from a range of difference fabrics using the pollen from two species. The numbers of pollen grains retrieved are given in Table 10.10.

Table 10.10. The number of pollen grains retrieved from four different fabrics (denim, fleece, viscose, and polycotton) using pollen from two species (*Phleum pratense* and *Ambrosia artemisiifolia*).

	Denim		Fleece		Viscose		Polycotton	
	Number of pollen grains	Rank	Number of pollen grains	Rank	Number of pollen grains	Rank	Number of pollen grains	Rank
A. artemisiifolia	8998	30	6670	25	6551	24	4858	23
	3019	17	8012	27	7557	26	12099	31
	3839	20	12108	32	8311	29	4754	22
	3354	18	8097	28	3613	19	4555	21
$\Sigma r = R$		85		112		98		97
n		4		4		4		4
R/n		21.25		28.00		24.50		24.25
P. pratense	2886	16	2316	15	1002	10	628	7
	402	2	1804	14	712	9	1104	11
	529	5	1018	12	392	1	509	4
	554	6	1513	13	506	3	698	8
$\Sigma r = R$		29		54		23		30
n		4		4		4		4
R/n		7.25		13.5		5.75		7.5

as a result of the fabrics, nor was there a significant interaction between the two variables (fabrics and species). However, there was a highly significant difference between the species in the pollen retrieved from the fabric ($K = 23.27$, $p = 0.001$). Again, there are only two species but there are four fabrics so we are still interested in obtaining more information about this significant difference. To obtain this, we can use a multiple comparisons test.

10.6.1 Using this test

To use this test you:

1) Have met the criteria for a Kruskal–Wallis test.
2) Do not need equal numbers of observations in each category (sample).

To illustrate this multiple comparisons test, we will use Example 10.4. The data in this example meet the criteria for using a two-way non-parametric ANOVA in that we wish to test the effect of two variables (fabric and species). The design is orthogonal (every combination of one variable has been compared with the other variable). The data are non-parametric (numbers of pollen grains) and can be ranked. There are four observations in each category. There are no tied values so we do not expect to obtain a negative K value.

10.6.2 The calculation

This multiple comparisons test is outlined in Box 10.6 and the calculation table (Table 10.11).

BOX 10.6 How to carry out multiple comparisons after a two-way non-parametric ANOVA	
GENERAL DETAILS	**EXAMPLE 10.4**
	This calculation is given in full in the Online Resource Centre. All values have been rounded to five decimal places.

1. Hypotheses to be tested
    H_0: There is no difference between the means of the ranks of the samples.
 	
    H_1: There is a difference between the means of the ranks of the samples.
	1. Hypotheses to be tested
	As in Tukey's test and in the multiple comparisons test in 10.4, you could write specific hypotheses for each comparison of means, for example:
	
	H_0: There is no difference between the mean of the ranks for the number of pollen grains retrieved from *P. pratense* pollen on polycotton compared with *P. pratense* pollen on viscose.

BOX 10.6 Continued

	H_1: There is a difference between the mean of the ranks for the number of pollen grains retrieved from *P. pratense* pollen on polycotton compared with *P. pratense* pollen on viscose. However, common sense suggests that where you are making many comparisons using this test, then it is sensible to report the overall hypotheses only.
2. Have the criteria for using this test been met?	**2. Have the criteria for using this test been met?** Yes (10.6.1).
3. To work out $Q_{calculated}$ i. Calculate the difference between each pairwise combination of means using the mean of the ranks.	**3. To work out $Q_{calculated}$** i. For example the mean of ranks for the *P. pratense* pollen on polycotton is 7.5. The mean of ranks for *P. pratense* pollen on viscose is 5.25. The difference is therefore 2.25 (Table 10.11). In total, 28 comparisons between means can be made. Table 10.11 gives all the differences between these pairs of means.
ii. For each pair of means, calculate the standard error of the difference of the means (SE). a. If there are few tied ranks the calculation is: $SE = \sqrt{\left[\left(\frac{N(N+1)}{12}\right) \times \left(\frac{1}{n_a} + \frac{1}{n_b}\right)\right]}$ where N is the total number of observations, n_a is the number of observations in the first of the two categories being compared, and n_b is the number of observations in the second of the two categories being compared. If n is the same for all categories (samples), then this value will be the same for all pairwise combinations. b. If there are tied ranks then the calculation for the standard error is modified: SE = $\sqrt{\left[\left(\left(\frac{N(N+1)}{12}\right) - \left(\frac{T^{kw}}{12(N-1)}\right)\right) \times \left[\left(\frac{1}{n_a}\right) + \left(\frac{1}{n_b}\right)\right]\right]}$ where N is the total number of observations, n_a is the number of observations in the first of the two categories being compared, n_b is the number of observations in the second of the two categories being compared, and T^{kw} is the value calculated during the two-way non-parametric ANOVA (10.5) reflecting the numbers of tied values in the data. If n is the same for all categories (samples), then this value will be the same for all pairwise combinations.	ii. In our example, there are no tied ranks so we may complete the calculation in step ii (a). $N = 32$ $n_{P.\ pratense\ polycotton} = 4$ and $n_{P.\ pratense\ viscose} = 4$ All n values in this example are 4 as there are four observations in each category. $SE = \sqrt{\left[\left(\frac{32(32+1)}{12}\right) \times \left(\frac{1}{4} + \frac{1}{4}\right)\right]}$ $= \sqrt{(88.0 \times 0.5)} = \sqrt{44.0}$ $= 6.63325$

BOX 10.6 Continued

iii. Divide the difference in the means by its SE. This is the $Q_{calculated}$ value for the comparison between that particular pair of means.	iii. For the comparison between the means of the ranks from the *P. pratense* polycotton and *P. pratense* viscose, $Q_{calculated}$ = 2.25/6.63325 = 0.33920. The $Q_{calculated}$ values for the other comparison are given in Table 10.11.
4. To find $Q_{critical}$ See Appendix d, Table D15. $Q_{critical}$ may be found using *a*, the number of categories or samples and *p*. This critical value for Q will be the same for each comparison of means.	**4. To find $Q_{critical}$** In our example there are eight categories (*P. pratense* polycotton, *P. pratense* viscose . . . *A. artemisiifolia* denim), *a* = 8. $Q_{critical}$ = 3.124 at *p* = 0.05
5. The rule If the calculated value of Q is greater than the critical value of Q, then you may reject the null hypothesis.	**5. The rule** When comparing the means of ranks for the *P. pratense* pollen on polycotton and *P. pratense* pollen on viscose, $Q_{calculated}$ (0.339) is less than $Q_{critical}$ (3.124) so we do not reject the null hypothesis. If we examine the means of ranks for *A. artemisiifolia* fleece and *P. pratense* denim, then $Q_{calculated}$ (3.128) is just greater than $Q_{critical}$ (3.124) at *p* = 0.05 so we may reject the null hypothesis. There are only two significant outcomes from these pairwise comparisons (Table 10.11).
6. What does this mean in real terms?	**6. What does this mean in real terms?** There is a significant difference (Q = 3.128, *p* = 0.05) in recovery of pollen between *A. artemisiifolia* pollen on fleece and *P. pratense* pollen on denim, and between *A. artemisiifolia* pollen on fleece and *P. pratense* pollen on viscose (Q = 3.354, *p* = 0.05).
Further examples relating to the topic including how to use statistical software are included in the Online Resource Centre. @ **online resource centre**	

As in our one-way multiple comparisons test, the outcome allows our interpretation of the data to be more specific. In this example, although there was a significant difference between species in the retrieval of pollen, this was mainly due to a difference between the fleece material for *A. artemisiifolia* and the smoother materials (denim and viscose) for *P. pratense*. This information can now be used to develop further more specific experiments to examine pollen retrieval.

There are a number of methods that can be used for reporting results from statistical tests. We review these in Chapter 11 (11.8.3). Here, we have placed the significant values in Table 10.11 in bold and indicated the level of significance using the asterisk system.

Table 10.11 The difference between the means of ranks ($\bar{x}_{ranks}$) and the Q values for a multiple comparisons test between the number of pollen grains retrieved from four fabrics (denim, fleece, viscose, and polycotton) using pollen from two species (*Phleum pratense* and *Ambrosia artemisiifolia*)

		Denim	Fleece	Viscose	Polycotton	Denim	Fleece	Viscose	Polycotton
		A. artemisiifolia $\bar{x}_{ranks}$ = 21.25	*A. artemisiifolia* $\bar{x}_{ranks}$ = 28.00	*A. artemisiifolia* $\bar{x}_{ranks}$ = 24.50	*A. artemisiifolia* $\bar{x}_{ranks}$ = 24.25	*P. pratense* $\bar{x}_{ranks}$ = 7.25	*P. pratense* $\bar{x}_{ranks}$ = 13.5	*P. pratense* $\bar{x}_{ranks}$ = 5.75	*P. pratense* $\bar{x}_{ranks}$ = 7.5
Polycotton *P. pratense* $\bar{x}_{ranks}$ = 7.5	Difference in means								
	Q								
Viscose *P. pratense* $\bar{x}_{ranks}$ = 5.25	Difference in means								7.5 – 5.25 = 2.25
	Q								2.25/6.63325 = 0.33920
Fleece *P. pratense* (13.25)	Difference in means							13.25 – 5.75 = 7.5	13.25 – 7.5 = 5.75
	Q							7.5/6.63325 = 1.13067	5.75/6.63325 = 0.86685
Denim *P. pratense* $\bar{x}_{ranks}$ = 7.25	Difference in means						13.5 – 7.25 = 6.25	7.25 – 5.75 = 1.5	7.5 – 7.25 = 0.25
	Q						6.25/6.63325 = 0.94222	1.5/6.63325 = 0.22613	0.25/6.63325 = 0.03769

		Polycotton $\bar{x}_{ranks}=24.25$	Viscose $\bar{x}_{ranks}=24.50$	Fleece $\bar{x}_{ranks}=28.00$				
Polycotton *A. artemisiifolia* $\bar{x}_{ranks}=24.25$	Difference in means				$24.25-7.25$ $=17.0$	$24.25-13.5$ $=10.75$	$24.25-5.75$ $=18.5$	$24.25-7.5$ $=16.75$
	Q				$17.0/6.63325$ $=2.56285$	$10.75/6.63325$ $=1.62062$	$18.5/6.63325$ $=2.78898$	$16.75/6.63325$ $=2.52516$
Viscose *A. artemisiifolia* $\bar{x}_{ranks}=24.50$	Difference in mean ranks	$24.5-24.25$ $=0.25$			$24.5-7.25$ $=17.25$	$24.5-13.5$ $=11.0$	$24.5-5.75$ $=18.75$	$24.5-7.5$ $=17.0$
	Q	$0.25/6.63325$ $=0.03769$			$17.25/6.63325$ $=0.15076$	$11.0/6.63325$ $=1.65831$	$18.75/6.63325$ $=2.82667$	$17.0/6.63325$ $=2.56285$
Fleece *A. artemisiifolia* $\bar{x}_{ranks}=28.00$	Difference in means	$28.0-24.25$ $=3.75$	$28.0-24.5$ $=3.5$		$28.0-7.25$ $=20.75$	$28.0-13.5$ $=14.5$	$28.0-5.75=$ 22.25	$28.0-7.5$ $=20.5$
	Q	$3.75/6.63325$ $=0.56533$	$3.5/6.63325$ $=0.52764$		$20.75/6.63325$ $=3.12818*$	$14.5/6.63325$ $=2.18596$	$22.25/6.63325=$ $3.35431*$	$20.5/6.63325$ $=3.09049$
Denim *A. artemisiifolia* $\bar{x}_{ranks}=21.25$	Difference in means	$24.25-21.25$ $=3.0$	$24.5-21.25$ $=3.25$	$28.0-21.25$ $=6.75$	$21.25-7.25$ $=14.0$	$21.25-13.5$ $=7.75$	$21.25-5.75$ $=15.5$	$21.25-7.5$ $=13.75$
	Q	$3.0/6.63325$ $=0.45227$	$3.25/6.63325$ $=0.48996$	$6.75/6.63325$ $=1.01760$	$14.0/6.63325$ $=2.11058$	$7.75/6.63325$ $=1.16836$	$15.5/6.63325$ $=2.33671$	$13.75/6.63325$ $=2.07289$

$*p \leq 0.05.$

10.7 Scheirer–Ray–Hare test

Key points This test combines a parametric ANOVA with the use of rank values rather than observations. We illustrate this test in relation to testing the effect of two variables.

The ANOVAs we have described have a number of criteria for use. One of these relates to the numbers of tied ranks, especially in large sample sizes. Under these conditions, the one- and two-way ANOVAs may produce a negative test statistic, which indicates insufficient variation within the data for these tests to function adequately as tests of hypotheses.

An alternative method, the Scheirer–Ray–Hare test, may then be helpful. This is similar to a parametric ANOVA in that it tests general hypotheses only, but uses rank values rather than the actual observations (Scheirer *et al.*, 1976).

10.7.1 Using this test

To use this test you:

1) Wish to examine the effect of two treatment variables each with at least two categories.

2) Have an orthogonal design.

3) Have non-parametric data that can be ranked.

4) Have equal numbers of replicates in each category or sample.

To illustrate this test, we will use the data from Example 10.3. These data meet the criteria for this test in that there are two treatment variables (location and fertilizer treatment) and each has two categories. The design is orthogonal and the observations are numbers of daisies, which is a non-parametric measure that can be ranked. There are four observations in each category.

10.7.2 The calculation

The ranks have already been assigned and it is these that are used in the analysis (Table 10.12). The three pairs of hypotheses being tested are those given in Boxes 10.5(A–C).

A two-way parametric ANOVA as described in Box 9.10 is carried out on these data, generating the values shown in the ANOVA table (Table 10.13). For more computation details, see the Online Resource Centre. At this point, the analysis diverges from that described for a two-way

online resource centre

parametric ANOVA in Chapter 9. Firstly, a MS is calculated by adding all the sum of squares (SS) values together and dividing the total by the total degrees of freedom (Table 10.13). An H value is calculated instead of an F value as SS/MS_{total}. The critical values for H are found in a chi-squared table (Appendix d, Table D1) for the degrees of freedom associated with the sum of squares (SS) (Table 10.13). The rule is the same as that given for the parametric two-way ANOVA (9.9 and Box 9.10) in that if $H_{calculated}$ is greater than $H_{critical}$, then you may reject the null hypothesis. In our example, it can be seen that there is a significant difference ($H = 10.38$, $p = 0.01$) between the median density of daisies in the two locations, but there is no significant difference for the fertilizer treatment or the interaction. The outcome is therefore the same as that for the non-parametric ANOVA described in Boxes 10.5(A–C).

Table 10.12 Ranking values assigned to Example 10.3: The density of *Bellis perennis* from two locations ('cricket pitch' and 'lawn') treated or not treated with fertilizer, for the Scheirer–Ray–Hare test

	Cricket pitch	Lawn
Fertilizer	5.0	14.5
	6.5	12.0
	6.5	14.5
	9.5	16.0
No fertilizer	1.0	8.0
	2.5	9.5
	2.5	12.0
	4.0	12.0

Table 10.13 ANOVA table for Scheirer–Ray–Hare analysis of data from Example 10.3: The density of *Bellis perennis* (daisy) on the University of Worcester campus, 2003

Source of variation	SS	v	MS	$H_{calculated}$	$H_{critical}$
Variable 1 (location)	232.5625	1		232.5625/22.4 = 10.382	$p = 0.01$ $H_{critical} = 6.64$
Variable 2 (fertilizer)	68.0625	1		68.0625/22.4 = 3.0385	$p = 0.05$ $H_{critical} = 3.84$
Interaction	0.25	1		0.25/22.4 = 0.01116	$p = 0.05$ $H_{critical} = 3.84$
Within samples	35.125	12			
Total	336.0	15	336/15 = 22.4		

10.8 Experimental design and non-parametric tests of hypotheses

Key points In this chapter, we have considered a number of non-parametric tests that examine the effect of one or more variables each with two or more categories, with replicates. In general, the non-parametric tests are less flexible, so recognizing any constraint placed on your design by the statistical tests during your planning stage is even more important than in previous chapters.

In this chapter, we have covered the non-parametric equivalents of the parametric tests covered in Chapter 9. Therefore, the experimental designs suitable for using these tests have equivalents in Chapter 9. The difference is largely that the observations will be non-parametric measurements and some of the criteria for the use of these tests differ, e.g. sample sizes.

10.8.1 One variable

In designs that test one variable, there are only non-parametric equivalents to the matched and unmatched tests and not to the Latin square.

i. Unmatched data

The Mann–Whitney U test may be suitable, if all other criteria are met, for an experiment with the design shown in Table 9.27 with one variable and two categories. The sample size is determined by the criteria in 10.1.1. Each observation in a sample is a replicate, although not usually referred to as such. One of the two samples could be a control and this test will allow you to make a comparison between the variation in a control sample and that of a sample exposed to a particular treatment.

The one-way non-parametric ANOVA may also be appropriate for this design, particularly where more than two categories are present. If the one-way ANOVA is followed by a multiple comparisons test, you may test both specific and general hypotheses if the criteria are met (10.3 and 10.4). These tests are the non-parametric equivalent of the parametric one-way ANOVA followed by Tukey's test.

ii. Matched data

Wilcoxon's matched pairs test is the non-parametric equivalent of the matched t- and z-tests (Table 9.27). The hypotheses being tested relate to one variable and two samples where the observations are not independent but are made on the same item. One set of observations could be a control and the other as a result of a treatment on the same item, e.g. 'before' and 'after', but need not always be so. The number of samples is determined by the criteria given in 10.2.1.

10.8.2 **Two variables**

As we indicated in Chapter 9, with two variables you may need to consider several more elements about an experimental design including whether the design is orthogonal, whether the variables are fixed or random, and therefore whether the ANOVA is Model I, Model II, or a mixed model. With the non-parametric tests we have included, only orthogonal designs can be tested, and ideally where equal numbers of observations (replicates) are included in each category. This design is illustrated in Table 9.29. For non-parametric tests that equate to the nested design (Table 9.30), we refer you to Sokal & Rohlf (1994) and Zar (2010).

The two-way ANOVA analyses three pairs of hypotheses, one of which relates to the possible interaction between the variables, in addition to which specific pairwise comparisons may be made using a multiple comparisons test.

Two-way designs with replicates are organized into blocks, as are the two-way parametric equivalents. This can allow for a certain degree of randomization and replication within the experiment and so help to minimize the impact of confounding variables.

10.8.3 **Three variables**

The Scheirer–Ray–Hare test is the only non-parametric test we have included that may be extended to more complex designs, such as three variables. The comments relating to three-way parametric ANOVAs remain pertinent to this non-parametric test.

 We are designing an experiment with one treatment variable and three categories, one of which is a control. Which test should we consider using and can we make a comparison that will allow us to analyse the difference between the control and other treatments?

Summary of Chapter 10

- We consider several non-parametric tests in this chapter. The Mann–Whitney U test for unmatched data (10.1) and Wilcoxon's matched pairs test (10.2) for matched data may both be used to analyse data from experiments with one treatment variable, with two categories. The one-way non-parametric ANOVA (Kruskal–Wallis test, 10.3) is particularly appropriate for a design with one treatment variable but three or more categories, and the two-way non-parametric ANOVA (10.3–10.6) and Scheirer–Ray–Hare tests (10.7) may be appropriate for designs with two or more variables.

- The non-parametric ANOVAs followed by a multiple comparisons test allow you to test both general and specific hypotheses (10.4 and 10.6). This provides you with considerably more explicit information about the effects of your treatment variables and is a design that should be used whenever possible.

- In general, these non-parametric tests can be applied to more restricted experimental designs compared with the parametric equivalents considered in Chapter 9, especially when an experiment sets out to investigate the effect of two or more variables.

- These non-parametric tests are distribution free and therefore the criteria for their use are often easier to meet than for the parametric tests (Chapter 9). Most of the tests are ranking tests: central steps are the determination of ranks and sum of ranks as outlined in Chapter 5 (Box 5.3).

- In Chapter 9, we consider the parametric equivalents to the tests covered in this chapter. If your data are parametric (Box 5.2) or can be normalized by transforming them (5.9), you should refer to Chapter 9.

- The Online Resource Centre includes interactive exercises that test your understanding of this chapter with other topics, particularly those considered in Chapters 7–10.

 **online resource centre**

Answers to chapter questions

A1 You wish to test the effect of one variable (consumption of alcohol). The data are matched, with observations on a total of 16 volunteers before and after consumption of alcohol. The observations (time taken) are measured on a continuous scale and the data may therefore be parametric. Gender is a possible confounding variable. To decide which test to use and whether the experiment needs to be modified, you need to decide whether the data are parametric. You cannot confirm this until you have carried out the investigation or a pilot experiment. If the data are parametric, you may consider a t-test for matched pairs. If the data are not parametric, then you may consider using Wilcoxon's matched pairs test. It would be worth analysing the male and female results separately to obtain some information about this confounding variable.

A2 Yes. The data from Example 10.2 are suitable for analysis by a non-parametric ANOVA as the unit of measurement is 'number of plants', a non-parametric measure that can be ranked. There is one variable (location on campus), four categories (cricket pitch, lawn, quadrangle, and rugby pitch), and eight replicates (observations) in each category.

A3 With this design, if the data are non-parametric, we should consider using a non-parametric one-way ANOVA followed by a multiple comparisons test to test both general and specific hypotheses (10.3 and 10.4). This will allow us to ask specific questions about the control and other treatments. If the data are parametric, we should consider the parametric one-way ANOVA, which, if significant, could be followed by Tukey's test (9.5 and 9.6).

Reporting your results

Reporting your research   11

In a nutshell

There is little point in carrying out research unless you communicate the findings to other researchers. In this chapter, we take you through the most common format used when writing about your results: that of a scientific report, thesis, or research paper. These all have similar elements: Title, Introduction, Methods, Results, and Discussion. We explain what each section is for, approaches to writing each section, and common errors.

There is little point in carrying out research unless your findings are communicated. The most common way to do this is to write a report or paper. Generally, publications of practical research in science adhere to a specific format. This is the structure also followed by most bioscience

departments most of the time for their science reports. In this section, we include a guide to the general format common in scientific reports. The format outlined in this chapter should be used in conjunction with local guidelines, either those relating to the journal you wish to publish your article in or your course or department requirements. Additional helpful advice and comments on writing scientific reports are included in Johnson & Scott (2009).

When learning how to write a scientific report or improving your skills, it is best if you have a journal article to look at. Try to have one from your subject area to hand, either as hardcopy or online, as we will use it to illustrate our points. We include some links to online journals in our Online Resource Centre. If you work through this chapter and the questions, it should take about four hours. Unlike previous chapters, only some of the answers to the questions are at the end of the chapter. The remaining answers are integrated into the main body of text. All our examples are based on real undergraduate research projects.

**online
resource
centre**

11.1 General format

Key points The general format of a scientific report includes the main sections Introduction, Methods, Results, and Discussion, usually all written in the passive past tense.

In general, there are nine sections to a full journal article: Title, Abstract, Keywords, Acknowledgements, Introduction, Methods, Results, Discussion, and References. Of these, the abstract, keywords, and acknowledgements are invariably not required for most undergraduate practical reports. When writing an honours project report or thesis, you may be required to include an abstract, and most people choose to include acknowledgements. In addition, in unpublished reports such as honours projects and theses, you may include an appendix for additional material such as the names of suppliers of chemicals, details about how stock solutions were made up, etc.

 Q1 Look at the journal article you have selected. What sections are there? How do they differ from the list given? Why do you think this is?

Your whole report should usually be written in the passive, past tense. The passive tense means that you are not referring to anyone in particular so you do not use 'I', 'you', 'we', etc. The 'past tense' means that you

write about what was done and not what is being done or will be done. Have a look at the introduction and methods in the article you are examining and note the tense that is used.

 Q2 Rephrase the following sentences so that they are in the passive, past tense.
1. I added 5mg of sodium chloride to water and stirred until it was dissolved.
2. In my reading, I found that ice is less dense than water.

Your report must be referenced correctly throughout. This is very important. Failure to do so is called plagiarism and usually carries severe penalties. We discuss plagiarism under 'cheating' (11.6) and referencing systems are discussed in 11.10.

In larger studies, you may have carried out more than one investigation. Some small elements may be redundant to the main thrust of your report. If this is the case, then leave them out or include this detail in an appendix. If your research extends over several months or years, you may tend to write about your work in the historical order in which it was carried out. This chronology may not be relevant to the actual outcome from the research and can be detrimental to the structure of your report or paper.

11.2 **Title**

Key points Your title needs to be explicit and succinct and to convey the essential elements of your aim.

This is the first thing a reader looks at and it should capture their attention. After all your hard work, you want people to read about your investigation. The title should be concise and should relate to the aim and conclusion. If you are publishing in a journal, your title may be part of the information made available for electronic searches so you need to consider who you want to read your article and include the terms they are likely to use in an internet search.

Common errors

Too long   It is not necessary to write 'An investigation into …'. Editors of journals like to save ink and will not wish to include unnecessary words. Your title will be more forceful if you go right to the subject under investigation.

In undergraduate reports, the same is true. Your title will be much more dynamic if written in this way.

Too short   You are not writing tabloid headlines. Your title does need to be complete and informative.

 Q3   Read the title of the journal article you have. Does it do everything we recommend? If not, how could it be improved?

11.3 Abstract

Key points An abstract provides a useful overview of your work, providing the reader with an understanding of the method and key findings.

An abstract is defined in the *Concise Oxford Dictionary* as a 'brief statement of content'. Most journal publishers do not want the 'brief statement' to be an extract copied from part of the report, so you have to write an abstract, not cut and paste it. This is also usually the case for undergraduate work where you are asked to write an abstract.

The format for an abstract varies among publications. In some journals, the key findings are given as a list; in others, the abstract is written as prose with sentences organized into paragraphs. Whichever format is required, you need to include enough background information so that your aim, which you also include, can be understood. This is followed by a summary of your data and your conclusion. There needs to be enough detail in the abstract for your conclusion to be understood. You should not include any new information in the abstract that is not elsewhere in your report.

Common errors

The most common error in undergraduate reports where an abstract is required is insufficient detail. This can include forgetting to include the aim or the conclusion. The conclusion also forms part of the discussion and is often written incorrectly (11.9). Clearly, these errors can then carry through to the abstract.

 Q4   Do not look at the abstract of the journal article you are examining. Instead, read the paper and write your own abstract. How does it compare with the author's version? What are the differences? Which is the better version and why?

11.4 Keywords

Key points Keywords are used as search terms and therefore need to be chosen with care to ensure that your work is not overlooked. These terms are less critical now that most searching is done electronically and using the full text.

Keywords are given as a short list of single words or linked words but not long phrases. For example, linked words might be 'resource protection' or 'agri-environment schemes'. The keywords you select from your paper will depend in part on who you are writing for and how the keywords will be used. Most often, keywords are only required when publishing a paper and there will be guidelines concerning them in the journal's guide to authors. In this context, the keywords are used by a reader to see quickly whether the article is of interest to them and by indexing systems. Indexes of keywords are used by the publishers of your paper to draw attention to your paper. They are also used by bibliographic journals that provide a quick method of locating articles of interest across a wide number of journals. Increasingly, the use of internet searching facilities means that keywords are becoming less important as the whole paper is often subjected to a search for a specific term.

In the same way as the title, the keywords are your way of flagging up your work for a particular audience, so you need to ensure that the terms they are likely to use when carrying out a literature search are included in your list. To achieve this, you should consider using one or two broad terms and one or two specific terms. For example, if you had carried out an investigation into the levels of pesticide residue on organic and non-organic fruit, you might include 'organic' as one broad term and the name of one of the pesticides as a specific term. Keywords may include terms not in the title.

Q5 How many keywords are there in the paper you are examining? How many of these terms are broad and how many are specific? Do all these terms appear in the title?

11.5 Acknowledgements

Key points Acknowledgements may be included in research publications but should be appropriate.

Many people feel that they would like to include acknowledgements at the completion of a significant piece of work. Those who may be acknowledged

fall into three categories: the funding body, those who provided professional assistance, and those who provided personal support. It is usually a requirement as well as a courtesy to include an acknowledgement to any organization or person who has provided you with funding or support in kind (e.g. seed supplies, an item of equipment) that has enabled you to carry out the research you are now reporting. You may also have received professional assistance, such as being given permission to carry out your work in a particular location, help with technical aspects of the research, or help in identification of species. Again, it is important that the input of these people is acknowledged. In journal articles, it would be unusual to find personal acknowledgements, although in honours project reports and theses, it is common practice to thank friends and/or family for their support. This is personal to you, and most universities have regulations that allow the inclusion of these acknowledgements. It is not appropriate to make these acknowledgements too personal in either a positive or negative way, nor should your acknowledgements include tenuous supporters such as drinking friends and distant cousins!

11.6 Introduction

Key points An introduction provides the background information so that the aim can be understood. It does not usually include any specific details relating to your experiment other than the aim of your research. The format used in scientific publishing is not that commonly taught at school.

An introduction is the part of a report that sets the scene and leads up to a statement of your aim.

 Q6 Have a look at the format of the introduction in your journal article. How many paragraphs are there? How does this compare with the discussion? Where is the aim? Are there any figures or tables? What information is contained in brackets throughout the introduction?

This quick look at your journal article should have drawn your attention to several features of an introduction: (i) it is written as prose with fully referenced sentences and paragraphs; (ii) there are not usually any figures or tables; and (iii) the aim is at the end.

Writing an introduction is like telling a joke: you must not give the punchline first or no one will understand it. The punchline in an introduction is the aim. It will be uppermost in your mind so it is tempting to

place it first, but if you do this, your reader, when reading the aim, will not understand the reasons for the investigation or fully understand the aim itself. Writing the aim first is a format often used in schools, but you should now use the format current in the scientific literature instead. The aim therefore is something you lead up to and should be in the last paragraph at the end of the introduction.

Your introduction should cover enough background information for a reader to understand the aim. If you are not sure how to achieve this, first look at your aim and note down critical words. Your introduction should include details about all these words. Secondly, complete your literature review as notes. Be guided by the words you have identified in your aim and draw a mind map or flowchart that develops step by step the background to the research, linking all the parts together and ending in the statement of your aim. Use this flowchart to guide your writing.

Common errors

The introduction reads like an essay   An introduction is not an essay. It performs a very specific and different function. Even if you are asked to include an extensive literature review, it should normally be written so that each paragraph takes you one step closer to being able to state your aim. One feature that tends to be associated with 'essay' introductions is their length. If in doubt, count how many paragraphs you have in the introduction compared with the discussion. If the introduction is considerably longer than the discussion, then you have probably made this error.

The introduction lacks detail   This is a feature that has become more common in recent years with the advent of electronic literature searching facilities. Many publishers make abstracts from journal articles available online and there is a temptation to rely on these as the prime source of information when reading about a topic. Abstracts do not provide enough detail for anyone to be able to appreciate fully the value or rigour of the research or the detailed arguments presented within the paper. A reliance on abstracts carries this lack of critical detail through to your report, leading to a weak introduction (and discussion).

The introduction is aimless   It is surprising how often students forget to include their aim in their introduction. This leaves the reader in utmost confusion, usually for the remainder of their time reading the report. Given report formats, it is extremely difficult to fully comprehend what the aim of the research is unless you explicitly state it.

The introduction is not focused on the actual piece of research that has been carried out   This usually occurs because you are using sources of information that are looking at a slightly different topic and you can get swept along.

Sometimes you may find a lot of information on a small specific part of your study and so you tend to note down everything you have found about that topic at the expense of areas where information is harder to find. You can avoid this by using the flow diagram and planning your introduction before you start writing. For example, one of our undergraduates investigated the opinions of students in relation to over-the-counter genetic testing. She might have written an introduction all about genetic testing and never touched on 'over-the-counter' aspects or how people develop their opinions. All of these topics should be covered in the introduction, but as information is easiest to find on genetic tests, it would not be surprising if the student was led astray. (We are pleased to say that she wasn't.)

You use quotations   In science, we are rarely interested in how someone else has phrased a sentence; we are interested in the facts within the sentence. Therefore, unlike reports written for courses such as English, you will not see quotations being used. When you are reading around the subject, you should extract the facts in note form only. You then use these succinct notes to write your own introduction. This makes for a much more readable piece of work. It also avoids inadvertent plagiarism (see Cheating (plagiarism), below). The exception to the use of quotes tends to be if you are defining a word.

 Q7   Look at your journal article. How many quotations are included in the introduction?

Predictions   It appears to be becoming increasingly common for students to declare what they are expecting as an outcome from their investigation in their introduction. (We use the term 'expectation' here as it might be used in general spoken English. Elsewhere, we use the term in a statistical way, e.g. Chapters 7 and 8.) This is a really bad idea for two reasons. Firstly, you should approach any investigation with an enquiring and open mind. If you start out by predicting the outcome, than you are likely to bias your design, your execution of the investigation, and your evaluation of the data. This can lead to a flawed piece of research and/or completely incorrect conclusions. Secondly, if you include your prediction in your report, your readers will know that you have not come to this with an enquiring and open mind and be prejudiced against your work as it is more likely to be biased.

More about the 'punchline too early' phenomenon   Another error we come across is where detailed elements usually relating to the methods appear in the introduction. In your introduction, you need to keep yourself aloof from what you actually did. The nearest you get to referring to what you did is the aim. The introduction is for background information only.

Cheating (plagiarism)   The computer age has brought easier access to sources of information through electronic searching and the 'cut-and-paste' facility. It has also tempted some undergraduates to cheat, either by downloading whole reports or essays, or by cutting and pasting paragraphs from articles. Assessments, including reports of investigations, are supposed to be your own work, which means that they should be your own ideas and your own words. If the ideas or words are not your own, you should indicate the source by referencing them. Failure to do this is called plagiarism. Not only will you not learn anything by cheating but you will also be subject to your institution's or the journal's penalties. Scientists make their living by the facts they discover and the communication of these facts. This is why you must always acknowledge the sources of your information (11.10) and this is why there are severe penalties if you do not.

If you are still tempted, please bear in mind that if you can find these articles so can others and it is surprisingly easy to spot the use of essay bank work and content that has been cut and pasted from another source.

Be a critical reader   Just because research is published does not mean it is necessarily 'good' research. You need to be discriminating in your choice of sources of literature and to read each piece of published research carefully. You can convey alternative sides to a debate or concerns about the quality of research in your report. This shows that you have the capacity for independent critical thought, which will earn you marks as an undergraduate and is an important life skill.

11.7 Methods

Key points   The methods section should provide sufficient detail so that your work can be repeated. It is not usual for there to be a separate section for materials. The methods may include information about the site when this is relevant and the statistics used in the data analysis.

The method outlines the procedure(s) you have undertaken during your investigation. Methods can vary in their content and you should check your local regulations or guide to authors to confirm the content. A method section can include a site description, the method, and details about the statistical analyses used.

11.7.1 Site

This section is included where the site is an important part of the research and its evaluation. You may either have carried out your research at a particular

site or have sampled from a particular site, and there is a possibility that the site has given your sample or study unique characteristics. The site description is written in such a way that your site can be relocated and the features of the site that impinge on your investigation need to be reported.

For example, an investigation was carried out into the percentage cover of five plant species growing on and/or off anthills in grassland. In this study, the site is important because the surrounding habitat and its management will affect the flora found in the locality. Therefore, you need to include a description of the area and its geology, typical plant species, and details about the management, e.g. grazing, presence of rabbits, cutting regimen, spraying, etc.

The location of your site is given in terms of Ordnance Survey (OS) coordinates. These are found using an OS map of the area. The complete coordinates consist of two letters and six numbers, e.g. the University of Worcester is at SO835556. The letters are printed on the map itself and are large and pale blue, e.g. Worcester is in the 'SO' compartment. Some OS maps may include details from more than one compartment, each of which has the two blue letters printed within the relevant area on the map. The six numbers comprise two groups of three numbers. The first three numbers come from the x-axis (west to east) and the second set of three numbers refer to the y-axis (north to south) (Fig. 11.1). Along the x-axis of the map there are a series of numbers. Locate your site in relation to these numbers first. Then subdivide the single grid square into ten and this gives the third value for the x-axis. Repeat this for the y-axis. The full OS coordinates are the two letters followed by the three x-axis numbers and the three y-axis numbers.

 The site of a survey in Worcestershire is shown as the letter A on the map in Fig. 11.1. What are the Ordnance Survey coordinates for the site?

11.7.2 Methods

The methods section describes the processes you carried out to investigate your aim. There needs to be enough detail included so that someone else can repeat your work. There is some variation in how much detail is required. For example, in undergraduate work you do not usually have to indicate the name of the company that supplied your chemicals, but you may explain how you made up stock solutions that were then used to make other solutions. In journal articles, the reverse tends to be the case. For example, some substances (e.g. compost) vary from one supplier to another and in these instances details about the supplier and the full trade name of the substance should be given.

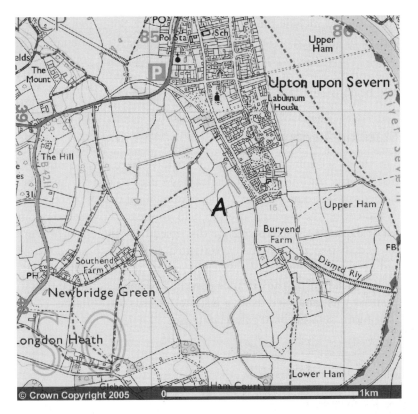

Fig. 11.1. Working out Ordnance Survey coordinates.

As this information about materials is an essential part of a method, this section may be titled Materials and Methods. However, the materials detail is usually integrated into the methods and is not given as a separate section in the methods.

> **Q9** Look at the methods section in the article you have been examining and imagine yourself repeating the experiment. Is there enough detail included for you to do this? If not, what more is needed?

The methods, as you will have noticed from examining the journal article (Q9), are written as paragraphs of prose usually in the passive, past tense. Often in school you are required to write a method as a list of steps, often with a separate section for materials. For undergraduate work, this format is no longer appropriate.

Some investigations are time sensitive. Most commonly, this is where you have collected samples from the field or carried out work outside the laboratory. If the outcome from your method is likely to be influenced by the

seasons, you should include the date in your method. Some investigations are influenced by the time of day, for example, the behaviour of animals. In this instance, you need to report the time at which you made your observations. Time and the effect of seasonal changes can be confounding variables and should be considered when planning your investigation (2.2.4).

11.7.3 Statistical analyses

In some methods sections where experiments have been carried out, you will also see a brief outline of the statistical tests used to test the hypotheses. This needs to be no more than a brief statement of the test used and a reference indicating the resource used to help you learn how to carry out this test. For example, in Chapter 2 we discussed how the effect of wind speed on seed dispersal might be investigated and we explained which test would be appropriate (2.2.7). When reporting this in a methods section, you would state that the distance travelled by the seed in relation to the wind speed was 'analysed using a Kruskal–Wallis test (Holmes *et al.*, 2010)'. For more details about referencing, see 11.10.

Common errors

Too little or too much information   Methods most often suffer from too little or too much information. Like Goldilocks, you need to get it 'just right'. To check whether you have included too little information, give your method to a friend and ask them to tell you how the investigation was carried out. This will quickly identify the gaps. Too much information can arise when you are using the 'materials and methods' format from school or you include unnecessary statements, e.g. 'The data were collected and summarized'. If you have developed a method during the research process and so in each use there are slight differences, you do not have to report the whole method each time. Reporting the full method once and indicating the subsequent revisions will provide sufficient detail for your work to be understood and repeatable.

It is difficult to find the correct balance when writing a method, but you can learn to do this by checking with your lecturers/supervisor and learning from the feedback you have received about earlier reports.

Symbols   As most reports are now word-processed, errors relating to the use of symbols have become common. Most versions of Microsoft Word have the following facilities that enable you to write symbols correctly:

- *Subscripts* (e.g. H_0) and *superscripts* (e.g. m^2) are generally located under Format–Font
- *Symbols* can usually be found either in Insert–Symbol (e.g. μ, σ, Σ) or as part of Equation Editor at Insert–Object–Microsoft Equation (e.g. Σ, $\bar{x}$).

When using symbols, be consistent in their use in terms of the font you choose and other text characteristics such as italics.

Cheating   If you are provided with a protocol for a method, this does not necessarily mean that it is written in the format appropriate for a report or that you necessarily have permission to report it verbatim in your report. You may also have made some revisions. Methods that you have been given need to be considered with as much care when you are preparing a report as a method you have developed yourself. Failure to acknowledge the source constitutes plagiarism; failure to follow the appropriate format or provide accurate details indicates learner incompetence.

 Q10   Read through the method in your article. Does it follow our guidelines? If not, in what way does it differ? Why do you think this is? Check your department's or the journal's guide to authors. What do they say about a method? How does it compare with the information in this chapter?

11.8 Results

Key points   The results section is where you present in writing the key trends seen in your data. Where relevant, you also include a statistical evaluation of hypotheses. The key trends are usually illustrated using figures and/or tables based on your data.

The whole point of a results section is that you draw to your reader's attention the main trends you have found in your data. You achieve this by first organizing your data in tables (11.8.1) or figures (11.8.2) and/or using summary statistics (5.3–5.7). By taking an overview of your data, you are more likely to be able to identify the apparent trends in your data. Secondly, if you have hypotheses to test (1.8), carry out the statistical analysis and draw conclusions from this (11.8.3). Finally, look at the raw data to see whether any specific further points can be made. This information is what needs to be communicated in your results section. It is not easy to write a results section: it requires practice. Like the introduction, it helps if you plan it out first.

 Q11   Have a look at the results section in your article. How many paragraphs of text are there? How many figures and how many tables?

A scientific report is written as words. This seems to be stating the obvious but it is amazing how many people forget to write any words in their

results section. Instead, they include a series of tables and/or figures and leave the reader to work out what is going on for themselves. To write a results section, look at the information you have about the trends in your data and prepare a flowchart/mind map that presents all the points you want to make in a coherent way. If you have a number of objectives, you will often find it useful to arrange the results under subheadings relating explicitly to each objective. When you have a suitable plan, you then write your results section using the flowchart. Only then should you consider whether to use any figures or tables. The role of figures and/or tables in a results section is a supporting one only. They allow you to include detail and to illustrate your trends. To review good practice in relation to figures and tables, see 11.8.1 and 11.8.2.

Common errors

'Therefore'   It can often feel very awkward when writing a results section to limit yourself to highlighting the trends seen in the data without referring to possible explanations for these results. If you find yourself including 'because' or 'therefore', this is probably what you are doing. The discussion section is the place for all these explanations. The reason there is a need to have separate sections is that explanations for the trends in your data are rarely straightforward and conclusive. If you embarked on lengthy discussions in the middle of presenting your data, the reader will lose any overall sense of your findings. It is better to keep all discussion separate from the presentation of your results.

Repetition   Many students suffer from overenthusiasm for tables and figures. The tendency, which again appears to come from early training in school, is to include a table of your data and then represent this with a figure. Such repetition shows that you have not thought through how best to support the points in your text. It is very rare indeed that a table and figure are needed to illustrate the same data. The exception to this would be a desire to demonstrate the overall trend using a figure but a need then to identify details in the data, which is best done using a table.

11.8.1 **Tables**

A long string of numbers does not help you to understand your raw data or help you to explain it to someone else (e.g. Example 2.3). Therefore, it is common practice to organize the raw data in a summary table. There are a number of different summary tables depending on the type of data you have. For discrete data (nominal or ordinal), the categories reflect those in which the observations naturally fall (e.g. Example 7.3). For continuous data (interval), you may construct a frequency table where

you select the categories within certain constraints. These divisions are often artificial in that they have no scientific significance or meaning. The categories have to demonstrate clearly and fairly the trend you wish to illustrate, but not overlap, and should normally all be the same size (e.g. Table 5.9). You then record the number of observations that fall within each category.

As you evaluate your data, you may also construct a calculation table; we have seen many examples of these in Chapters 7–10 (e.g. Tables 7.5 and 7.7). In scientific papers or reports, you may include a summary table and occasionally a calculation table. Which table will depend on the points you have made in your results section. You do not normally, however, include the calculation itself.

When writing for a particular journal or preparing other reports, there may be local requirements that you need to follow when producing a table. For example, it is usual when reporting summary statistics to include the mean (5.4.3), the standard error of the mean (5.6), and *n*. In doing so, you have provided sufficient detail for a reader if they wish to be able to calculate confidence intervals for your data.

A table should always:

- *Have a title* This should indicate clearly what the data are. For most publishers, the title should have enough detail so that the table can be taken out of context and still be understood. In most scientific journals, the title for a table is placed above the table. The title may be followed by additional details about the contents of the table.

- *Be numbered* Numbering tables makes it much easier for you to explain which table you are referring to. Table numbers should be sequential.

- *Be labelled (rows and columns)* Most tables are organized into rows and columns. It is important to make it clear what information each row and each column contains.

- *Indicate the units of measurement* If a row or column includes data, you should indicate the units of measurement, e.g. mm, °C, etc., in the column and/or row headings. Units should also be included in the table title (if they apply to the entire table) or in column or row headings. Having indicated the units in the title or column and row heading, you do not then include units in the central body of the table.

- *Distinguish between zero and missing data* Sometimes investigations do not go according to plan and you have 'missing data'. You may then need to distinguish between a zero measurement and missing data (see Table 11.1). If you are going to analyse your data using computer software, this is also very important. If you enter a zero for missing data on a spreadsheet, the software assumes this is a real value.

EXAMPLE 11.1 The response of tobacco explants to auxin

In an undergraduate investigation into the effect of auxin on the growth of tobacco in tissue culture, the relative increase in diameter (mm) of leaf explants was recorded after 2 weeks. Some of the samples e.g. explant 1, auxin 1, showed no increase after 2 weeks. Two of the samples became contaminated and died. These missing data are indicated by a dash (–) in Table 11.1.

Table 11.1. The relative increase in diameter of explants of tobacco after treatment with auxins

Explant	Relative increase in diameter (mm) after 2 weeks' tissue culture	
	Auxin 1	**Auxin 2**
1	0	1
2	2	0
3	–	1
4	1	–

- *Be simple* If the table is complicated, break it down into several different tables.
- *Use appropriate classes in a frequency table* Where data are measured on a continuous scale, 'classes' must be used. Two common errors relating to the use of frequency tables are choosing the wrong number of size classes and choosing size classes that overlap.

In Example 2.3, we described some research where the natural variation in growth of rye seedlings was examined and the researcher recorded heights (mm) after 3 weeks for 15 seedlings. In her work book, the researcher recorded the following:

12.0, 12.0, 11.5, 18.0, 14.0, 11.0, 14.5, 11.5,
10.0, 10.0, 19.5, 19.0, 21.0, 15.5, 14.5.

This is continuous data and is therefore best summarized in a frequency table with size classes of your choosing. We illustrate three possible sets of size class for this data so that you can see which class sizes best illustrate the trend in the data (Tables 11.2–11.4).

It is also a common failing to choose size classes that overlap. If you were planning a frequency table for the rye data (Example 2.3), you may have chosen classes 10.5–11.0mm, 11.0–11.5mm, etc. With these classes there is an overlap. If you had a value of 11.0mm,

Table 11.2. Frequency distribution of leaf length (mm) in rye seedlings after 3 weeks' growth

	Leaf length (mm)											
	10.0–10.4	10.5–10.9	11.0–11.4	11.5–11.9	12.0–12.4	12.5–12.9	13.0–13.4	13.5–13.9	14.0–14.4	14.5–14.9	15.0–15.4	15.5–15.9
Number of individuals	2	0	1	2	2	0	0	0	1	2	0	1
	16.0–16.4	16.5–16.9	17.0–17.4	17.5–17.9	18.0–18.4	18.5–18.9	19.0–19.4	19.5–19.9	20.0–20.4	20.5–20.9	21.0–21.4	
Number of individuals	0	0	0	0	1	0	1	1	0	0	1	

Too many size classes. It is not possible to see any overall trend in this data.

Table 11.3. Frequency distribution of leaf length (mm) in rye seedlings after 3 weeks' growth

	Leaf length (mm)	
	10.0–19.9	20.0–29.9
Number of individuals	12	1

Too few size classes. This summary does not allow you to see any detail.

Table 11.4. Frequency distribution of leaf length (mm) in rye seedlings after 3 weeks' growth

	Leaf length (mm)					
	10.0–11.9	12.0–13.9	14.0–15.9	16.0–17.9	18.0–19.9	20.0–21.9
Number of individuals	5	2	4	0	3	2

Just about right. This summary table allows you to see some detail (compared with Table 11.3) whilst allowing the trend to be evident (which it was not in Table 11.1).

which class would you place it in? The classes in Table 11.4 do not overlap and are therefore correct.

 Q12 Examine the tables in your paper. Do they follow all these guidelines? Find a table you have prepared. Are there any improvements you could make?

- *Rounding up* When carrying out a calculation you should use the complete value known to you including all decimal places. Rounding up where the value is abbreviated introduces errors to your calculation. In our chapters of choosing and using statistical tests, all calculations were carried out using the full values but for the purposes of reporting the values in this book they were rounded up to five decimal places.

When reporting numerical values, it is common and sensible practice to round these values up. As a general rule of thumb, if you are reporting a mean or variance for data, you round the values up to the same number of decimal places used in your observations. When reporting calculated statistical values, these would be rounded up to the same number of decimal places as the critical values.

11.8.2 **Figures**

If you believe that a figure best illustrates the points you are making in your results section, then you need to decide which is the correct figure to use. The five most common types of figures for data are a scatter plot, pie chart, bar chart, line graph, and histogram. For a review of more figure types, see Crothers (1981) and Hawkins (2005). The type of figure you use depends on the point you are making, as some styles will illustrate the points you are interested in more clearly than others.

Choosing the correct format for a figure also depends on the type of data (5.1) you have. Table 11.5 gives a guide to choosing the correct figure for certain types of data.

Table 11.5. Guide to the correct figures

Type of data	Pie chart	Bar chart	Line graph	Histogram	Scatter plot
Nominal	✓	✓			
Ordinal	✓	✓			
Interval, one variable. Few points within the variable are observed. The variable is usually under the control of the investigator.			✓		
Interval, one variable. Many points within the variable are observed. The variable is not usually under the control of the investigator.				✓	
Interval, two variables					✓

EXAMPLE 11.2 The behaviour of captive orang-utans

An undergraduate examined the behaviour of orang-utans at a zoo in the UK. She categorized the various behaviours, e.g. grooming, playing, solitary, eating, and showing aggression, and recorded the time spent on each.

If the data from Example 11.2 were to be reported and the key trends illustrated using a figure, these data would best be presented as a **pie chart** (Fig. 11.2) or **bar chart** (Fig. 11.3) because there is one variable (behaviour), the data collected are nominal, each type of behaviour is a discrete category independent of the others, and the types of behaviour cannot be arranged in any particular order.

Pie chart A circle is divided up to reflect the relative proportions (or percentages) of observations in each category or sample

Bar chart Each category from a nominal or ordinal scale is represented as a discrete (separate) bar on the figure

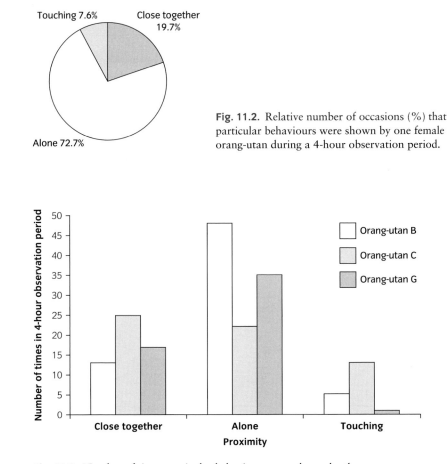

Fig. 11.2. Relative number of occasions (%) that particular behaviours were shown by one female orang-utan during a 4-hour observation period.

Fig. 11.3. Number of times particular behaviours were shown by three orang-utans (B, C, and G) during a 4-hour observation period.

> **EXAMPLE 11.3 The effect of petrol on the growth of *Lolium perenne***
>
> As part of a study of a polluted site, an undergraduate investigated the effect of petrol on the relative increase in leaf length of a grass (*Lolium perenne*) over a 6-week period. Three concentrations were examined.

In Example 11.3, the effect of only one variable (concentration of petrol) is being examined. Concentration (ml/g soil) is measured on a continuous scale. However, only a few (three) discrete points along this scale have been examined. The variable (concentration of petrol) was determined by the investigator. Therefore, the results are best shown as a **line graph** (Fig. 11.4).

Line graph Each observation or mean from a group of observations for a given *x* value is plotted as a point on the figure and may be joined together with a trend line

 Q13   Represent the data relating to the growth of rye seedlings (Example 2.3) by an appropriate figure. You may wish to repeat this exercise using a computer package. How does the output compare?

If you are using a figure to illustrate your data, it is common practice to include confidence limits (5.7). These are added as vertical lines about the point or bar on your figure (e.g. Fig. 11.4). For regression analyses, confidence limits for the regression coefficient may also be calculated (Sokal & Rohlf, 1994).

For some studies, e.g. microbiology, molecular biology, and ecology, you may have photographs that illustrate your results. These used to be

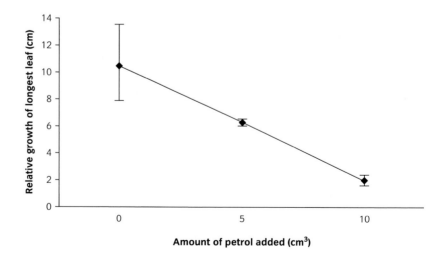

Fig. 11.4. Relative growth of longest leaf (cm) of Lolium perenne exposed to petrol in the soil.

called plates but more recently have tended to be included in with the figures. If they are called plates, then the numbering system should be sequential for all plates included.

More errors in figures are appearing with the use of computer software, often, although not always, through user inexperience. If you are going to use a software package, it is worth roughing out the figure beforehand to ensure that you are quite clear in your own mind what is required. As with tables, you are likely to have local regulations about how you present figures in your reports, papers for publication, etc. The main aim is always to make figures explicit and appropriate for the data. Therefore, a figure should:

- *Include a title* Most journals require figure titles to contain enough detail so that they can be taken out of context and still be understood. You do not need to include such things as 'A figure to show ...'. It is clear that this is a figure, so you do not need to include this in your title. In most scientific reports the standard practice is to place the title below the figure. You may also add brief additional information about the figure.

- *Be numbered* This enables you to explain clearly which figure you are referring to at any one time. The numbering should be sequential.

- *Have clearly labelled axes, stating the units in use* If you fail to label your axes, then your figure will convey no useful information at all.

- *Provide a key if necessary* You may be using your figure to allow easy comparison between several samples or treatments. Figure 11.3 shows the behaviours of several orang-utans. Each column is shaded in a different manner and a key is given so that you can identify which individual is which. Where shading or colour is used to identify particular samples or treatments, this coding should be consistent for all figures. Thus, if sample 1 is shaded blue in the first figure, it should be shaded blue in all subsequent figures.

- *Use the bar chart or histogram correctly* Both bar charts and **histograms** are drawn as bars on a figure. It is therefore tempting to think that they are the same, but bar charts have a gap between the bars whereas histograms do not. The gaps indicate, on a bar chart, that the data are measured on a discrete scale (e.g. blood groups). If the data are nominal then the order of bars on a bar chart is imposed by the reporter and does not reflect any meaningful order. A histogram with no gaps shows that the data are measured on a continuous scale (e.g. height in mm) and the order of the bars is derived from the data and is meaningful. For example, Fig. 11.3 has a gap between each 'behaviour' to indicate that this is not a continuous scale, whereas in Fig. 5.1 the scale of measurement is centimetres and is therefore a continuous interval scale. In this representation of

Histogram Each class is represented as a bar and each bar abuts the next, indicating that the scale of measurement is continuous not discrete

the data, there are no gaps between the bars so that the continuous nature of this scale is shown diagrammatically. Figure 5.1 is therefore a histogram.

- *Make it clear if you have dependent and independent variables* For some data it is possible to identify a dependent and an independent variable. These terms are used by different authors in different ways (see Glossary). When we use these terms, we mean that the independent variable is taken without sampling error, is usually under the investigator's control, and that there is probably an association between the dependent and independent variables.

In Chapter 8, we considered two examples: Example 8.2: Heavy metal contamination of soil under electricity pylons, and Example 8.4: Lower arm (cm) and lower leg (cm) length of a small cohort of female undergraduates. In both these investigations, two variables were being considered. In Example 8.2, soil samples were taken at regular points from the electricity pylon. This treatment variable (distance from pylon) is under the investigator's control and is the independent variable. The level of lead contamination is not under the investigator's control and is said to be the dependent variable. In Example 8.4, however, neither parameter (arm length or leg length) is being manipulated within the experiment and neither could sensibly be said to be dependent on, or independent of, the other.

Where you have a dependent and an independent variable, the dependent variable is plotted on the *y*-axis and the independent variable is plotted on the *x*-axis (Fig. 8.12). Where there is no clear case for the variables being classed as dependent and independent then either variable may be plotted on the *x*-or *y*-axis (Fig. 8.15).

 Q14 Cows in calf were fed known amounts of supplementary corn. The calves were weighed when born. The investigator wanted to test whether there was any association between the amount of corn the cows were fed and the calves' birth weight. Is there a dependent and an independent variable? If so, which is which?

- *Use trend lines correctly* Lines may be fitted to data in two ways: either by joining the points on a graph together as a trend line or as the result of a statistical analysis, e.g. regression (Chapter 8). Trend lines are very problematic. Before using one you should ask yourself whether it is sensible to draw a line between two points on your graph and read off an intervening value. It is usually appropriate to draw trend lines on line and scatter plots but not on histograms and bar charts.

In Example 11.3, only three concentrations of petrol contamination of soil were investigated. The data are measured on a continuous scale (amount of petrol added (ml)) and the data are plotted as a line graph. A trend line that links each point on this line graph makes sense: you could read off an intervening point and it would be meaningful (Fig. 11.4). If you have continuous data that are organized into classes and plotted as a histogram, it is not appropriate to link these classes together with a trend line. The imposition of classes has effectively taken away the continuous nature of the scale when visualized as a figure (e.g. Fig. 5.4).

In Example 11.2, the various behaviours of an orang-utan were recorded. When these data are plotted on a bar chart (Fig. 11.3), it is quite clear that any line joining the tops of these bars would be meaningless. This is particularly so because the data are nominal and the categories (in this case different behaviours) have no inherent order.

Beware: computer software is very keen to draw in trend lines. Do not assume this is correct. Where it is not correct, do not leave the trend line in your figure.

- *Distinguish between a trend line and a regression line* If a line is drawn, you must indicate clearly whether this is a trend line or one derived from statistical analysis. Some statistical tests (e.g. regression analysis, 8.4) enable you to work out a mathematical equation that describes the line that best fits your data (Figs 8.14 and 8.15). If you have a regression equation, you should add this to your figure; if you do not, then you must make it clear that the line is drawn in only to indicate the possible trend.

- *Have a restricted line* If you are drawing a line on your figure, do not extend it beyond the range of your data. You do not know what happens outside this range and you could be suggesting a trend that is quite wrong. If you need to know what is happening outside the data set you have, you need to repeat your experiment with more observations in the area you are interested in. Chapter 8 (Q3) illustrates this graphically.

Q15 Examine the figures in your paper. Do they fulfil all the guidance on good practice? If not, how might they be improved? If there are no figures included in the paper, do you think it would have been useful to include some? Why? To avoid overlap with data already represented in a table, would any tables need modification? In what way?

11.8.3 **Reporting statistics**

When reporting the outcomes from your analysis, there is very little detail that is required but what is needed is critical to making sense of your analysis. In each box in Chapters 7–10 where we have illustrated worked examples for a number of statistical tests, we have included one way of reporting our results. However, there are several conventions that may be used. These conventions vary depending on the statistical test you used, the p value, and the local requirements of the journal or your department. The following, therefore, needs to be used in conjunction with these other guides.

i. When p = 0.05

Throughout Chapters 7–10, we have indicated the usual format for reporting the results from statistical analyses. This reporting requires the inclusion of details from your hypothesis, the calculated value of your test statistic, an indication as to whether you reject or do not reject the null hypothesis, and the level of probability (p) at which this decision has been made. In each box in Chapters 7–10, we have written this information formally so that you become used to the inclusion of these various elements. For example, in Box 7.2, the outcome from the analysis is reported as 'There is no significant difference ($\chi^2_{calulated} = 4.40$, $p = 0.05$) between the observed lengths of ladybirds compared with that expected if the data are normally distributed'. This formal phrasing includes all the information required. However, in most scientific papers, a looser phrasing is used.

ii. When p < 0.05

We explained in Chapter 5 how the p value you initially use in hypothesis testing is $p = 0.05$, but that if you may reject the null hypothesis at $p = 0.05$, you should go further and identify the smallest p value at which the null hypothesis is rejected. For example, in Box 7.1, the study examined the distribution of holly leaf miners on a single tree using a chi-squared goodness-of-fit test. It was found that $\chi^2_{calculated}$ (155.47) is greater than $\chi^2_{critical}$ (5.99) at $p = 0.05$ and therefore we reject the null hypothesis. Looking again at the chi-squared table of critical values, you can see that, in fact, at $p = 0.001$, $\chi^2_{critical}$ is 13.82. Therefore, we can reject the null hypothesis at this higher level of significance and have a greater degree of certainty that the decision we have made is correct. It is important to convey this increased confidence and this can be done using one of two conventions.

You may emphasize how significant the difference is between your samples by using particular phrases. These are a widely adopted standard and should be adhered to.

p value at which null hypothesis is rejected	Phrasing
0.05	Significant difference
0.01	Highly significant difference
0.001	Very highly significant difference

In our example from Box 7.1, we would therefore write: 'There is a very highly significant difference ($\chi^2_{calculated}$ = 155.47, $p<0.001$) between the numbers of holly leaf miners found at three different levels on the tree'.

As an alternative to using these particular phrases, a system of asterisks (*) can be used to indicate the level of significance at which you rejected your null hypothesis. If you are using a table that includes the data from your investigation, then it is common practice to incorporate the * symbol within the table. The number of asterisks used is again a standardized convention that you should follow.

p	Symbol
>0.05	NS (not significant)
≤0.05	*
≤0.01	**
≤0.001	***

These symbols tend to be used when reporting the outcome from an ANOVA and the ANOVA table is included in your results section (e.g. Table 10.5).

iii. Statistical software and the exact p value

The results from the analysis of data using statistical software usually return the exact p value. Clearly, you are then able to use this value rather than the selected values for $p = 0.05$, etc. The phrasing and use of an asterisk can be adapted for the exact p values. For example, if the significance of the hypothesis testing was reported as $p = 0.0037$, you would use the phrasing relating to $p = 0.01$. This is because the significance is less than $p = 0.01$ but is not significant at $p = 0.001$.

iv. Regression analysis

If you have carried out a regression analysis (Chapter 8), the data are usually represented as a **scatter plot** and the regression equation and level of significance on the regression line of the figure is added (e.g. Figs 8.13

Scatter plot When observations for two or more variables are made for each item and the observations from each item are marked as a point on the figure

and 8.15). In 11.8.2, we discuss good practice in relation to drawing lines on figures including regression lines.

v. Tukey's test

The outcome from a multiple comparisons test such as Tukey's test is often reported in a visual way, where lines are used to link non-significant values. Examples of this can be seen in Chapter 9 (9.6). An alternative method is the use of superscripts where the same letter is used to identify those means that are significantly different from each other (9.6.2).

vi. Confidence limits

An estimate of the probability that the population mean lies within a particular range around our sample estimate is known as a confidence limit (5.7). This statistic is usually represented in a figure as a vertical line (Fig. 11.4). Confidence limits can be added to most figure types as long as you have appropriate data to summarize.

Where data are summarized in a table, you may indicate the variation around a mean value using either the standard deviation (s) from the data or the standard error of the mean (SEM). The first of these provides a valuable indication of the variation in the data in the same units as the original observations. In this case, you would report $\bar{x} \pm s$. If you use the mean and standard error of the mean, this allows a reader to calculate confidence limits for the mean value. In this case, you would be reporting $\bar{x} \pm SE$. Variation around a mean is more usually indicated using SEM. Either way, you must make it clear in table headings which terms you are reporting.

 Q16   Look in the journal article you have been reading. How have the statistics been reported?

11.9 Discussion

Key points In the discussion, you evaluate the key trends identified in the results section in the context of other studies and in relation to weaknesses in your own experiment. A concluding paragraph that presents the main findings of your research may be included at the end.

In the results section, you have *reported* your results, the key trends identified in your data and any outcomes from testing hypotheses. In your discussion, you *evaluate* these findings. This is achieved by starting with the

points you have identified in your data and one by one providing an explanation and comment about each point. Each part of this discussion would usually draw on other reported scientific literature and here we remind you not to rely on abstract sources only and to be a critical reader (11.6). You need to have an overall plan for your discussion, so a mind map or flow chart devised before you start writing will help to ensure that you address all the trends identified in your results in a logical and coherent way.

If the structure of the results section is based on a number of objectives, you may use these as a basis for the early part of the discussion. However, you do not want to fragment your discussion unnecessarily. The discussion is where you pull aspects of your study together.

As part of the evaluation, you need to be open about any assumptions you have made or any possible bias arising from your design (2.2.9). You also need to demonstrate your awareness of faults and limitations (2.1.3) within your investigation. There may be some weaknesses in the report as a result of practical requirements; however, careful planning before carrying out the research should ensure that these are minimal.

Recommendations may also form part of your discussion section. For example, you may have recommendations arising from the development of a site management plan, or you may have ideas about how to take your research forward or strengthen it. This information needs to be integrated into your overall discussion and should not be tacked on at the end so that it appears to be an afterthought.

At the end of your discussion, you should include a conclusion. Some report formats place the conclusion in its own section; however, in most journal articles the conclusion is an integral and final part of the discussion. Conclusions are one of the key parts of your report as this is what the reader will 'take away' with them and it will remind them of the contents of your paper or report. It is worth thinking carefully about the conclusion and making sure that it refers only to what you found in *your* study and is supported by *your* data analysis. A conclusion should not include any new material: sometimes people wrongly embark on another discussion in the middle of a conclusion. Often there is a desire to make a great flourish at the end of a report, especially if you are writing a lengthy report for an honours project or thesis. This flourish needs to be avoided as it tends to lead you to overemphasize your findings and to make them out to be more than they are.

Q17 Read the discussion in the journal article you have been examining. Note how the discussion is constructed, how the information from other studies is linked to the findings from this study, and the location and construction of the conclusion. How might you improve your own discussion section? Read your institution's or department's guidance. Does it differ? How?

Common errors

Essays   A discussion can suffer from the same problems that the introduction section is prey to (11.6). Your discussion should be structured and should progress in a logical manner, leading to the conclusion. A common error is that a discussion can become essay-like and lose its focus, which should be your results. This can be evident if you do not explicitly link information from other sources to a debate about your results. If you have free-standing facts, then you have probably made this error. Planning your discussion before you start writing it and making sure you cover each point identified in your results section in a balanced and systematic manner should produce a balanced discussion.

No results   For many students having just written their results section, there is a sense that the results have been dealt with and the discussion then plunges directly and exclusively into a consideration of other people's work with little or no reference to your own results. To avoid this, make a list of all the points arising from your results section and tick them off as these are covered in your discussion.

11.10 References

Key points   It is critical to acknowledge the sources of the facts you include in your report. This is achieved by including a brief 'tag' in your report at the point where you use each particular fact and providing a list with complete reference details at the end of your report. A reference list only includes material you have explicitly referred to and is not therefore a more general bibliography.

Correctly referencing your work is essential (11.6). There are two parts to referencing a report or paper: you must acknowledge your sources of information in the body of the report and then include full details of these sources at the end. There are several systems for referencing. In the biosciences, the most widely used is the Harvard system and this is what is covered here. You should check with the journal's guide to authors or your department's regulations to confirm that they are using the Harvard system. Within the Harvard system, you will find some slight variations in the use of punctuation, etc., and again you should check which version is in use by referring to your local regulations. The Harvard system does not cover referencing electronic sources and here we draw on best practice from a number of different higher education institutions.

The essence of the Harvard system is to identify in your text the source of the information using the author and date of publication as a 'tag'. This tag allows you to cross-reference to the reference list at the end of your paper or report where the full details of the source of information

are given. Clearly, there is a wide range of sources that may be used, from journal articles to videos and television. In Tables 11.6 and 11.7, we summarize the Harvard system and current best practice in referencing.

11.10.1 Referencing in the text

You must acknowledge the sources of information at the point at which you incorporate them into your report. If one sentence contains information from one source and the next sentence contains information from another source, you need to reference at the end of each sentence. If all the facts in a paragraph have been derived from a single source then the reference can be placed at the end of the paragraph. The reference in this context is the author and date of publication or, where this is not possible, some other unique tag that allows unambiguous cross-referencing to the full information in the reference list.

There are many possible scenarios for referencing. For example, the type of source may be a journal article or a video or television programme, there may be one or many authors, you may be using a table or figure taken directly from a source or facts taken from the source but presented in your own words. It is neither feasible nor helpful to cover all possible combinations of these but the guidance in Table 11.6 is constructed so that you may use one or more rows to produce an appropriate reference tag.

Q18 Examine the journal article. Find ten examples of referencing in the introduction and/or discussion. How do the format and positioning of these various references differ?

11.10.2 Reference list

You may, at the end of a report or paper, be asked to prepare a bibliography or a reference list. A bibliography is usually a record of all sources of information you have read, whereas a reference list is those sources of information you have included directly in your report (i.e. you have referred to them). You will usually be asked for one or the other and need to ensure that you include the appropriate list of sources. In most cases, you will be asked to compile a reference list and it is to this that the Harvard referencing system really applies, although the principles can also be used when constructing a bibliography.

There are two aims for a reference list. Firstly, the tag from the text (11.10.1) must link uniquely to one reference, and secondly, sufficient detail must be provided in the reference list for the reader to be able to obtain the reference themselves. All the references in a reference list need

Table 11.6. Summary of formats used in one version of the Harvard referencing system in the main body of text in a report

Context	Referencing in the text	Example
You refer to the author in the sentence.	Only the date is added in brackets after the author's name.	In an excellent study, Herbert (2005) identified …
You refer to the facts in the sentence.	The author (or other, see below) and date of publication are given in brackets.	A rare moth was recorded in Shropshire during a study of *A. moschatellina* (Holmes, 2005).
You are using a table or figure taken from another source.*	You should make it clear that this is taken from another source, and include the author (or other) of that work and date of publication.	Fig. 11.1. The variation in lead concentration in a soil sample from Hartlebury Common (taken from Hiles, 2008).
You are using more than one paper by the same author from the same year.	To distinguish between several papers add a, b, etc., after the date.	(Weaver, 2005a; Weaver, 2005b; etc.)
You use information that is shown to have come from somewhere else but you have not read this original source.	You give details of both in the text.	(Sokal & Rohlf, 1981, as cited in Holmes *et al.*, 2006)
You need to indicate particular parts of a document.	The additional information (usually a page number) should be given after the year, but within the brackets.	(Ruxton & Colegrave, 2003, p. 64)
There are two authors for a journal article, book, or conference paper.	You include both authors' surnames and the date of the publication. The authors' names are given in the same order as that used in the source.	(Trueman & Mortimore, 1999)
There are more than two authors for a journal article, book or conference paper.	Only include the first author's surname and then add *et al.* and the date of publication. *et al.* is Latin and is generally placed in italics. This is a less common practice now than formerly.	(Alma et al., 2004) or (Alma *et al.*, 2004)
Contributor to a book.	Include the surname(s) of the contributor and date of publication.	(James, 2001)
The work is anonymous or the author not given in the source.	Anonymous may be used in place of the author's surname, or use details of the source.	(Anonymous, 2004) (Defra, 2004) (*The Independent*, 2004)
You have personally been provided with unpublished information.	Indicate that this is a personal communication, from whom, and include the date you received the information.	(J. Huffer, personal communication, 7.4.97)
e-journals	Author's surname and date of publication.	(Davis, 2005)
Web material other than e-journals.	Author's surname (or other) and date of publication (if known) or date accessed.	(The Bat Society, accessed 10.7.09)
Television programme or film	Give the title and year of publication or date screened.	(The Boy with the Incredible Brain, 23.5.05)
CD ROM and DVDs	Use the author's surname (or other) and year of publication.	(Joseph, 1994)

*If you are preparing a paper for publication (rather than a student essay or internal report), you need to obtain permission from the author and copyright holder to reproduce previously published material. Student work is usually covered by the institution's copyright licence.

to be arranged in a consistent manner to enable the reader to be able to find the full reference details quickly. The reference list can be organized in a number of ways but most often it is in alphabetical order using the author's surname (or other) first. If you need to, you can also use the second author's name and the date of publication for ordering the references. The earliest paper by the same author(s) will come first at that point in the list. Where you have also used a, b, etc. in the tag, then these references are ordered by these letters.

 Examine the reference list in the journal article you have been scrutinizing. Are the references organized alphabetically by author? Are there any papers with very similar authorship? What criteria have been used to order these papers?

As we outlined in 11.10.1, there can be many types of information and therefore many possible types of references. Again, we give a guide that, by combining one or more rows, will provide you with the requisite detail for dealing with most types of sources of information you will encounter (Table 11.7).

 The Harvard system is used extensively in bioscience publications, although there is considerable variation between journal styles especially in relation to the type of fonts, case, and punctuation used. Examine your journal article and note how the system differs from that described in Tables 11.6 and 11.7. Find your local regulations relating to referencing. How do these differ from the ones we have outlined.

Common errors

Not recording the sources of information   One of the most common errors in relation to referencing is a failure to record the sources of information at the time when you are taking notes. Clearly, making good this omission takes a lot of unnecessary work and can be easily avoided.

Referencing in the wrong place   If you are discussing your ideas and your results, you need to check that any referencing in that sentence is in the right place. Often referencing is misplaced within a sentence and appears to imply that your own ideas or your own results are in fact those of another author. For example, an undergraduate wrote 'It is clear that the organic content of the soil in this study (Holmes *et al.*, 1999) should be higher.' This implies that Holmes *et al.* carried out the study and not the student. What the undergraduate should have written was 'It is clear from other studies (Holmes *et al.*, 1999) that the organic content of the soil in this study should be higher.'

Table 11.7. Format of information in a reference list following one version of the Harvard system

Context	Format in reference list	Example
Article in a journal, one author	Author's surname and initials, (the year of publication). The title of the article. The title of the journal (in italics) Volume and (part number) where known: page numbers of article.	Holmes, D. S. (2005). Sexual reproduction in British populations of *Adoxa moschatellina* L. *Watsonia* 25(3):265–273.
Article in a journal with two authors	As above but list both authors. Use 'and' or '&' between the two authors.	Trueman, l. & Mortimore, D. (1999). Nutrition and school meals. *Schools Today* 35(5): 21–24.
Article in a journal with more than two authors	As above, listing all authors' names and use 'and' or '&' before the last author.	Dutton, J., Pip, D. & Hannah, M. (2004). The population genetics of tepal colour in *Allium schoenoprasum. Journal of Ecology and Genetics* 10: 54–63.
An authored book	Author's surname(s) and initials (the year of publication). The title of the book (in italics). Edition (if not a first edition). Publisher, place of publication.	Ruxton, G. D. & Colegrave, N. (2005). *Experimental Design for the Life Sciences*, 2nd ed. Oxford University Press, Oxford.
Contribution in a book	Contributor's surname and initials (year of publication). Title of contribution. Initials and surname of the author(s) or editor(s) of the book, if the latter include ed. or eds. The title of the book (in italics). The publisher and place of publication, page number(s) of contribution.	James, R. (2001). Bactericide properties of herbs. D. Christopher, ed. *Herbs Today* Worcester University Press, Worcester, p. 35.
Conference paper	Contributor's surname and initials (year of publication). Title of contribution. Initials and surname of editor(s) of conference proceedings. Title of conference proceedings and date and place of conference. Publisher and place of publication, page number(s) of contribution.	Joseph, J. (1992). Art and the biosciences. C. Joseph, ed. American Society of Bioscience meeting 24.3.94. New Orleans. Houston Press, Houston, p. 13.
Thesis	Author's surname and initials (the year of publication). Title of thesis. Award (Ph.D., M.Sc., etc.). Name of institution to which work submitted.	Holmes, D. S. (1986). Selection and population dynamics in *Allium schoenoprasum*. D.Phil., University of York.
Publication from a corporate body	Name of body (year of publication). Title of publication. Publisher, place of publication, report number (if any).	Botanical Society of the British Isles (1999). Code of conduct for the conservation and enjoyment of wild plants. Botanical Society of the British Isles, London.
Newspaper article	Author's surname and initials (if known) or title of newspaper (year of publication). Title of article. Title of newspaper. Day and month, page number(s) and column number.	Verkaik, R. (2005). Police investigate retaliation attacks. *The Independent*. 9 July, p. 21.
Personal communications	Do not include in the reference list.	–
e-journals	Author's surname and initials (year of publication). Title of article. Title of e-journal. Volume or part if known. Publisher. URL [date accessed].	Davis, G. (2005). How to keep active during your retirement. Lifelong Learning 40. Gill Press. www.gillpress.com [24.5.05]
Web material other than e-journals	Authors or editors surmane if known and initials., (year of publication, last updated), url [date accessed].	Bat Conservation Trust. (last updated 2.7.05), www.bats.org.uk [6.7.05].

Table 11.7. Continued

Context	Format in reference list	Example
Television programme, film	Title (year of production). Type of material (video, TV, etc.) Directors surname and initials. Production details—place, organization. [date screened or seen].	The boy with the incredible brain (2005). TV documentary. Oxford Scientific, C4. [23.5.05].
CD ROMs and DVDs	Author's surname and initials (year of publication). Title {type of medium}, (Edition). Publisher and place of publication. Any identifier number. [Date accessed]	Joseph, A. (1992). The wild child. {CD ROM}. Houston Press, Houston. [14.3.92].

Referencing in the wrong place can also lead to the reverse implication, i.e. that you carried out work that was in fact the result of someone else's efforts. Although this will have occurred in error, it is plagiarism (see cheating).

Not a bibliography   You are usually asked to include a reference list. There-fore you should not include sources of information that have not been referred to in the text. Your reference list should be complete, and all reference tags in the text should link to full details in the reference list.

Only include the sources you have read   Books are a frequently used source of information. These are invariably derived from other (primary) sources that are referenced in the book. If you have not read these primary sources you need to make this clear in your referencing (Table 11.6). This is par-ticularly important, as the authors of the book may have misunderstood the original report and misrepresented the findings. If you have not read the original source, you will not know this, but the error will appear to be yours and not that of the author of the book, unless you make it clear that you have read only the book and not the primary source.

Cheating   A failure to reference correctly, both in the main body of the text and by not providing a reference list when asked to, is plagiarism. This is considered to be a form of cheating (11.6). Cheating in this way is increasing with the ease of accessing electronic forms of information. Make sure you avoid this by referencing correctly throughout all written work.

11.11 Appendix

Key points An appendix may be used as a repository of useful but not critical information. You should not include any information in an appendix that must be read to understand the main points in your report. However, you may wish to include raw data, details of equipment, basic methods, and evidence of compli-ance with the law, for example.

An appendix is not part of a published paper, but may form part of undergraduate and graduate reports, including theses. There is considerable variation in what may be included in an appendix, which reflects not only differences in practice between departments but also variation in research areas. The golden rule when considering whether to use an appendix is never to include anything that a reader needs to refer to in order for them to understand your report. Therefore, you should not include critical tables or figures in the appendix of your report.

For many reports, it would be appropriate to include evidence supporting your compliance with the law (Chapter 4), such as your risk assessment, ethics approval, confirmation of permission for working in an area, etc. If you have needed to use a consent form, an unsigned copy may be incorporated into the appendix. However, you should not include anything that might compromise the confidentiality of any volunteers, such as signed consent forms or completed questionnaires, if these could be used to identify individuals.

The appendix can also be useful as a repository for raw data, but only if this is not required in the methods or results section. Including this type of information in an appendix can be useful both for the institution and for yourself; however, before you commit yourself to including pages of numbers, check with your supervisor. The data on which most of the worked examples in this book and web site are based come (with permission) from undergraduate projects and for many of these the raw data were included in the appendix.

If you have been carrying out an experiment that required the use of stock solutions (e.g. 0.5M hydrochloric acid), you may wish to include details on how the stock solution was made. Similarly, it is usually more appropriate to include details about how well-known solutions or compounds (e.g. Feulgen stain) were made in the appendix rather than clutter your methods with these details. Details about suppliers can also be an important piece of information that will allow someone to repeat your experiment. These can also be included in the appendix. If you are in doubt as to what information your institution or department prefers in an appendix you should discuss your draft methods and appendix with your supervisor.

11.12 Your approach to writing a report

When writing a scientific report, you need to consider how to use words, figures, and tables and the scientific report format to best convey the findings from your research. The tendency is to write the report

in the format order, i.e. Title, Introduction, Method, Results, Discussion and Conclusion, References, and Abstract. In fact, a much more efficient approach and one that will enable you to avoid some of the problems we have outlined in this chapter (e.g. essay-like introductions) is to write your draft report as follows: Method (and site if relevant), Results, Discussion, and Introduction, and then on reviewing the draft add the Conclusions, Abstract, and finally the Title. From our experience, this order will require the least number of revisions, give you greater mastery of the overall report, and lead to a better, more balanced report at the end.

The reason for this order is that the methods section is usually the most straightforward section and is a matter of reporting from (hopefully) a clear set of notes. Completing one section such as the methods gives you a sense of satisfaction and progress, which can give you a much-needed lift. The results section is the key to your report as it is here that you present the information you will go on to discuss. The results section usually takes the longest time to construct because all the data need to be examined, summarized, and analysed before you begin to consider communication, and the communication itself can involve figures and tables that can take some time to select and construct properly. Therefore, we suggest that the results is the second section to tackle. With your results in mind, it is then usually easiest to move on to the discussion. One advantage in writing the discussion before the introduction is that by drafting the discussion you will have a clearer idea as to how you need to approach the introduction to avoid unnecessary repetition. The introduction is then the last main section to write in draft. By keeping this until towards the end of the writing process, you will have a much clearer idea of what background material is needed in the Introduction to properly ground the rest of the report and to allow you to introduce the points you have developed throughout your report.

Writing, even using a structured format, is a creative process and you may find yourself uncomfortable with tackling one section and more comfortable about tackling another. If you feel strongly about which section you wish to work on, then you will probably be more efficient if you go with the flow.

Q21 Make a list of the areas where your report writing could be improved. Identify three points from those on your list that you will tackle when writing your next report.

Summary of Chapter 11

- The aim of this chapter is to encourage you to review and improve your writing skills in relation to writing scientific reports in the format usually used in science journals.

- We examine each section of a scientific report, look at its construction, and discuss approaches towards identifying and addressing common errors.

- This evaluation is achieved by examining published research papers.

- The Online Resource Centre includes interactive exercises that test your understanding of this chapter with other topics considered earlier in this book.

 online resource centre

Answers to chapter questions

A2
  a. 5mg of sodium chloride were added to water and stirred until it was dissolved.

  b. It is clear from the literature that ice is less dense than water.

A8 The coordinates are SO853395.

A13 An interval scale of measurement (length of leaves (mm)) has been used but the points along this scale of measurement have not been determined by the investigator and are scattered. To impose some order on the representation of this data, a histogram with classes (Table 11.4) is more useful than a line graph (Fig. 11.5).

A14 Yes. The amount of corn given to the cows is under the investigator's control and is the independent variable. The weight of the calves at birth is the responding dependent variable.

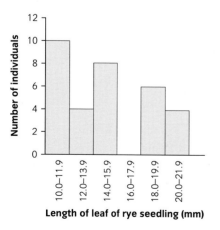

Fig. 11.5. Frequency distribution of leaf length in rye seedlings after 3 weeks' growth.

Appendix a. How to choose a research project

Many undergraduates are expected to carry out a piece of independent research during their final year of a degree course. For some this may be literature-based. In some institutions, the topics are listed and you indicate a preference. Some students, however, are asked to choose and design their own research project. This checklist is designed to help the last group of students in particular, but some points are also relevant for students choosing topics from a prescribed list.

Your honours project in your final year can be used to develop skills or contacts that will be valuable for obtaining employment, or the topic may be chosen because it particularly interests you. Above all, it must be a project you are enthusiastic about as, of all your degree work, it is this research project that demonstrates your abilities as a scientist in your chosen field and as a graduate to be independent in your learning.

Most institutions do not require you, as an undergraduate, to carry out original work. You will, however, be expected to show some initiative. This means that, although you must not copy other people's work, it can usually be adapted. For example, you might apply the same method to a different location or species. However, you must check your local regulations to confirm this.

The checklist that follows is a series of prompts that will help you come to a better understanding of yourself and your interests and therefore help you to focus in on a topic of research that will most suit you.

A1 How to choose a research project

Here is a checklist to help you choose a topic for your independent research.

1) Do you already have an idea?
   YES (go to 14)   NO (go to 2)

2) What do you wish to do when you complete your programme of study?
   You can often use your honours research project to strengthen links with future potential employers, develop skills relevant to your planned career, or provide you with school experience.

3) When would you like to carry out the work?
   Some research is seasonal and it may not be practical for you to carry out research at a particular time of year e.g. fieldwork, observations of reproductive behaviour in vertebrates or invertebrates, or a survey of student opinion.

4) Where would you like to carry out the work?
   You may prefer to work from home or you have contacts abroad, etc.

5) Do you have any potentially useful outside contacts?
   Undergraduates may be able to set up projects with schools' sports clubs, support groups, businesses, charities, or industry. Speak to your supervisor/ personal tutor about this as there will be institutional regulations that relate to such work.

6) Do you prefer fieldwork or laboratory work?
   Most people have a preference for one or other of these and this can guide you when choosing a project.

7) Which part of your course have you most enjoyed?

8) Has there been a particular subject in your course that you would like to investigate further?

9) Often lecturers and PhD students have small research projects they would like to see carried out. Ask.

10) Organizations such as English Nature and the Wildlife Trusts invariably have a list of projects they would like undergraduates to carry out. Get in touch with them or ask a member of staff who is likely to have contacts with the organization that interests you.

11) Go to the library and look through a journal in your area of interest, e.g. *Journal of Biological Education, Journal of Ecology, Heredity*, etc.
    See what other research has been done. Can you extend or repeat part of this work?

12) Look at projects that have been carried out by students in the past. Research reports such as these are often held in the Institution's library.

13) Now do you have an idea?
    YES (go to 14)   NO (see your supervisor/personal tutor)

14) Check that your institution has the equipment you will need and that you can be supervised by someone who works in a similar area. Contact the lecturer whose research area is most similar to the topic you are interested in. Ask them: is this topic suitable?

A2 Common problems

From our experience, we are aware that students' choices of project often suffer from problems. The common ones we encounter most often are listed below. Check your ideas to make sure your project does not suffer from any of these.

A2.1 The topic is too broad

Often students are enthusiastic and find it really hard to narrow down their topic and to remain focused. One approach is to use a flowchart or mind map to show how different sections interlink. Then select from this a central narrow area. You can always expand later if you have time.

A2.2 **The research topic is vague**

Some of the proposals we have seen from our students are very vague, such as 'something on fish'! Vague choices usually reflect a lack of thought and planning. They can also occur when no thought is given to the justification for the research. It may well be possible to examine something, but do you have a scientific reason for doing so? To resolve this make sure you write an aim and objectives at the outset as this will help to guide your planning. Draw a flowchart of the investigation and the thinking behind it, as this can also help you to see whether your choice makes sense.

A2.3 **Time management**

For nearly all researchers from undergraduates in their final year to research scientists, there are time constraints that can have an impact on your research. These fall into two groups:

i. Not appreciating the seasonality of some work

We have referred to this in the checklist in A1, in that there are often time constraints on the work you may wish to carry out. For example, if you wished to examine the association between asthma and pollen, the work would have to be carried out when the pollen was present.

ii. Not thinking about how long it will take to complete the work

When carrying out research in a new area, this is always a point to remember. For example, in Chapter 2 (Example 2.2) we described an undergraduate project in which the student investigated the relative effectiveness of tea-tree oil and triclosan as antibactericides in a GP's practice. To collect the swabs from the volunteers and to inoculate the plates took 13 hours. The student had carried out a pilot study and so was aware that this part of the procedure would take a long time and she was able to make suitable arrangements before she started.

When planning your work, it is beneficial to draw up a detailed timetable and, if you are not familiar with the techniques, to have a trial run to enable you to judge how long it will take to complete each part of the investigation. If your work is to be reported, then ensure you also leave enough time for panicking, computer downtime, drawing figures, printing, and binding. A general rule of thumb for all time management decisions is to work out carefully how long you think it should take and then double the time you have allowed.

A2.4 **Independent learners**

Undergraduate and graduate research projects are usually intended to allow you to show your abilities as an independent researcher. Therefore, they are invariably less formally structured and you will not be chased up by staff. Attending regular meetings with your supervisor and drawing up and sticking to a timetable are essential. Although research is meant to demonstrate your own abilities, this does not mean you should work in a vacuum. Students who do this nearly always perform badly.

This is not what is intended by the words 'independent learners'. Although your peers are not necessarily having the same problems that you are, use their support but also use supervisory support and that of other established researchers.

A2.5 **Trials and tribulations**

Research is almost always unpredictable and things go wrong. If you are getting upset, see your supervisor and if they do not seem to be sympathetic, go to another lecturer or a counsellor. Try to keep things in perspective.

A2.6 **Non-significant results**

Most practical work carried out by undergraduates is designed to illustrate differences between treatments; therefore, you get into the habit of expecting significant differences when testing hypotheses. As a result, a non-significant result is often automatically assumed to be a mistake and dismissed as such. You must be confident that your results are a true reflection of the real underlying biology. Be critical of your work but do not undervalue the importance of non-significant results.

A2.7 **Cheating**

There are many temptations in research to cheat. Perhaps you have an outlier and would like to ignore it to simplify your analysis. You may have missing data because an organism died or a treatment did not work. Failure to acknowledge all your real data or creating data is cheating and will carry significant penalties both as an undergraduate and as a professional scientist. You will not be penalized for being honest.

Appendix b. Planning your experiment

Decision web

The decision web takes you through the process of designing an experiment. If you have more than one objective, you should complete a decision web for each. This decision web may not suit all designs so you may need to be creative. Relevant sections in the book are indicated throughout.

This decision web is available on the Online Resource Centre written using a standard Microsoft application. Using the electronic version, you can expand boxes, move them around and delete them if not relevant. You can also delete the notes on each page.

If you need help choosing a project, you should refer to Appendix a for some suggestions.

online resource centre

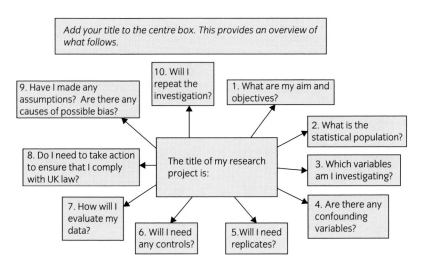

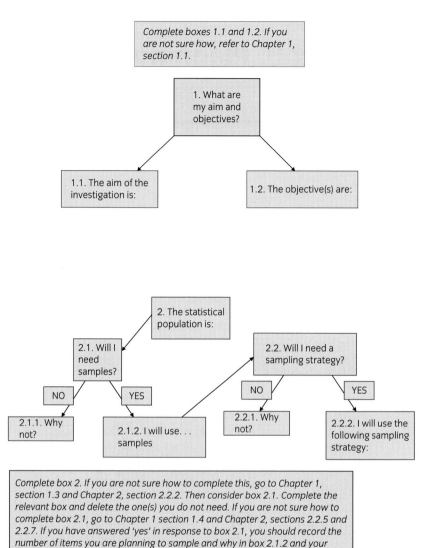

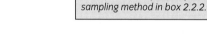

Complete box 2. If you are not sure how to complete this, go to Chapter 1, section 1.3 and Chapter 2, section 2.2.2. Then consider box 2.1. Complete the relevant box and delete the one(s) you do not need. If you are not sure how to complete box 2.1, go to Chapter 1 section 1.4 and Chapter 2, sections 2.2.5 and 2.2.7. If you have answered 'yes' in response to box 2.1, you should record the number of items you are planning to sample and why in box 2.1.2 and your sampling method in box 2.2.2.

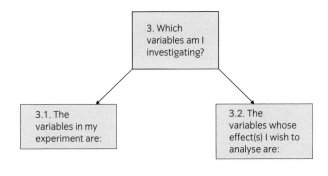

In box 3.1, list all the variables you are interested in. Then consider whether you are investigating the effect of all of them or a subset. Record the variables whose effects you are examining in box 3.2. If you are not sure, you can refer to Chapter 1, sections 1.6 and 1.7 and Chapter 2, section 2.2.3. If you are using a questionnaire etc. to gather data, you should refer to Chapter 3.

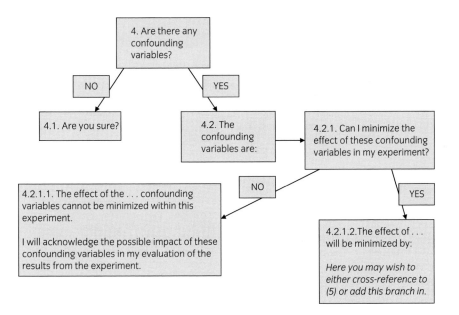

Complete box 4.1 or 4.2. Confounding variables are first considered in Chapter 1, section 1.7 and again in Chapter 2, sections 2.2.4 and 2.2.5. If you have completed box 4.2, then proceed to box 4.2.1. Look at each of the confounding variables in turn and decide whether they should be placed in 4.2.1.1 or 4.2.1.2. If they are in 4.2.1.2, you need to explain what you are going to do to minimize the effect of the confounding variable, e.g. are you keeping the temperature constant throughout, are you going to randomize your sampling etc.

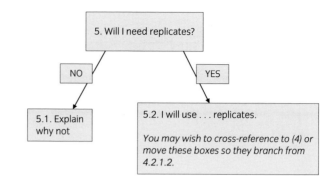

5. Will I need replicates?

NO

YES

5.1. Explain
why not

5.2. I will use . . . replicates.

You may wish to cross-reference to (4) or move these boxes so they branch from 4.2.1.2.

Complete box 5.1 or 5.2. If you are completing box 5.2, then you need to explain what your replicates will be, how many there will be and why you have used this number. If you are not sure, read Chapter 2, section 2.2.5. Remember, replication is not the same as a repeated experiment. Also take care to avoid pseudoreplication.

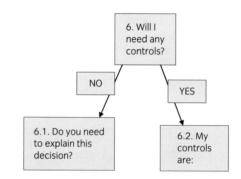

6. Will I
need any
controls?

NO

YES

6.1. Do you need
to explain this
decision?

6.2. My
controls
are:

Complete box 6.1 or 6.2. If you are completing box 6.2, then you need to explain what controls you will have and what they are for. If you are not sure, read Chapter 2, section 2.2.6.

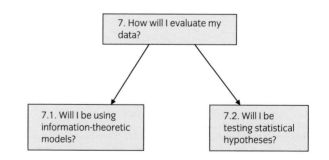

7. How will I evaluate my
data?

7.1. Will I be using
information-theoretic
models?

7.2. Will I be
testing statistical
hypotheses?

Here you need to decide whether you are starting out with a hypothesis to test or not. This is explained in Chapter 1, section 1.8.

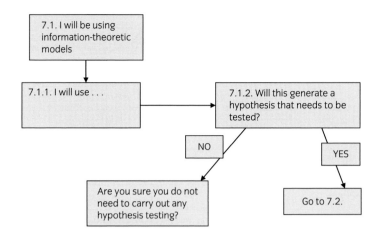

We are not able to explain in this book how to use the range of information-theoretic models that are available so you will need to find other texts to help. You can add notes on the model you will use in box 7.1.1. Often such models generate hypotheses that then need to be tested. You may therefore need to go on to box 7.2. In addition, these models may require you to sample in a specific way or to collect a certain number of observations. This information will help your planning.

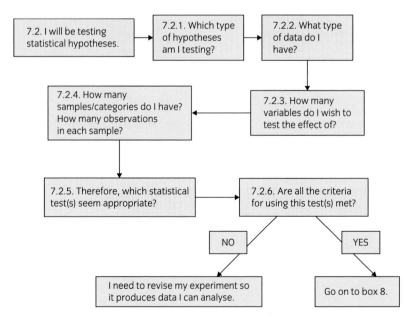

An explanation of these critical steps is on page 406.

Completing this part of your planning often seems to be the hardest step. It is critical that you do this before you start any research or you may generate data that cannot be easily analysed. This step can also often show you that you need to revise your sample size or number of samples.

The first step is to note down something about the hypotheses you are planning to test (boxes 7.2.1 and 7.2.2). You then need to record information in box 7.2.2 about the scale you are using in your measurements (i.e. interval, nominal, or ordinal), whether the data are matched, and whether the data are likely to have a normal distribution and may therefore be considered to be parametric. If you are not sure how to do this, then refer to the following sections in the book:

Decision web box number	Relevant section in book
7.2.1	1.8, 2.2.7, 6.1
7.2.2	5.1, 2.2.7, 9.3
7.2.3	1.7, 2.2.3
7.2.4	1.2, 1.4, 2.2.7

The information gathered in steps 7.2.1–7.2.4 makes it easier for you to select the correct statistical test. We have summarized one method for doing this in Appendix c and an example is given in Chapter 2, section 2.2.8. There are many more examples in the Online Resource Centre with practice questions. Record your decision in box 7.2.5 along with a note about how you came to this conclusion.

Now look up the specific statistical test and check in the text that the criteria for using this test are met, as far as you can tell. Some of the criteria can only be checked when you have your data so you may need to return to this having carried out the experiment. If the criteria are not met, then you need to revise your experiment or review your choice of statistical test. This step can lead you to revise your experiment and you may need to amend earlier parts of the decision tree as a result.

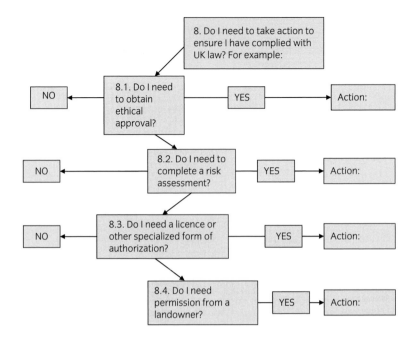

This is not an exhaustive list of the legislation you may need to consider Chapter 2, section 2.2.8 and Chapter 4 which provide further details. You should also discuss this with your supervisor and check your institute's regulations.

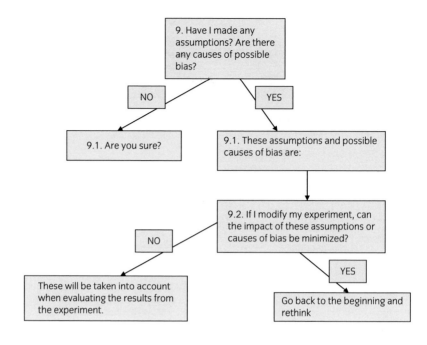

These will be taken into account when evaluating the results from the experiment.

Go back to the beginning and rethink

Occasionally, you consciously or inadvertently make assumptions or introduce possible causes of bias into designs. You need to minimize these if at all possible (Chapter 2, section 2.2.9).

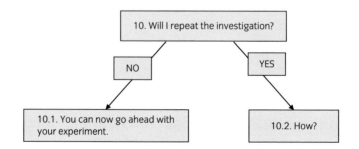

If you are able to repeat the experiment, you need to add any additional details to box 10.1. For an example, see Chapter 2, section 2.2.10.

Appendix c. Which statistical test should I choose?

This book is about the process of designing and reporting research in the biosciences. As an integral part of this process, you will often use statistics. You must make the decisions concerning your statistics before you carry out your research as the statistical tests determine features such as sample size and the number of samples. As statistics is an integral part of research design, we touch on choosing statistical tests in virtually every chapter in this book. Here, we therefore briefly draw these elements together.

Which type of hypotheses am I testing?

There are three types of hypotheses that you need to choose between:

1) Do the data match an expected ratio? or
2) Is there an association between two or more variables? or
3) Do samples come from the same or different populations?

If you are not sure which type of hypotheses you will be testing, read the information in C1–C3 before deciding. For more information about hypotheses and hypothesis testing, read 1.8 and 6.1.

C1 Do the data match an expected ratio?

C1.1 What is meant by 'expectation'?

The term 'expectation' can be used in two ways, only one of which is correct in this context. You may have reason to expect that your observations will follow a particular ratio or mathematical relationship defined by biological or mathematical principles (an *a priori* expectation). For example, you may expect your data to be normally distributed and so to fit a Gaussian equation, or you may expect your data to fit a 1:1 gender ratio. This is the correct way to use the term 'expected'. The alternative is the notion of you having a personal hunch about the likely outcome. This is not an *a priori* expectation and plays no part in science at this level of your

education. When we use the term 'expectation', it is the first of these meanings we are using.

C1.2 When might we have an expectation?

There are three types of investigation where you may have an *a priori* expectation:

- You can reasonably argue that all samples or observations should have the same value. For example, if you sampled the numbers of beetles falling into different-coloured pitfall traps, you might expect that there should be the same number of beetles collected in each trap.

- You will be carrying out a genetic cross or sample in a population and have reason to expect a particular segregation ratio, population genetic ratio, or sex ratio. For example, you may use the Hardy–Weinberg theorem to predict the allele frequencies in the F_1 generation of a population of *Drosophila melanogaster* exposed to a known selection pressure.

- You wish to confirm that your data have a particular distribution such as a normal or a Poisson distribution (5.2).

C1.3 How to choose the correct test

Your choice is determined by the number of variables and number of categories for each variable. This table directs you to the most likely statistical test. If you are not sure, then go to the specific sections indicated and look at the examples to see whether they are similar to the work you are planning. Each test has additional criteria that need to be met. These are given in the sections indicated.

You wish to examine the effect of one variable and have only one sample. You have an *a priori* reason for expecting certain outcomes from your investigation. The variable has more than two categories. The data are counts or frequencies.	Chi-squared goodness-of-fit test (7.1) or *G* goodness-of-fit test (7.5.1).
You wish to examine the effect of one variable and have only one sample. You have an *a priori* reason for expecting certain outcomes from your investigation. The variable has only two categories. The data are counts or frequencies.	Chi-squared goodness-of-fit test with Yates's correction (7.4.1) or *G* goodness-of-fit test (7.5.1).
You wish to examine the effect of one variable with two or more categories. You have more than two samples in your data set and wish to know if the samples are similar or different from each other. The data are counts or frequencies.	Chi-squared test for heterogeneity (7.2).

One of the criteria for using a chi-squared test is that the expected values are greater than 5. Having calculated your expected values, if you find any that are less than 5, you should refer to 7.6.

C2 **Is there an association between two or more variables?**

C2.1 **What is meant by 'association'?**

By association, we mean where one variable is found to change in a similar manner to another variable. For example, is there a significant association between the hardness of shells of eggs laid by pullets and the amount of food supplement they have eaten? Tests for an association can allow you to model the association, make predictions, and test the significance of the association.

C2.2 **How to choose the correct test**

Your choice is determined primarily by whether you expect your data to be parametric (interval) or non-parametric (interval, ordinal, or nominal) and whether you wish to model the association or make predications as well as test the significance of the association. To tell whether your data are parametric, refer to Box 5.2. This table directs you to the most likely statistical test. If you are not sure, go to the specific sections indicated and look at the examples to see whether they are similar to the work you are planning. Each test has additional criteria that need to be met. These are given in the sections indicated.

You wish to test the significance of an association between two variables. The data are counts or frequencies organized into discrete categories or classes. The categories can be derived from ordinal or nominal data or can be imposed on interval data. The distribution does not need to be linear.	Chi-squared or *G* test for association (Chapter 7).
You wish to test the significance of an association between two variables. The distribution appears to be linear. The data are non-parametric.*	Spearman's rank correlation (8.2).
You wish to test the significance of an association between two variables. The distribution appears to be linear. The data are parametric.	Pearson's product moment correlation (8.3).
You have two treatment variables, one of which is usually under the control of the investigator so that the observations are not subject to sampling error. Both variables are measured on interval scales. There is only one *y* value for each value of *x*. The distribution appears to be linear. You wish to test the significance of an association and/or you wish to model the association and/or predict *y* values for given *x* values within the range of your observations.	Simple linear regression (8.5).

You have two treatment variables, one of which is usually under the control of the investigator so that the observations are not subject to sampling error. Both variables are measured on interval scales and the *y* values are parametric. There is more than one *y* value for each value of *x*, with equal replication in all categories. The association appears to be linear. You wish to test the significance of an association and/or you wish to model the association and/or predict *y* values for given *x* values within the range of your observations.	One-way ANOVA to test for a linear regression: more than one *y* for each *x* (equal replicates) (8.6).
You have two treatment variables and neither is under the control of the investigator. The two variables are measured on the same interval scale, e.g. mm. You wish to model the association and/or predict *y* values for given *x* values within the range of your observations.	Principal axis regression (8.7).
You have two treatment variables and neither is under the control of the investigator. The two variables are measured on different scales but both scales are interval, e.g. grams and centimetres. You wish to model the association and/or predict *y* values for given *x* values within the range of your observations.	Ranged principal axis regression (8.8).
None of these tests seems to be right for your data. For example, you have three or more treatment variables or you do not have a linear distribution.	Grafen & Hails (2002); Legendre & Legendre (1998); Sokal & Rohlf (1994).

*Parametric tests are more powerful; therefore, if you have parametric data you should always use a parametric test. If you have non-parametric data, you should consider transforming it (5.9).

C3 Do samples come from the same or different populations?

These are the types of hypotheses that are most frequently investigated by undergraduates. For example, is there is a significant difference between the shell heights (mm) of periwinkles from the lower and mid-shore at Aberystwyth? There are many tests that will test this type of hypothesis. These tests fall into parametric tests (Chapter 9) to be used when you have normally distributed data, and non-parametric tests (Chapter 10) when your data are not normally distributed or the distribution is not known. To tell whether your data are normally distributed, refer to Box 5.2.

C3.1 Parametric tests

These are largely selected on the basis of the experimental design: How many variables? How many categories in each variable? How many replicates in each category?

This table directs you to the most likely statistical test. It is assumed that you have parametric data. The terms 'matched' and 'unmatched' are explained in 9.3 and in the glossary.

If you are not clear about the criteria, refer to the specific section indicated and examine the examples to see whether they are similar to the work you are planning. Each test has additional criteria that need to be met. These are given in the sections indicated.

You wish to examine the effect of one variable with two categories or samples. The data are unmatched.	t- or z-test for unmatched data (9.1 or 9.2).
You wish to examine the effect of one variable with two categories or samples. The data are matched.	t- or z-test for matched data (9.3).
You wish to examine the effect of one variable with more than two categories or samples. There are equal replicates in each category.	One-way ANOVA for equal replicates (9.5). If the outcome from this test is significant, you may follow it with Tukey's multiple comparisons test (9.6).
You wish to examine the effect of one variable with more than two categories or samples. There are unequal replicates in each category.	One-way ANOVA for unequal replicates (9.7). If the outcome from this test is significant, you may follow it with a Tukey–Kramer multiple comparisons test (9.8).
You wish to examine the effect of one variable with more than two categories or samples. There are no replicates. There is a notable confounding variable or the data are repeated measures or matched.	Two-way ANOVA with no replicates (9.12).
You wish to examine the effect of two variables with more than two categories or samples. There are no replicates and all categories from one variable are combined with all categories from the second variable (orthogonal).	Two-way ANOVA with no replicates (9.12).
You wish to examine the effect of two variables. Each variable has at least two categories or samples and all categories from one variable are combined with all categories from the second variable (orthogonal). There are equal replicates in each category.	Two-way ANOVA with equal replicates (9.9) If the outcome from this test is significant, you may follow it with Tukey's multiple comparisons test (9.10).
You wish to examine the effect of two variables. Each variable has at least two categories or samples and all categories from one variable are combined with all categories from the second variable (orthogonal). There is missing data or unequal replicates.	Two-way ANOVA for unequal replicates (9.11).

You wish to examine the effect of two variables. Each variable has at least two categories. One variable is randomized or nested with regard to the second variable. There are equal replicates in each category.	Two-way nested ANOVA for equal replicates (9.13). If the outcome from this test is significant, you may follow it with Tukey's multiple comparisons test (9.13.3).
You wish to examine the effect of three variables. Each variable has at least two categories and all categories from each variable are combined with all other categories from the other variables (orthogonal). There are no replicates.	Three-way factorial ANOVA with no replicates (9.14).
You wish to examine the effect of three treatment variables. Each variable has at least two categories and all categories from each variable are combined with all other categories from the other variables (orthogonal). There are equal replicates in each category.	Three-way ANOVA with equal replicates (9.15).
None of the above.	See Chapter 10, and Sokal & Rohlf (1994) and Zar (2010).

C3.2 Non-parametric tests

The choice of these tests is similar to choosing the parametric tests. Your selection depends on how many treatment variables you are planning to examine, how many categories are in each variable, and how many replicates are in each category. It is assumed that you have non-parametric data. The terms 'matched' and 'unmatched' are explained at the beginning of section 10.2 and in the glossary.

This table directs you to the most likely statistical test. If you are not sure, look at the specific section indicated for the test and examine the examples to see whether they are similar to the work you are planning. Each test has additional criteria that need to be met. These are given in the sections indicated.

You wish to examine the effect of one variable. You are going to compare two categories or samples. The data are unmatched.	Mann–Whitney U test (10.1).
You wish to examine the effect of one variable. You are going to compare two categories or samples. The data are unmatched. The data are measured on a continuous scale and you have more than 30 observations in each sample.	z-test for unmatched data (9.1).
You wish to examine the effect of one variable. You are going to compare two categories or samples. The data are matched. You have fewer than 30 pairs of observations.	Wilcoxon's rank paired test (10.2).

You wish to examine the effect of one variable. You are going to compare two samples. The data are matched. You have more than 30 pairs of observations.	z-test for matched data (9.3).
You wish to examine the effect of one variable. You are going to compare three or more categories or samples.	One-way ANOVA (Kruskal–Wallis test) (10.3). If the outcome from this test is significant, you may follow it with a multiple comparisons test (10.4).
You wish to examine the effect of more than one variable. You are going to compare two or more categories or samples. You have equal numbers of observations in each category.	Two-way non-parametric ANOVA (10.5). If the outcome from this test is significant, you may follow it with a multiple comparisons test (10.6).
You wish to examine the effect of more than one variable. Each variable has two or more categories.	Scheirer–Ray–Hare test (10.7).
You wish to examine the effect of two variables each with two or more categories. The design is orthogonal. There is only one observation in each category. These may be repeated measures.	Friedman's test; see also Sokal & Rohlf (1994) and Zar (2010).

C4 Are you going to report your results?

With most research, you will wish (or be expected) to report your findings. We consider this in Chapter 11. When reporting your findings, you may use summary statistics and you may illustrate the trends in your data using a figure or table. These topics can be found in the following sections:

Summary statistics	Central tendency	Mean, median, mode, skew, and kurtosis	5.4
	Variation	Range, interquartile range, percentiles, variance, standard deviation, standard error of the mean	5.5
		Coefficient of variation	5.6
		Confidence limits	5.7
Tables			11.8.1
Figures			11.8.2
Reporting statistics			11.8.3

Appendix d. Tables of critical values for statistical tests

D1. Critical values for the chi-squared test (χ^2) between $p = 0.05$ and $p = 0.001$, where v is the degrees of freedom and p is the probability

D2. Critical values for Spearman's rank correlation (r_s) between $p = 0.10$ and $p = 0.001$ for a two-tailed test, where n is the number of pairs of observations

D3. Critical values for the Pearson's correlation (r) between $p = 0.10$ and $p = 0.001$ for a two-tailed test, where the degrees of freedom (v) are the number of pairs of observations (n) – 2 and p is the probability

D4. Critical values for the F-test that is carried out to confirm that the variances are homogeneous before a z- or t-test, where $p = 0.05$ and where v_1 is the degrees of freedom for the larger variance (numerator) and v_2 is the degrees of freedom for the smaller variance (denominator)

D5. Critical values for a two-tailed z-test

D6. Critical values for the t-tests between $p = 0.10$ and $p = 0.001$ for a two-tailed test, where v is the degrees of freedom and p is the probability

D7. Critical values for an F_{max} test that is carried out to confirm that the variances are homogeneous before a parametric analysis of variance, where $p = 0.05$, a is the number of samples or treatments, and v is the degrees of freedom

D8. Critical values for an F-test for a parametric ANOVA, where $p = 0.05$, v_1 is the degrees of freedom for the numerator, and v_2 is the degrees of freedom for the denominator

D9. Critical values for an F-test for a parametric ANOVA, where $p = 0.01$, v_1 is the degrees of freedom for the numerator, and v_2 is the degrees of freedom for the denominator

D10. Critical values for an F-test for a parametric ANOVA, where $p = 0.001$, v_1 is the degrees of freedom for the numerator, and v_2 is the degrees of freedom for the denominator

D11. q values for Tukey's test at $p = 0.05$, $p = 0.01$, and $p = 0.001$, where a is the number of samples and v is the degrees of freedom for MS_{within} from the ANOVA calculation

D12. Critical values of U for the Mann–Whitney U test at $p = 0.05$ for a two-tailed test, where n_1 is the number of observations in sample 1 and n_2 is the number of observations in sample 2

D13. Critical values for Wilcoxon's matched pairs test (T) between $p = 0.1$ and $p = 0.002$ for a two-tailed test, where N is the number of pairs of observations used to provide ranks for the calculation (i.e. not those where $d = 0$)

D14. Critical values for H for the Kruskal–Wallis test, between $p = 0.10$ and $p = 0.01$ for a two-tailed test, where a is the number of categories (samples) and n is the number of observations in a category

D15. Critical values for Q for the multiple comparisons test to follow a significant non-parametric ANOVA. The Q statistic is given between $p = 0.10$ and $p = 0.001$ for a two-tailed test, where a is the number of categories (samples)

Table D1 Critical values for the chi-squared test (χ^2) between $p = 0.05$ and $p = 0.001$, where v is the degrees of freedom and p is the probability.

v	p			v	p		
	0.05	0.01	0.001		0.05	0.01	0.001
1	3.84	6.64	10.83	29	42.56	49.59	58.30
2	5.99	9.21	13.82	30	43.77	50.89	59.70
3	7.81	11.34	16.27	31	44.99	52.19	61.10
4	9.49	13.28	18.47	32	46.19	53.47	62.49
5	11.07	15.09	20.51	33	47.40	54.78	63.87
6	12.59	16.81	22.46	34	48.60	56.06	65.25
7	14.07	18.47	24.32	35	49.80	57.34	66.62
8	15.51	20.09	26.12	36	51.00	58.62	67.99
9	16.92	21.67	27.88	37	52.19	59.89	69.35
10	18.31	23.21	29.59	38	53.38	61.16	70.70
11	19.67	24.72	31.26	39	54.57	62.43	72.06
12	21.03	26.22	32.91	40	55.76	63.69	73.40
13	22.36	27.69	34.53	41	56.94	64.95	74.75
14	23.68	29.14	36.12	42	58.12	66.21	76.08
15	25.00	30.58	37.70	43	59.30	67.46	77.42
16	26.30	32.00	39.25	44	60.48	68.71	78.75
17	27.59	33.41	40.79	45	61.66	69.96	80.08
18	28.87	34.80	42.31	46	62.83	71.20	81.40
19	30.14	36.19	43.82	47	64.00	72.44	82.72
20	31.41	37.57	45.31	48	65.17	73.68	84.04
21	32.67	38.93	46.80	49	66.34	74.92	85.35
22	33.92	40.29	48.27	50	67.51	76.15	86.66
23	35.17	41.64	49.73	60	79.08	88.38	99.61
24	36.41	42.98	51.18	70	90.53	100.43	112.32
25	37.65	44.31	52.62	80	101.88	112.33	124.84
26	38.88	45.64	54.05	90	113.145	124.12	137.21
27	40.11	46.96	55.48	100	124.34	140.17	149.45
28	41.34	48.28	56.89				

Table D2 Critical values for Spearman's rank correlation (r_s) between $p = 0.10$ and $p = 0.001$ for a two-tailed test, where n is the number of pairs of observations and p is the probability

(To find the critical values for a one-tailed test, divide the p value in half. For example, the critical values for a two-tailed test when p = 0.1 will be the critical values for p = 0.05 for a one-tailed test.)

n	p				
	0.10	0.05	0.02	0.01	0.001
5	0.900	1.000	1.000	–	–
6	0.829	0.886	0.943	1.000	–
7	0.714	0.786	0.893	0.929	1.000
8	0.643	0.738	0.833	0.881	0.976
9	0.600	0.683	0.783	0.833	0.933
10	0.564	0.648	0.745	0.794	0.903
11	0.523	0.623	0.709	0.818	0.873
12	0.497	0.591	0.678	0.780	0.846
13	0.475	0.566	0.648	0.745	0.824
14	0.457	0.545	0.626	0.716	0.802
15	0.441	0.525	0.604	0.689	0.779
16	0.425	0.507	0.582	0.666	0.762
17	0.412	0.490	0.564	0.645	0.748
18	0.399	0.476	0.549	0.625	0.728
19	0.388	0.462	0.534	0.608	0.712
20	0.377	0.450	0.521	0.591	0.696
21	0.368	0.438	0.508	0.576	0.681
22	0.359	0.428	0.496	0.562	0.667
23	0.351	0.418	0.485	0.549	0.654
24	0.343	0.409	0.475	0.537	0.642
25	0.336	0.400	0.465	0.526	0.630
26	0.329	0.392	0.456	0.515	0.619
27	0.323	0.385	0.448	0.505	0.608
28	0.317	0.377	0.440	0.496	0.598
29	0.311	0.370	0.432	0.487	0.589
30	0.305	0.364	0.425	0.478	0.580
35	0.283	0.335	0.394	0.433	0.539
40	0.264	0.313	0.368	0.405	0.507
45	0.248	0.294	0.347	0.382	0.479
50	0.235	0.279	0.329	0.363	0.456

Table D3 Critical values for the Pearson's correlation (r) between p = 0.10 and p = 0.001 for a two-tailed test, where the degrees of freedom (v) is the number of pairs of observations (n) − 2 and p is the probability®

(To find the critical values for a one-tailed test, divide the p value in half. For example, the critical values for a two-tailed test when p = 0.10 will be the critical values for p = 0.05 in a one-tailed test.)

v	p 0.05	0.01	v	p 0.05	0.01
1	0.997	1.000	26	0.374	0.479
2	0.950	0.990	27	0.367	0.471
3	0.878	0.959	28	0.361	0.463
4	0.811	0.917	29	0.355	0.456
5	0.754	0.874	30	0.349	0.449
6	0.707	0.834	31	0.344	0.443
7	0.666	0.798	32	0.339	0.436
8	0.632	0.765	33	0.334	0.430
9	0.602	0.735	34	0.329	0.424
10	0.576	0.708	35	0.325	0.418
11	0.553	0.684	36	0.320	0.413
12	0.532	0.661	37	0.316	0.408
13	0.514	0.641	38	0.312	0.403
14	0.497	0.623	39	0.308	0.398
15	0.482	0.606	40	0.304	0.393
16	0.468	0.590	41	0.300	0.389
17	0.456	0.575	42	0.297	0.384
18	0.444	0.561	43	0.294	0.380
19	0.433	0.549	44	0.291	0.376
20	0.423	0.537	45	0.288	0.372
21	0.413	0.526	46	0.284	0.368
22	0.404	0.515	47	0.281	0.365
23	0.396	0.505	48	0.279	0.361
24	0.388	0.496	49	0.276	0.358
25	0.381	0.487	50	0.273	0.354

Table D4 Critical values for the F-test that is carried out to confirm that the variances are homogeneous before a z- or t-test, where $p = 0.05$ and v_1 is the degrees of freedom for the larger variance (numerator) and v_2 is the degrees of freedom for the smaller variance (denominator)

v_2	v_1																		
	1	2	3	4	5	6	7	8	9	10	12	15	20	24	30	40	60	120	∞
1	647.8	799.5	864.2	899.6	921.8	937.1	948.2	956.7	963.3	968.6	976.7	984.9	993.1	997.2	1001	1006	1010	1014	1018
2	38.51	39.00	39.17	39.25	39.30	39.33	39.36	39.57	39.39	39.40	39.41	39.43	39.45	39.46	39.46	39.47	39.48	39.49	39.50
3	17.44	16.04	15.44	15.10	14.88	14.73	14.62	14.54	14.47	14.42	14.34	14.25	14.17	14.12	14.08	14.04	13.99	13.95	13.90
4	12.22	10.65	9.98	9.60	9.36	9.20	9.07	8.98	8.90	8.84	8.75	8.66	8.56	8.51	8.46	8.41	8.36	8.31	8.26
5	10.01	8.43	7.76	7.39	7.15	6.98	6.85	6.76	6.68	6.62	6.52	6.43	6.33	6.28	6.23	6.18	6.12	6.07	6.02
6	8.81	7.26	6.60	6.23	5.99	5.82	5.70	5.60	5.52	5.46	5.37	5.27	5.17	5.12	5.07	5.01	4.96	4.90	4.85
7	8.07	6.54	5.89	5.52	5.29	5.12	4.99	4.90	4.82	4.76	4.67	4.57	4.47	4.42	4.36	4.31	4.25	4.20	4.14
8	7.57	6.06	5.42	5.05	4.82	4.65	4.53	4.43	4.36	4.30	4.20	4.10	4.00	3.95	3.89	3.84	3.78	3.73	3.67
9	7.21	5.71	5.08	4.72	4.48	4.32	4.20	4.10	4.03	3.96	3.87	3.77	3.67	3.61	3.56	3.51	3.45	3.39	3.33
10	6.94	5.46	4.83	4.47	4.24	4.07	3.95	3.85	3.78	3.72	3.62	3.52	3.42	3.37	3.31	3.26	3.20	3.14	3.08
11	6.72	5.26	4.63	4.28	4.04	3.88	3.76	3.66	3.59	3.53	3.43	3.33	3.23	3.17	3.12	3.06	3.00	2.94	2.88
12	6.55	5.10	4.47	4.12	3.89	3.73	3.61	3.51	3.44	3.37	3.28	3.18	3.07	3.02	2.96	2.91	2.85	2.79	2.72
13	6.41	4.97	4.35	4.00	3.77	3.60	3.48	3.39	3.31	3.25	3.15	3.05	2.95	2.89	2.84	2.78	2.72	2.66	2.60
14	6.30	4.86	4.24	3.89	3.66	3.50	3.38	3.29	3.21	3.15	3.05	2.95	2.84	2.79	2.73	2.67	2.61	2.55	2.49
15	6.20	4.77	4.15	3.80	3.68	3.41	3.29	3.20	3.12	3.06	2.96	2.86	2.76	2.70	2.64	2.59	2.52	2.46	2.40
16	6.12	4.69	4.08	3.73	3.50	3.34	3.22	3.12	3.05	2.99	2.89	2.79	2.68	2.63	2.57	2.51	2.45	2.38	2.32
17	6.04	4.62	4.01	3.66	3.44	3.28	3.16	3.06	2.98	2.92	2.82	2.72	2.62	2.56	2.50	2.44	2.38	2.32	2.25
18	5.98	4.56	3.95	3.61	3.38	3.22	3.10	3.01	2.93	2.87	2.77	2.67	2.56	2.50	2.44	2.38	2.32	2.26	2.19

ν_2	\multicolumn{19}{c}{ν_1}
	1	2	3	4	5	6	7	8	9	10	12	15	20	24	30	40	60	120	∞
19	5.92	4.51	3.90	3.56	3.33	3.17	3.05	2.96	2.88	2.82	2.72	2.62	2.51	2.45	2.39	2.33	2.27	2.20	2.13
20	5.87	4.46	3.86	3.51	3.29	3.13	3.01	2.91	2.84	2.77	2.68	2.57	2.46	2.41	2.35	2.29	2.22	2.16	2.09
21	5.83	4.42	3.82	3.48	3.25	3.09	2.97	2.87	2.80	2.73	2.64	2.53	2.42	2.37	2.31	2.25	2.18	2.11	2.04
22	5.79	4.38	3.78	3.44	3.22	3.05	2.93	2.84	2.76	2.70	2.60	2.50	2.39	2.33	2.27	2.21	2.14	2.08	2.00
23	5.75	4.35	3.75	3.41	3.18	3.02	2.90	2.81	2.73	2.67	2.57	2.47	2.36	2.30	2.24	2.18	2.11	2.04	1.97
24	5.72	4.32	3.72	3.38	3.15	2.99	2.87	2.78	2.70	2.64	2.54	2.44	2.33	2.27	2.21	2.15	2.08	2.01	1.94
25	5.69	4.29	3.69	3.35	3.13	2.97	2.85	2.75	2.68	2.61	2.51	2.41	2.30	2.24	2.18	2.12	2.05	1.98	1.91
26	5.66	4.27	3.67	3.33	3.10	2.94	2.82	2.73	2.65	2.59	2.49	2.39	2.28	2.22	2.16	2.09	2.03	1.95	1.88
27	5.63	4.24	3.65	3.31	3.08	2.92	2.80	2.71	2.63	2.57	2.47	2.36	2.25	2.19	2.13	2.07	2.00	1.93	1.85
28	5.61	4.22	3.63	3.29	3.06	2.90	2.78	2.69	2.61	2.55	2.45	2.34	2.23	2.17	2.11	2.05	1.98	1.91	1.83
29	5.59	4.20	3.61	3.27	3.04	2.88	2.76	2.67	2.59	2.53	2.43	2.32	2.21	2.15	2.09	2.03	1.96	1.89	1.81
30	5.57	4.18	3.59	3.25	3.03	2.87	2.75	2.65	2.57	2.51	2.41	2.31	2.20	2.14	2.07	2.01	1.94	1.87	1.79
40	5.42	4.05	3.46	3.13	2.90	2.74	2.62	2.53	2.45	2.39	2.29	2.18	2.07	2.01	1.94	1.88	1.80	1.72	1.64
60	5.29	3.93	3.34	3.01	2.79	2.63	2.51	2.41	2.33	2.27	2.17	2.06	1.94	1.88	1.82	1.74	1.67	1.58	1.48
120	5.15	3.80	3.23	2.89	2.67	2.52	2.39	2.30	2.22	2.16	2.05	1.94	1.82	1.76	1.69	1.61	1.53	1.43	1.31
∞	5.02	3.69	3.12	2.79	2.57	2.41	2.29	2.19	2.11	2.05	1.94	1.83	1.71	1.64	1.57	1.48	1.39	1.27	1.00

Table D5 Critical values for a two-tailed z-test

(To find the critical values for a one-tailed test, divide the p value in half. For example, the critical values for a two-tailed test when p = 0.1 will be the critical values for p = 0.05 for a one-tailed test.)

Probability value	z
0.10	1.647
0.05	1.960
0.01	2.576
0.02	2.326
0.002	3.100
0.001	3.291

Table D6 Critical values for the t-test between p = 0.10 and p = 0.001 for a two-tailed test, where ν is the degrees of freedom, and p is the probability

(To find the critical values for a one-tailed test, divide the p value in half. For example, the critical values for a two-tailed test when p = 0.1 will be the critical values for p = 0.05 for a one-tailed test.)

ν	p					
	0.100	0.050	0.025	0.010	0.005	0.001
1	6.314	12.706	25.452	63.657	127.320	636.620
2	2.920	4.303	6.205	9.925	14.089	31.598
3	2.353	3.182	4.176	5.841	7.453	12.941
4	2.132	2.776	3.495	4.604	5.598	8.610
5	2.015	2.571	3.163	4.032	4.773	6.859
6	1.943	2.447	2.969	3.707	4.317	5.959
7	1.895	2.365	2.841	3.499	4.029	5.405
8	1.860	2.306	2.752	3.355	3.832	5.041
9	1.833	2.262	2.685	3.250	3.690	4.781
10	1.812	2.228	2.634	3.169	3.581	4.587
11	1.796	2.201	2.593	3.106	3.497	4.437
12	1.782	2.179	2.560	3.055	3.428	4.318
13	1.771	2.160	2.533	3.012	3.372	4.221
14	1.761	2.145	2.510	2.977	3.326	4.140
15	1.753	2.131	2.490	2.947	3.286	4.073
16	1.746	2.120	2.473	2.921	3.252	4.015
17	1.740	2.110	2.458	2.898	3.222	3.965
18	1.734	2.101	2.445	2.878	3.197	3.922
19	1.729	2.093	2.433	2.861	3.174	3.883
20	1.725	2.086	2.423	2.845	3.153	3.850
21	1.721	2.080	2.414	2.831	3.135	3.819

v	p					
	0.100	0.050	0.025	0.010	0.005	0.001
22	1.717	2.074	2.406	2.819	3.119	3.792
23	1.714	2.069	2.398	2.807	3.104	3.767
24	1.711	2.064	2.391	2.797	3.090	3.745
25	1.708	2.060	2.385	2.787	3.078	3.725
26	1.706	2.056	2.379	2.779	3.067	3.707
27	1.703	2.052	2.373	2.771	3.056	3.690
28	1.701	2.048	2.368	2.763	3.047	3.674
29	1.699	2.045	2.364	2.756	3.038	3.659
30	1.697	2.042	2.360	2.750	3.030	3.646
50	1.676	2.008	2.310	2.678	2.937	3.496
100	1.661	1.982	2.276	2.625	2.871	3.390
∞	1.645	1.960	2.241	2.576	2.807	3.290

Table D7 Critical values for an F_{max} test that is carried out to confirm that the variances are homogeneous before a parametric ANOVA, where $p = 0.05$, a is the number of samples or treatments, and v is the degrees of freedom

v	a										
	2	3	4	5	6	7	8	9	10	11	12
2	39.0	87.5	142	202	266	333	403	475	550	626	704.0
3	15.4	27.8	39.2	50.7	62.0	72.9	83.5	93.9	104	114	124.0
4	9.60	15.5	20.6	25.2	29.5	33.6	37.5	41.1	44.6	48.0	51.4
5	7.15	10.8	13.7	16.3	18.7	20.8	22.9	24.7	26.5	28.2	29.9
6	5.82	8.38	10.4	12.1	13.7	15.0	16.3	17.5	18.6	19.7	20.7
7	4.99	6.94	8.44	9.70	10.8	11.8	12.7	13.5	14.3	15.1	15.8
8	4.43	6.00	7.18	8.12	9.03	9.78	10.5	11.1	11.7	12.2	12.7
9	4.03	5.34	6.31	7.11	7.80	8.41	8.95	9.45	9.91	10.3	10.7
10	3.72	4.85	5.67	6.34	6.92	7.42	7.87	8.28	8.66	9.01	9.34
12	3.28	4.16	4.79	5.30	5.72	6.09	6.42	6.72	7.00	7.25	7.48
15	2.86	3.54	4.01	4.37	4.68	4.95	5.19	5.40	5.59	5.77	5.93
20	2.46	2.95	3.29	3.54	3.76	3.94	4.10	4.24	4.37	4.49	4.59
30	2.07	2.40	2.61	2.78	2.91	3.02	3.12	3.21	3.29	3.36	3.39
60	1.67	1.85	1.96	2.04	2.11	2.17	2.22	2.26	2.30	2.33	2.36
∞	1.00	1.00	1.00	1.00	1.00	1.00	1.00	1.00	1.00	1.00	1.00

Table D8 Critical values for an *F*-test for a parametric ANOVA, where $p = 0.05$, v_1 is the degrees of freedom for the numerator, and v_2 is the degrees of freedom for the denominator

v_2	v_1									
	1	**2**	**3**	**4**	**5**	**6**	**8**	**10**	**20**	**∞**
1	161.4	199.5	215.7	224.6	230.2	234.0	238.9	242.0	249.0	254.3
2	18.51	19.00	19.16	19.25	19.30	19.33	19.37	19.40	19.45	19.50
3	10.13	9.55	9.28	9.12	9.01	8.94	8.84	8.79	8.66	8.53
4	7.71	6.94	6.59	6.39	6.26	6.16	6.04	5.96	5.80	5.63
5	6.61	5.79	5.41	5.19	5.05	4.95	4.82	4.74	4.56	4.36
6	5.99	5.14	4.76	4.53	4.39	4.28	4.15	4.06	3.87	3.67
7	5.59	4.74	4.35	4.12	3.97	3.87	3.73	3.64	3.44	3.23
8	5.32	4.46	4.07	3.84	3.69	3.58	3.44	3.35	3.15	2.93
9	5.12	4.26	3.86	3.63	3.48	3.37	3.23	3.14	2.94	2.71
10	4.96	4.10	3.71	3.48	3.33	3.22	3.07	2.98	2.77	2.54
11	4.85	3.98	3.59	3.36	3.20	3.09	2.95	2.85	2.65	2.40
12	4.75	3.88	3.49	3.26	3.11	3.00	2.85	2.75	2.54	2.30
13	4.67	3.80	3.41	3.18	3.02	2.92	2.77	2.67	2.46	2.21
14	4.60	3.74	3.34	3.11	2.96	2.85	2.70	2.60	2.39	2.13
15	4.54	3.68	3.29	3.06	2.90	2.79	2.64	2.54	2.33	2.07
16	4.49	3.63	3.24	3.01	2.85	2.74	2.59	2.49	2.28	2.01
17	4.45	3.59	3.20	2.96	2.81	2.70	2.55	2.45	2.23	1.96
18	4.41	3.55	3.16	2.93	2.77	2.66	2.51	2.41	2.19	1.92
19	4.38	3.52	3.13	2.90	2.74	2.63	2.48	2.38	2.16	1.88
20	4.35	3.49	3.10	2.87	2.71	2.60	2.45	2.35	2.12	1.84
21	4.32	3.47	3.07	2.84	2.68	2.57	2.42	2.32	2.10	1.81
22	4.30	3.44	3.05	2.82	2.66	2.55	2.40	2.30	2.07	1.78
23	4.28	3.42	3.03	2.80	2.64	2.53	2.38	2.27	2.05	1.76
24	4.26	3.40	3.01	2.78	2.62	2.51	2.36	2.25	2.03	1.73
25	4.24	3.38	2.99	2.76	2.60	2.49	2.34	2.24	2.01	1.71
26	4.22	3.37	2.98	2.74	2.59	2.47	2.32	2.22	1.99	1.69
27	4.21	3.35	2.96	2.73	2.57	2.46	2.30	2.20	1.97	1.67
28	4.20	3.34	2.95	2.71	2.56	2.44	2.29	2.19	1.94	1.65
29	4.18	3.33	2.93	2.70	2.54	2.43	2.28	2.18	1.93	1.64
30	4.17	3.32	2.92	2.69	2.53	2.42	2.27	2.16	1.88	1.62
40	4.08	3.23	2.84	2.61	2.45	2.34	2.18	2.08	1.84	1.51
50	4.03	3.18	2.79	2.56	2.40	2.29	2.13	2.03	1.78	1.44
60	4.00	3.15	2.76	2.52	2.37	2.25	2.10	1.99	1.75	1.39
70	3.98	3.13	2.74	2.50	2.35	2.23	2.07	1.97	1.72	1.35
80	3.96	3.11	2.72	2.49	2.33	2.21	2.06	1.95	1.70	1.31
90	3.95	3.10	2.71	2.17	2.32	2.20	2.04	1.94	1.69	1.28
100	3.94	3.09	2.70	2.46	2.30	2.19	2.03	1.93	1.68	1.26
∞	3.84	2.99	2.60	2.37	2.21	2.10	1.94	1.83	1.57	1.00

Table D9 Critical values for an *F*-test for a parametric ANOVA, where $p = 0.01$, v_1 is the degrees of freedom for numerator, and v_2 is the degrees of freedom for the denominator

	1	2	3	4	5	6	8	10	20	∞
1	4052	4999	5403	5625	5764	5859	5982	6106	6234	6366
2	98.50	99.00	99.17	99.25	99.30	99.33	99.37	99.42	99.40	99.50
3	34.12	30.82	29.46	28.71	28.24	27.91	27.49	27.20	26.70	26.12
4	21.20	18.00	16.69	15.98	15.52	15.21	14.80	14.50	14.0	13.46
5	16.26	13.27	12.06	11.39	10.97	10.67	10.29	10.01	9.55	9.02
6	13.74	10.92	9.78	9.15	8.75	8.47	8.10	7.87	7.40	6.88
7	12.25	9.55	8.45	7.85	7.46	7.19	6.84	6.62	6.16	5.65
8	11.26	8.65	7.59	7.01	6.63	6.37	6.03	5.81	5.36	4.86
9	10.56	8.02	6.99	6.42	6.06	5.80	5.47	5.26	4.81	4.31
10	10.04	7.56	6.55	5.99	5.64	5.39	5.06	4.85	4.41	3.91
11	9.65	7.20	6.22	5.67	5.32	5.07	4.74	4.54	4.10	3.60
12	9.33	6.93	5.95	5.41	5.06	4.82	4.50	4.30	3.86	3.36
13	9.07	6.70	5.74	5.20	4.86	4.62	4.30	4.10	3.66	3.16
14	8.86	6.51	5.56	5.03	4.69	4.46	4.14	3.94	3.51	3.00
15	8.68	6.36	5.42	4.89	4.56	4.32	4.00	3.80	3.37	2.87
16	8.53	6.23	5.29	4.77	4.44	4.20	3.89	3.69	3.26	2.75
17	8.40	6.11	5.18	4.67	4.34	4.10	3.79	3.59	3.16	2.65
18	8.28	6.01	5.09	4.58	4.25	4.01	3.71	3.51	3.08	2.57
19	8.18	5.93	5.01	4.50	4.17	3.94	3.63	3.43	3.00	2.49
20	8.10	5.85	4.94	4.43	4.10	3.87	3.56	3.37	2.94	2.42
21	8.02	5.78	4.87	4.37	4.04	3.81	3.51	3.31	2.88	2.36
22	7.94	5.72	4.82	4.31	3.99	3.76	3.45	3.26	2.83	2.31
23	7.88	5.66	4.76	4.26	3.94	3.71	3.41	3.21	2.78	2.26
24	7.82	5.61	4.72	4.22	3.90	3.67	3.36	3.17	2.74	2.21
25	7.77	5.57	4.68	4.18	3.86	3.63	3.32	3.13	2.70	2.17
26	7.72	5.53	4.64	4.14	3.82	3.59	3.29	3.09	2.66	2.13
27	7.68	5.49	4.60	4.11	3.78	3.56	3.26	3.06	2.63	2.10
28	7.64	5.45	4.57	4.07	3.75	3.53	3.23	3.03	2.60	2.06
29	7.60	5.42	4.54	4.04	3.73	3.50	3.20	3.00	2.57	2.03
30	7.56	5.39	4.51	4.02	3.70	3.47	3.17	2.98	2.55	2.01
40	7.31	5.18	4.31	3.83	3.51	3.29	2.99	2.80	2.37	1.80
50	7.17	5.06	4.20	3.72	3.41	3.19	2.89	2.70	2.27	1.68
60	7.08	4.98	4.13	3.65	3.34	3.12	2.82	2.63	2.20	1.60
70	7.01	4.92	4.07	3.60	3.29	3.07	2.78	2.59	2.15	1.53
80	6.96	4.88	4.04	3.56	3.26	3.04	2.74	2.55	2.12	1.47
90	6.92	4.85	4.01	3.53	3.23	3.01	2.72	2.52	2.09	1.43
100	6.90	4.82	3.98	3.51	3.21	2.99	2.69	2.50	2.07	1.39
∞	6.64	4.60	3.78	3.32	3.02	2.80	2.51	2.32	1.88	1.00

Table D10 Critical values for an *F*-test for a parametric ANOVA, where $p = 0.001$, v_1 is the degrees of freedom for the numerator, and v_2 is the degrees of freedom for the denominator

	1	2	3	4	5	6	8	10	20	∞
1										
2	998	999	999	999	999	999	999	999	999	999
3	167.0	148.5	141.1	137.1	134.6	132.8	130.6	129.00	126.00	123.5
4	74.13	61.24	56.18	53.48	51.71	50.52	49.00	48.10	46.10	44.05
5	47.04	36.61.1	33.28	31.09	29.75	28.84	27.64	26.90	25.40	23.78
6	35.51	27.00	23.70	21.90	20.81	20.03	19.03	18.40	17.10	15.75
7	29.25	21.69	18.77	17.20	16.21	15.52	14.63	14.10	12.90	11.70
8	25.42	18.49	15.83	14.39	13.49	12.86	12.05	11.50	10.50	9.34
9	22.86	16.59	13.98	12.50	11.71	11.13	10.37	9.89	8.90	7.81
10	21.04	14.91	12.55	11.28	10.48	9.93	9.20	8.75	7.80	6.76
11	19.68	13.81	11.56	10.35	9.58	9.05	8.35	7.29	7.01	6.00
12	18.64	12.97	10.81	9.63	8.89	8.38	7.71	7.29	6.40	5.42
13	17.81	12.31	10.21	9.07	8.35	7.86	7.21	6.80	5.93	4.97
14	17.14	11.78	9.73	8.62	7.92	7.43	6.80	6.40	5.56	4.60
15	16.58	11.34	9.34	8.25	7.57	7.09	6.47	6.08	5.25	4.31
16	16.12	10.97	9.00	7.94	7.27	6.81	6.19	5.81	4.99	4.06
17	15.72	10.66	8.73	7.68	7.02	6.56	5.96	5.58	4.78	3.85
18	15.38	10.39	8.49	7.46	6.81	6.35	5.76	5.39	4.59	3.67
19	15.08	10.16	8.28	7.27	6.61	6.18	5.59	5.22	4.43	3.51
20	14.82	9.95	8.10	7.10	6.46	6.02	5.44	5.08	4.29	3.38
21	14.59	9.77	7.94	6.95	6.32	5.88	5.31	4.95	4.17	3.26
22	14.38	9.61	7.80	6.81	6.19	5.76	5.19	4.83	4.06	3.15
23	14.19	9.47	7.67	6.70	6.08	5.65	5.09	4.73	3.96	3.05
24	14.03	9.34	7.55	6.59	5.98	5.55	4.99	4.64	3.87	2.97
25	13.87	9.22	7.45	6.49	5.89	5.46	4.91	4.56	3.79	2.89
26	13.74	9.12	7.36	6.41	5.80	5.38	4.83	4.48	3.72	2.82
27	13.61	9.02	7.27	6.33	5.73	5.31	4.76	4.41	3.66	2.75
28	13.50	8.93	7.19	6.25	5.66	5.24	4.69	4.35	3.60	2.69
29	13.39	8.85	7.12	6.19	5.59	5.18	4.65	4.29	3.54	2.64
30	13.29	8.77	7.05	6.12	5.53	5.12	4.58	4.24	3.49	2.59
40	12.61	8.25	6.60	5.70	5.13	4.73	4.21	3.87	3.14	2.23
50	12.22	7.96	6.34	5.46	4.90	4.51	4.00	3.67	2.95	2.02
60	11.97	7.77	6.17	5.31	4.76	4.37	3.87	3.54	2.83	1.89
70	11.80	7.64	6.06	5.20	4.66	4.28	3.77	3.45	2.74	1.78
80	11.67	7.54	5.97	5.12	4.58	4.20	3.70	3.39	2.68	1.72
90	11.57	7.47	5.91	5.06	4.53	4.15	3.87	3.34	2.63	1.66
100	11.50	7.41	5.86	5.02	4.48	4.11	3.61	3.30	2.59	1.61
∞	10.83	6.91	5.42	4.62	4.10	3.74	3.27	2.96	2.27	1.00

Table D11 q values for a Tukey's test at $p = 0.05$, $p = 0.01$, and $p = 0.001$, where a is the number of samples and v is the degrees of freedom for MS_{within} from the ANOVA calculation

v	p	2	3	4	5	6	7	8	9	10	11	12	13	14	15	16	17	18	19	20
									a										
1	0.05	17.97	26.98	32.82	37.08	40.41	43.12	45.40	47.36	49.07	50.59	51.96	53.30	54.33	55.36	56.32	57.22	58.04	58.83	59.56
	0.01	90.03	135.0	164.3	185.6	202.2	215.8	227.0	237.0	245.6	253.2	260.0	266.2	271.8	277.0	281.8	286.3	290.4	294.3	298.0
	0.001	90.3	1351	1643	1856	2022	2158	2272	2370	2455	2532	2600	2662	2718	2770	2818	2863	2904	2934	2980
2	0.05	6.09	8.33	9.80	10.90	11.74	12.44	13.03	13.54	13.99	14.39	14.75	15.08	15.38	15.65	15.91	16.14	16.37	16.57	16.77
	0.01	14.04	19.02	22.29	24.72	26.63	28.20	29.53	30.68	31.69	32.59	33.40	34.13	34.81	35.43	36.00	36.53	37.03	37.50	37.95
	0.001	44.69	60.42	70.77	78.43	84.49	89.46	93.67	97.30	100.5	103.3	105.9	108.2	110.4	112.3	114.2	115.9	117.4	118.9	120.3
3	0.05	4.50	5.91	6.83	7.50	8.04	8.48	8.85	9.18	9.46	9.72	9.95	10.15	10.35	10.53	10.69	10.84	10.98	11.11	11.24
	0.01	8.26	10.62	12.17	13.33	14.24	15.00	15.64	16.20	16.69	17.13	17.53	17.89	18.22	18.52	18.81	19.07	19.32	19.55	19.77
	0.001	18.28	23.32	26.65	29.13	31.11	32.74	34.12	35.33	36.39	37.34	38.20	38.98	39.69	40.35	40.97	41.54	42.07	42.58	43.05
4	0.05	3.93	5.04	5.76	6.29	6.71	7.05	7.35	7.60	7.83	8.03	8.21	8.37	8.53	8.66	8.79	8.91	9.03	9.13	9.23
	0.01	6.51	8.12	9.17	9.96	10.58	11.10	11.55	11.93	12.27	12.57	12.84	13.09	13.32	13.53	13.73	13.91	14.08	14.24	14.40
	0.001	12.18	14.99	16.84	18.23	19.34	20.26	21.04	21.73	22.33	22.87	23.36	23.81	24.21	24.59	24.94	25.27	25.58	25.87	26.14
5	0.05	3.64	4.60	5.22	5.67	6.03	6.33	6.58	6.80	6.99	7.17	7.32	7.47	7.60	7.72	7.83	7.93	8.03	8.12	8.21
	0.01	5.70	6.97	7.80	8.42	8.91	9.32	9.67	9.97	10.24	10.48	10.70	10.89	11.08	11.24	11.40	11.55	11.68	11.81	11.93
	0.001	9.71	11.67	12.96	13.93	14.71	15.35	15.90	16.38	16.81	17.18	17.53	17.85	18.13	18.41	18.66	18.89	19.10	19.31	19.51
6	0.05	3.46	4.34	4.90	5.31	5.63	5.89	6.12	6.32	6.49	6.65	6.79	6.92	7.03	7.14	7.24	7.34	7.43	7.51	7.59
	0.01	5.24	6.33	7.03	7.56	7.97	8.32	8.61	8.87	9.10	9.30	9.49	9.65	9.81	9.95	10.08	10.21	10.32	10.43	10.54
	0.001	8.43	9.96	10.97	11.72	12.32	12.83	13.26	13.63	13.97	14.27	14.54	14.79	15.01	15.22	15.42	15.60	15.78	15.94	17.97
7	.05	3.34	4.16	4.68	5.06	5.36	5.61	5.82	6.00	6.16	6.30	6.43	6.55	6.66	6.76	6.85	6.94	7.02	7.09	7.17
	.01	4.95	5.92	6.54	7.01	7.37	7.68	7.94	8.17	8.37	8.55	8.71	8.86	9.00	9.12	9.24	9.35	9.46	9.55	9.65
	.001	7.65	8.93	9.76	10.40	10.90	11.32	11.68	11.99	12.27	12.52	12.74	12.95	13.14	13.32	13.48	13.64	13.78	13.92	14.04
8	0.05	3.26	4.04	4.53	4.89	5.17	5.40	5.60	5.77	5.92	6.05	6.18	6.29	6.39	6.48	6.57	6.65	6.73	6.80	6.87
	0.01	4.74	5.63	6.20	6.63	6.96	7.24	7.47	7.68	7.87	8.03	8.18	8.31	8.44	8.55	8.66	8.76	8.85	8.94	9.03
	0.001	7.13	8.25	9.00	9.52	10.00	10.32	10.64	10.91	11.15	11.36	11.56	11.74	11.91	12.06	12.21	12.34	12.47	12.59	12.70
9	0.05	3.20	3.95	4.42	4.76	5.02	5.24	5.43	5.60	5.74	5.87	5.98	6.09	6.19	6.28	6.36	6.44	6.51	6.58	6.64
	0.01	4.60	5.43	5.96	6.35	6.66	6.91	7.13	7.32	7.49	7.65	7.78	7.91	8.03	8.13	8.23	8.32	8.41	8.49	8.57
	0.001	6.76	7.77	8.42	8.91	9.30	9.62	9.90	10.14	10.36	10.55	10.73	10.89	11.03	11.18	11.30	11.42	11.54	11.64	11.75

v	p	2	3	4	5	6	7	8	9	10	11	12	13	14	15	16	17	18	19	20
																				a
10	0.05	3.15	3.88	4.33	4.65	4.91	5.12	5.30	5.46	5.60	5.72	5.83	5.93	6.03	6.11	6.20	6.27	6.34	6.40	6.47
	0.01	4.48	5.27	5.77	6.14	6.43	6.67	6.87	7.05	7.21	7.36	7.48	7.60	7.71	7.81	7.91	7.99	8.07	8.15	8.22
	0.001	6.49	7.41	8.01	8.45	8.80	9.10	9.35	9.57	9.77	9.95	10.11	10.25	10.39	10.52	10.64	10.75	10.85	10.95	11.03
11	0.05	3.11	3.82	4.26	4.57	4.82	5.03	5.20	5.35	5.49	5.61	5.71	5.81	5.90	5.99	6.06	6.14	6.20	6.26	6.33
	0.01	4.39	5.14	5.62	5.97	6.25	6.48	6.67	6.84	6.99	7.13	7.25	7.36	7.46	7.56	7.65	7.73	7.81	7.88	7.95
	0.001	6.28	7.14	7.69	8.10	8.43	8.70	8.93	9.14	9.32	9.48	9.63	9.77	9.89	10.01	10.12	10.22	10.31	10.41	10.49
12	0.05	3.08	3.77	4.20	4.51	4.75	4.95	5.12	5.27	5.40	5.51	5.62	5.71	5.80	5.88	5.95	6.03	6.09	6.15	6.21
	0.01	4.32	5.04	5.50	5.84	6.10	6.32	6.51	6.67	6.81	6.94	7.06	7.17	7.26	7.36	7.44	7.52	7.59	7.66	7.73
	0.001	6.11	6.92	7.44	782	8.13	8.38	8.60	8.79	8.96	9.11	9.25	9.38	9.50	9.61	9.71	9.80	9.89	9.98	10.06
13	0.05	3.06	3.73	4.15	4.45	4.69	4.88	5.05	5.19	5.32	5.43	5.53	5.63	5.71	5.79	5.86	5.93	6.00	6.05	6.11
	0.01	4.26	4.96	5.40	5.73	5.98	6.19	6.37	6.53	6.67	6.79	6.90	7.01	7.10	7.19	7.27	7.34	7.42	7.48	7.55
	0.001	5.97	6.74	7.23	7.60	7.89	8.13	8.33	8.51	8.67	8.82	8.95	9.07	9.18	9.28	9.38	9.46	9.55	9.63	9.70
14	0.05	3.03	3.70	4.11	4.41	4.64	4.83	4.99	5.13	5.25	5.36	5.46	5.55	5.64	5.72	5.79	5.85	5.92	5.97	6.03
	0.01	4.21	4.89	5.32	5.63	5.88	6.08	6.26	6.41	6.54	6.66	6.77	6.87	6.96	7.05	7.12	7.20	7.27	7.33	7.39
	0.001	5.86	6.59	7.06	7.41	7.69	7.92	8.11	8.28	8.43	8.51	8.70	8.81	8.91	9.01	9.10	9.19	9.27	9.34	9.42
15	0.05	3.01	3.67	4.08	4.37	4.60	4.78	4.94	5.08	5.20	5.31	5.40	5.49	5.58	5.65	5.72	5.79	5.85	5.90	5.96
	0.01	4.17	4.83	5.25	5.56	5.80	5.99	6.16	6.31	6.44	6.55	6.66	6.76	6.84	6.93	7.00	7.07	7.14	7.20	7.26
	0.001	5.76	6.47	6.92	7.25	7.52	7.74	7.93	8.09	8.23	8.37	8.48	8.59	8.69	8.79	8.87	8.95	9.03	9.10	9.17
16	0.05	3.00	3.65	4.05	4.33	4.56	4.74	4.90	5.03	5.15	5.26	5.35	5.44	5.52	5.59	5.66	5.72	5.79	5.84	5.90
	0.01	4.13	4.78	5.19	5.49	5.72	5.92	6.08	6.22	6.35	6.46	6.56	6.66	6.74	6.82	6.90	6.97	7.03	7.09	7.15
	0.001	5.68	6.37	6.80	7.12	7.37	7.59	7.77	7.92	8.06	8.19	8.30	8.41	8.50	8.59	8.68	8.76	8.83	8.90	8.96
17	0.05	2.98	3.63	4.02	4.30	4.52	4.71	4.86	4.99	5.11	5.21	5.31	5.39	5.47	5.55	5.61	5.68	5.74	5.79	5.84
	0.01	4.10	4.74	5.14	5.43	5.66	5.85	6.01	6.15	6.27	6.38	6.48	6.57	6.66	6.73	6.80	6.87	6.94	7.00	7.05
	0.001	5.61	6.28	6.70	7.01	7.25	7.45	7.63	7.78	7.92	8.04	8.15	8.25	8.34	8.43	8.51	8.58	8.65	8.72	8.78
18	0.05	2.97	3.61	4.00	4.28	4.49	4.67	4.82	4.96	5.07	5.17	5.27	5.35	5.43	5.50	5.57	5.63	5.69	5.74	5.79
	0.01	4.07	4.70	5.09	5.38	5.60	5.79	5.94	6.08	6.20	6.31	6.41	6.50	6.58	6.65	6.72	6.79	6.85	6.91	6.96
	0.001	5.55	6.20	6.60	6.91	7.14	7.34	7.51	7.66	7.79	7.91	8.01	8.11	8.20	8.23	8.36	8.43	8.50	8.57	8.63

v	p																			a
		2	3	4	5	6	7	8	9	10	11	12	13	14	15	16	17	18	19	20
19	0.05	2.96	3.59	3.98	4.25	4.47	4.65	4.79	4.92	5.04	5.14	5.23	5.32	5.39	5.46	5.53	5.59	5.65	5.70	5.75
	0.01	4.05	4.67	5.05	5.33	5.55	5.73	5.89	6.02	6.14	6.25	6.34	6.43	6.51	6.58	6.65	6.72	6.78	6.84	6.89
	0.001	5.49	6.13	6.53	6.82	7.05	7.24	7.41	7.55	7.68	7.79	7.89	8.00	8.08	8.16	8.23	8.30	8.37	8.43	8.49
20	0.05	2.95	3.58	3.96	4.23	4.45	4.62	4.77	4.90	5.01	5.11	5.20	5.28	5.36	5.43	5.49	5.55	5.61	5.66	5.71
	0.01	4.02	4.64	5.02	5.29	5.51	5.69	5.84	5.97	6.09	6.19	6.29	6.37	6.45	6.52	6.59	6.65	6.71	6.76	6.82
	0.001	5.44	6.07	6.45	6.74	7.00	7.15	7.31	7.45	7.58	7.69	7.79	7.88	8.00	8.04	8.12	8.19	8.25	8.32	8.37
24	0.05	2.92	3.53	3.90	4.17	4.37	4.54	4.68	4.81	4.92	5.01	5.10	5.18	5.25	5.32	5.38	5.44	5.50	5.54	5.59
	0.01	3.96	4.54	4.91	5.17	5.37	5.54	5.69	5.81	5.92	6.02	6.11	6.19	6.26	6.33	6.39	6.45	6.51	6.56	6.61
	0.001	5.30	5.90	6.24	6.50	6.71	6.88	7.03	7.16	7.27	7.37	7.47	7.55	7.63	7.70	7.77	7.83	7.89	7.95	8.00
30	0.05	2.89	3.49	3.84	4.10	4.30	4.46	4.60	4.72	4.83	4.92	5.00	5.08	5.15	5.21	5.27	5.33	5.38	5.43	5.48
	0.01	3.89	4.45	4.80	5.05	5.24	5.40	5.54	5.65	5.76	5.85	5.93	6.01	6.08	6.14	6.20	6.26	6.31	6.36	6.41
	0.001	5.16	5.70	6.03	6.28	6.47	6.63	6.76	6.88	6.98	7.10	7.16	7.24	7.31	7.38	7.44	7.49	7.55	7.60	7.65
40	0.05	2.86	3.44	3.79	4.04	4.23	4.39	4.52	4.63	4.74	4.82	4.91	4.98	5.05	5.11	5.16	5.22	5.27	5.31	5.36
	0.01	3.82	4.37	4.70	4.93	5.11	5.27	5.39	5.50	5.60	5.69	5.77	5.84	5.90	5.96	6.02	6.07	6.12	6.17	6.21
	0.001	5.02	5.53	5.84	6.06	6.24	6.39	6.51	6.62	6.71	6.80	6.87	6.94	7.01	7.07	7.12	7.17	7.22	7.27	7.32
60	0.05	2.83	3.40	3.74	3.98	4.16	4.31	4.44	4.55	4.65	4.73	4.81	4.88	4.94	5.00	5.06	5.11	5.16	5.20	5.24
	0.01	3.76	4.28	4.60	4.82	4.99	5.13	5.25	5.36	5.45	5.53	5.60	5.67	5.73	5.79	5.84	5.89	5.93	5.98	6.02
	0.001	4.89	5.37	5.65	5.86	6.02	6.16	6.27	6.37	6.45	6.53	6.60	6.66	6.72	6.77	6.82	6.87	6.91	6.96	7.00
120	0.05	2.80	3.36	3.69	3.92	4.10	4.24	4.36	4.48	4.56	4.64	4.72	4.78	4.84	4.90	4.95	5.00	5.05	5.09	5.13
	0.01	3.70	4.20	4.50	4.71	4.87	5.01	5.12	5.21	5.30	5.38	5.44	5.51	5.56	5.61	5.66	5.71	5.75	5.79	5.83
	0.001	4.77	5.21	5.48	5.67	5.82	5.94	6.04	6.13	6.21	6.28	6.34	6.40	6.45	6.50	6.54	6.58	6.62	6.66	6.70
∞	0.05	2.77	3.31	3.63	3.86	4.03	4.17	4.29	4.39	4.47	4.55	4.62	4.68	4.74	4.80	4.85	4.89	4.93	4.97	5.01
	0.01	3.64	4.12	4.40	4.60	4.76	4.88	4.99	5.08	5.16	5.23	5.29	5.35	5.40	5.45	5.49	5.54	5.57	5.61	5.65
	0.001	4.65	5.06	5.31	5.48	5.62	5.73	5.82	5.90	5.97	6.04	6.09	6.14	6.19	6.23	6.27	6.31	6.35	6.38	6.41

Table D12 Critical values of U for the Mann–Whitney U test at $p = 0.05$ for a two-tailed test, where n_1 is the number of observations in sample 1 and n_2 is the number of observations in sample 2

(To find the critical values for a one-tailed test, divide the p value in half. For example, the critical values for a two-tailed test when $p = 0.1$ will be the critical values for $p = 0.05$ for a one-tailed test.)

n_1 \ n_2	2	3	4	5	6	7	8	9	10	11	12	13	14	15	16	17	18	19	20
2							0	0	0	0	1	1	1	1	1	2	2	2	2
3				0	1	1	2	2	3	3	4	4	5	5	6	6	7	7	8
4			0	1	2	3	4	4	5	6	7	8	9	10	11	11	12	13	13
5		0	1	2	3	5	6	7	8	9	11	12	13	14	15	17	18	19	20
6		1	2	3	5	6	8	10	11	13	14	16	17	19	21	22	24	25	27
7		1	3	5	6	8	10	12	14	16	18	20	22	24	26	28	30	32	34
8	0	2	4	6	8	10	13	15	17	19	22	24	26	29	31	34	36	38	41
9	0	2	4	7	10	12	15	17	20	23	26	28	31	34	37	39	42	45	48
10	0	3	5	8	11	14	17	20	23	26	29	33	36	39	42	45	48	52	55
11	0	3	6	9	13	16	19	23	26	30	33	37	40	44	47	51	55	58	62
12	1	4	7	11	14	18	22	26	29	33	37	41	45	49	53	57	61	65	69
13	1	4	8	12	16	20	24	28	33	37	41	45	50	54	59	63	67	72	76
14	1	5	9	13	17	22	26	31	36	40	45	50	55	59	64	67	74	78	83
15	1	5	10	14	19	24	29	34	39	44	49	54	59	64	70	75	80	85	90
16	1	6	11	15	21	26	31	37	42	47	53	59	64	70	75	81	86	92	98
17	2	6	11	17	22	28	34	39	45	51	57	63	67	75	81	87	93	99	105
18	2	7	12	18	24	30	36	42	48	55	61	67	74	80	86	93	99	106	112
19	2	7	13	19	25	32	38	45	52	58	65	72	78	85	92	99	106	113	119
20	2	8	13	20	27	34	41	48	55	62	69	76	83	90	98	105	112	119	127

Table D13 Critical values for *T* for Wilcoxon's matched pairs test between *p* = 0.1 and *p* = 0.001 for a two-tailed test, where *N* is the number of pairs of observations used to provide ranks for the calculation (i.e. not those where *d* = 0)

(To find the critical values for a one-tailed test, divide the p value in half. For example, the critical values for a two-tailed test when p = 0.1 will be the critical values for p = 0.05 for a one-tailed test.)

N	p			
	0.1	0.05	0.01	0.001
5	0			
6	2	0		
7	3	2		
8	5	3	0	
9	8	5	1	
10	10	8	3	
11	13	10	5	0
12	17	13	7	1
13	21	17	9	2
14	25	21	12	4
15	30	25	15	6
16	35	29	19	8
17	41	34	23	11
18	47	40	27	14
19	53	46	32	18
20	60	52	37	21
21	67	58	42	25
22	75	65	48	30
23	83	73	54	35
24	91	81	61	40
25	100	89	68	45
26	110	98	75	51
27	119	107	83	57
28	130	116	91	64
29	140	126	100	71
30	151	137	109	78
31	163	147	118	86
32	175	159	128	94
33	187	170	138	102
34	200	182	148	111
35	213	195	159	120

Table D14 Critical values for H for the Kruskal–Wallis test, between $p = 0.10$ and $p = 0.01$ for a two-tailed test, where a is the number of categories (samples) and n is the number of observations in a category

	a											
	3			4			5			6		
n \ p	0.10	0.05	0.01	0.10	0.05	0.01	0.10	0.05	0.01	0.10	0.05	0.01
2	4.571	–	–		6.167	6.667	6.982	7.418	8.291	8.154	8.846	9.846
3	4.622	5.600	7.200	5.667	7.000	8.538	7.333	8.333	10.200	8.620	9.789	11.820
4	4.654	5.692	7.654	6.026	7.235	9.287	7.457	8.685	11.070	8.800	10.140	12.720
5	4.560	5.780	8.00	6.088	7.377	9.789	7.532	8.876	11.570	8.902	10.360	13.260
6	4.643	5.801	8.222	6.120	7.453	10.090	7.557	9.002	11.910	8.958	10.500	13.600
7	4.594	5.819	8.378	6.127	7.501	10.250	7.600	9.080	12.140	8.992	10.590	13.840
8	4.595	5.805	8.465	6.141	7.534	10.420	7.624	9.126	12.290	9.037	10.660	13.990
9	4.586	5.831	8.529	6.148	7.557	10.530	7.637	9.166	12.410	9.057	10.710	14.130
10	4.581	5.853	8.607	6.161	7.586	10.620	7.650	9.200	12.500	9.078	10.750	14.240
11	4.587	5.885	8.648	6.167	7.623	10.690	7.660	9.242	12.580	9.093	10.760	14.320
12	4.578	5.872	8.712	6.163	7.629	10.750	7.675	9.274	12.630	9.105	10.790	14.380
13	4.601	5.901	8.735	6.185	7.645	10.800	7.685	9.303	12.690	9.115	10.830	14.440
14	4.592	5.896	8.754	6.191	7.658	10.840	7.695	9.307	12.740	9.125	10.840	14.490
15	4.591	5.902	8.821	6.198	7.676	10.870	7.701	9.302	12.770	9.133	10.860	14.530
16	4.595	5.909	8.822	6.201	7.678	10.900	7.705	9.313	12.790	9.140	10.880	14.560
17	4.593	5.915	8.856	6.205	7.682	10.920	7.709	9.325	12.830	9.144	10.880	14.600
18	4.596	5.932	8.865	6.206	7.698	10.950	7.714	9.334	12.850	9.149	10.890	14.630
19	4.598	5.923	8.887	6.212	7.701	10.980	7.717	9.342	12.870	9.156	10.900	14.640
20	4.594	5.926	8.905	6.212	7.703	10.980	7.719	9.353	12.910	9.159	10.920	14.670
21	4.597	5.930	8.918	6.216	7.709	11.010	7.723	9.356	12.920	9.164	10.930	14.700
22	4.597	5.932	8.928	6.218	7.714	11.030	7.724	9.362	12.920	9.168	10.940	14.720
23	4.598	5.937	8.947	6.215	7.719	11.030	7.727	9.368	12.940	9.171	10.930	14.740
24	4.598	5.936	8.964	6.220	7.724	11.060	7.729	9.375	12.960	9.170	10.930	14.740
25	4.605	5.942	8.975	6.222	7.727	11.070	7.730	9.377	12.960	9.177	10.940	14.770

Table D15 Critical values for Q for the multiple comparisons test to follow a significant non-parametric ANOVA. The Q statistic is given between $p = 0.10$ and $p = 0.001$ for a two tailed test, where a is the number of categories (samples)

a	p			
	0.10	0.05	0.01	0.001
2	1.645	1.960	2.576	3.291
3	2.128	2.394	2.936	3.588
4	2.394	2.639	3.144	3.765
5	2.576	2.807	3.291	3.891
6	2.713	2.936	3.403	3.988
7	2.823	3.038	3.494	4.067
8	2.914	3.124	3.570	4.134
9	2.992	3.197	3.635	4.191
10	3.059	3.261	3.692	4.241
11	3.119	3.317	3.743	4.286
12	3.172	3.368	3.789	4.326
13	3.220	3.414	3.830	4.363
14	3.264	3.456	3.878	4.397
15	3.304	3.494	3.902	4.428
16	3.342	3.529	3.935	4.456
17	3.376	3.562	3.965	4.483
18	3.409	3.593	3.993	4.508
19	3.439	3.662	4.019	4.532
20	3.467	3.649	4.044	4.554
21	3.494	3.675	4.067	4.575
22	3.519	3.699	4.089	4.595
23	3.543	3.722	4.110	4.614
24	3.566	3.744	4.130	4.632
25	3.588	3.765	4.149	4.649

Appendix e. Maths and statistics

online resource centre

In this book, we come across a number of symbols and mathematical procedures. For those of you who are not familiar with these symbols or are not confident using them and the maths, we give further explanations and worked examples. There are questions at the end of each section to test your understanding. The answers are at the end of this appendix. If you work through all the questions, it will take about two hours. In addition, in the Online Resource Centre, we include the full step-by-step calculations for each of the worked examples that we have included in the book.

E1 You and your calculator

In this appendix, we encourage you to use some functions already integrated into your calculator. This requires you to own a statistical calculator. There are three types of calculators: ones that will carry out basic steps in maths such as addition and subtraction and no more; a statistical calculator; and a programmable and/or graphical calculator. The first of these does not offer you enough built-in functions to be useful and the last has considerably more than you will ever need and is unnecessarily difficult to master. For graphics and more complex statistics, use a computer package. Statistical calculators are not expensive. Look for one that calculates n, Σx, Σx^2 and the sample variance or standard deviation. Other useful functions have been highlighted below.

E2 Add, subtract, divide, multiply, equals, more than, and less than

In this book, we use a number of symbols. These are a way of giving instructions to you, telling you what type of calculation to carry out.

You will be familiar with + (add), − (subtract), × (multiply), and ÷ (divide). Some of these processes can be written in other ways.

Multiply can be indicated as ab (i.e. a × b) or if you have more complex sums you may use brackets $(1 + 2)(2 + 3)$. This tells you to multiply the result from the sum inside the first set of brackets by the result from the sum inside the second set of brackets. In this case, this would be $3 \times 5 = 15$.

Division can be written as ÷, as in 4 ÷ 2, but it is more usual to write 4/2. If the calculations are more complex, division may be written for example as $(6 \times 3)/(4 - 2)$.

The top part of an equation where you are asked to carry out a division is called the **numerator**. The lower part is called the denominator.

You will also be familiar with the symbol = (equals) as in 4 ÷ 2 = 2, or 4/2 = 2. There are a number of other symbols that can also be used here. For example, we use ≈, which means 'approximately equals'. Two other symbols used in place of the equals sign are < and >. The < symbol means less than, and the > symbol means more than. For example, $p < 0.05$ means that p is less than 0.05. If $p > 0.05$, then p is greater than 0.05. These two symbols can be combined to set limits to a value. For example, $0.05 > p > 0.01$ means that 0.05 is greater than p which is greater than 0.01, i.e. p is a value somewhere between 0.05 and 0.01. We also use these symbols when writing out specific hypotheses, e.g. A > B > C would mean that the median of sample A is greater than the median of sample B which is greater than the median of sample C.

Numerator The value that is being divided into

Denominator The value that is being used to divide

Q1 Write in English your understanding of the following expressions:

a. 2/4

b. *xy*

c. *a/b*

d. $p < 0.001$

e. $12.2/2 \approx 6$

E3 Brackets and absolute values

The next group of symbols is those that are placed around parts of a sum. These are collectively called brackets. The first set of brackets used is usually the round brackets ( ). If you need to surround round brackets with another set of brackets, then you usually use square brackets [ ]. For example, $[(6 \times 2) - 2]/2$.

You carry out the calculation inside the brackets first. So for this example you first work out 6 × 2 = 12 and subtract 2, i.e. 12 − 2 = 10, before carrying out the division, which will be 10/2 = 5.

$[(5 - 1) \times (4 - 2)] - 3$ is an example of using two sets of brackets. Here, you work out the calculations for the inside brackets first, so (5 − 1) = 4 and (4 − 2) = 2. You now complete the sum inside the square brackets [4 × 2] = 8 and finally complete the remaining steps in the sum, 8 − 3 = 5.

Another symbol that can enclose parts of a calculation are the straight brackets | |. These are used to tell you that when the calculation within the brackets is completed, the sign of the answer is positive, even if the calculation produces a negative number. These are called absolute values. For the example |3 − 1| = 2, the sign is already positive, so being an absolute value has made no difference. But for |5 − 6| = 1 the actual sum of 5 − 6 would normally equal −1. As this is an absolute value, we ignore the negative sign, so the answer is +1.

Q2 Complete the following calculations:

a. $(10 - 2)/[(3 \times 2) - 2]$

b. $2(x - 1)$ where $x = 2$

c. $(6 + 3) / |-3|$

E4 N, n, x, y, Σx, Σy, and Σxy

To show you how the terms n, N, x, and y can be used, we have included a table from Chapter 9 (Table 9.2). In this study, periwinkles from the lower and mid-shore at Porthcawl have been collected and the shell heights recorded.

Table 9.2. Height of shells in two putative species of periwinkles from the mid- and lower shore at Porthcawl, 2002

Shell height (mm)			
Periwinkles on the lower shore		Periwinkles on the mid-shore	
5.5	4.0	3.3	6.7
8.4	5.0	6.3	5.7
5.0	6.2	6.1	4.2
5.0	5.0	8.0	6.3
5.6	7.7	13.5	6.0
4.8	6.0	5.3	7.2
8.4		6.7	

We have two columns of data. Either to simplify our explanations of calculations using this data or when explaining about choosing and constructing figures, we need to identify one column of data. For this we use the letters x and y. In this example the periwinkles on the lower shore are the x observations and the periwinkles on the mid-shore are the y observations. There are 13 x observations, which we would record as $n_x = 13$. There are also 13 y observations, so for this sample $n_y = 13$. In many calculations, you need to work out $n - 1$, which for both of these columns of data in this example would be $n - 1 = 13 - 1 = 12$. If we needed to know the total number of observations in our data, this would be $N = n_x + n_y = 13 + 13 = 26$.

In some calculations (e.g. Chapter 9), you need to know the totals by adding all x observations together or all the y observations together. The symbol used to indicate this is Σ. In this example, $\Sigma x = 5.5 + 8.4 + 5.0 + \ldots + 5.0 + 7.7 + 6.0 = 76.6$. The same can be calculated for Σy.

Another example of the use of x and y can be seen in data from Chapter 8. In this example, 13 pullets have been examined and the amount of feed each has eaten and the hardness of the egg each produced has been recorded (Table 8.1). In this example, there are 13 items (pullets) so for each column of data $n = 13$.

Table 8.1. The hardness of eggshells produced by 13 Maren pullets and their consumption of a food supplement

Pullet	Amount of food supplement (g)	Hardness of shells
1	19.5	7.1
2	11.2	3.4
3	14.0	4.5
4	15.1	5.1
5	9.5	2.1
6	7.0	1.2
7	9.8	2.1
8	11.6	3.4
9	17.5	6.1
10	11.2	3.0
11	8.2	1.7
12	12.4	3.4
13	14.2	4.2

In this example, instead of two separate samples as seen in Table 9.2, two observations have been recorded for each pullet: the amount of food supplement (g) and the hardness of shells. Although these are not independent samples, we still use x and y to indicate which column of data we are referring to. In this example, the amount of food eaten is x and the hardness of egg shells is y. Unlike the previous example, one of these columns of data is said to be dependent and this is always called y; the other is said to be independent and this is always called x. We explain these terms in Chapter 8.

As we showed you in relation to Table 9.2, you may need to calculate the totals for each column, i.e. Σx and Σy. The idea of a sum (Σ) can be extended to the product xy. The symbol xy means that an x observation has been multiplied by its y observation $= xy$. We show this for the pullet example in Table 8.3. If all these 13 xy values are added together, this is $\Sigma xy = 661.0$. (In Table 8.3. the original data (Table 8.1) has been reorganized into numerical order to simplify other parts of a calculation which we consider in Chapter 8.)

a. What is Σy for the data in Table 9.2?

b. What is Σx for the data in Table 8.1?

c. What is Σy for the data in Table 8.1?

Table 8.3. Calculating Pearson's product moment correlation for Example 8.1 between the hardness of shells and food consumption in Maren pullets

Amount of food supplement (g) (x)	x^2	Hardness of shells (y)	y^2	$x \times y$
7.00	49.00	1.20	1.44	8.40
8.20	67.24	1.70	2.89	13.94
9.50	90.25	2.10	4.41	19.95
9.80	96.04	2.10	4.41	20.58
11.20	125.44	3.00	9.00	33.60
11.20	125.44	3.40	11.56	38.08
11.60	134.56	3.40	11.56	39.44
12.40	153.76	3.40	11.56	42.16
14.00	196.00	4.50	20.25	63.00
14.20	201.64	4.20	17.64	59.64
15.10	228.01	5.10	26.01	77.01
17.50	306.25	6.10	37.21	106.75
19.50	380.25	7.10	50.41	138.46
$\Sigma x = 161.20$	$\Sigma(x)^2 = 2153.88$	$\Sigma y = 47.30$	$\Sigma(y)^2 = 208.35$	$\Sigma xy = 661.00$
$\Sigma(x)^2 = 25985.44$		$\Sigma(y)^2 = 2237.29$		
$\bar{x} = 12.40$		$\bar{y} = 3.63846$		
$n = 13$		$n = 13$		

E5 Powers (x^n) and roots ($^n\sqrt{}$)

You will come across x and y with a superscript such as x^2 or y^2. This indicates that you must square the x or y observation by multiplying it by itself. For example, if the x observation was 2, then $x^2 = 2^2 = 2 \times 2 = 4$. If a different superscript is used, you multiply the number by itself as many times as the superscript indicates. So if $x = 2$, then $x^3 = 2 \times 2 \times 2 = 8$. This is described as x cubed. Although squaring most whole numbers is a straightforward process, which you can work out in your head, for more complex numbers, use the x^2 button on your calculator.

This process can be reversed by taking the root of a value. In this book, we only take a square root, and the symbol that denotes this is $\sqrt{}$. If you were to take the cube root, the symbol would be $^3\sqrt{}$. You may learn square roots for some numbers such as $\sqrt{4} = 2$, $\sqrt{9} = 3$, $\sqrt{81} = 9$. For other numbers you can use the $\sqrt{}$ button on your calculator.

Taking squares and square roots is an important step in the calculation of a variance and standard deviation (e.g. Box 5.1), as the variance is the square of the standard deviation and therefore the standard deviation is the square root of the variance.

When calculating the variance or standard deviation, two other important squared terms are used: $(\Sigma x)^2$ and Σx^2. These appear to be the same apart from the brackets but the brackets are especially important here. Remember, whenever you have brackets like this you first carry out whatever calculation is inside the brackets. Therefore, $(\Sigma x)^2$ is the sum of all the x values (Σx), which is then squared.

This compares with Σx^2, which means first square each x observation and then sum the squares.

These two terms can be calculated for the x observations in Table 9.2. We have already calculated $\Sigma x = 76.6$ so $(\Sigma x)^2 = (76.6)^2 = 76.6 \times 76.6 = 5867.56$. The term Σx^2 is calculated as $5.5^2 + 8.4^2 + 5.0^2 + \ldots + 5.0^2 + 7.7^2 + 6.0^2 = 475.5$. Clearly $(\Sigma x)^2$ and Σx^2 are very different calculations with very different results, so it is important to make sure you do not get them mixed up.

 Q4
a. What is $(\Sigma y)^2$ for the data from Table 9.2?
b. What is Σy^2 for the data from Table 9.2?
c. What is $(\Sigma x)^2$ for the data from Table 8.1?
d. What is Σx^2 for the data from Table 8.1?
e. These calculations can also be used for the column of $x \times y$ values in Table 8.3. What is $(\Sigma xy)^2$?
f. What is $\Sigma(xy)^2$ for the data from Table 8.3?

When you write y^2, this is sometimes referred to as y to the power 2. Here y is multiplied by itself so the calculation is $y \times y$: there are two y values in the calculation. This can be extended. For example, y could be multiplied by itself three times, $y^3 = y \times y \times y$. If $y = 2$, then $y^3 = 2 \times 2 \times 2 = 8$. The power is 3. Powers may sometimes be shown using symbols instead of numbers.

The binomial equation is written as:

$$y = \frac{n!}{x!(n-x)!} \times p^x \times q^{(n-x)}$$

Here there are two powers: p^x and $q^{(n-x)}$. If you encounter these, you first need to identify what number the power symbol is; for example, what is x or what is $n - x$? Having found these values, and if you know the numerical value for p or q, you can work out the numerical value for p^x and $q^{(n-x)}$ using the powers button on your calculator. For example, if $p = 0.5$ and $x = 4$, then $p^x = 0.5^4 = 0.0625$. If $q = 6$, $n = 8$, and $x = 4$, then $q^{(n-x)} = 6^{(8-4)} = 6^4 = 1296$.

Powers can be negative, for example x^{-2}. In this case, the sign is indicating that this calculation must be inverted (turned upside down). For example, if $x = 2$, and $x^2 = 2 \times 2$, then:

$$x^{-2} = \frac{1}{2 \times 2} = \frac{1}{4} = 0.25$$

You can use the power button (x^y) and the change sign button ($^+/_-$) on your calculator to work out positive and negative powers.

 Q5   Use your calculator to work out the following powers:
a. 6^5
b. 6^{-5}
c. 0.5^2
d. 0.5^{-2}

E6 Factorials (x!)

In the binomial equation (E5) another symbol (!) was used. Maths revolves around number patterns. These often occur so frequently that they are given a particular symbol. Factorials are one of these patterns that appear in this book. A factorial is the value that results from multiplying a whole number (integer) by all the integers less than itself. A factorial of $3! = 3 \times 2 \times 1 = 6$, a factorial of $4! = 4 \times 3 \times 2 \times 1 = 24$ and so on. In the binomial equation (5.2.2), two factorials are included: $x!$ and $(n - x)!$. To work these factorials out, you first need to know what x is, or what $n - x$ is. You could then work this sum out by hand, but it is simpler to calculate factorials using the factorial button on your calculator.

 Q6   Use a calculator to work out the following:
a. 10!
b. $x!$, where $x = 6$
c. $(n - x)!$, where $x = 6$ and $n = 13$

E7 Constants: pi (π) and exponential (e)

Patterns in maths can also include constant numbers or constants. They are specific numbers that have been found to be needed in many calculations for the mathematical relationships to work. The two constants that appear in this book are pi (π) ≈ 3.14 and exponential (e) ≈ 2.72. π can usually be found on your calculator. e is invariably used as a power as we see in the Gaussian equation (5.2.1). On a calculator, this button is indicated as e^x. To find the value of e itself, press the e^x button, 1, equals.

 Q7   a. What is π to five decimal places?
b. What is e to five decimal places?
c. What is e^2?
d. What is e^{-2}?
e. What is e^x when $x = 2(6 - 4)$?

E8 Scales of measurement (%, log, ln, and × 10^n)

We use many scales of measurements. The one most familiar to you is our numbering system based on 10, where we have units (numbers 1–9), tens (numbers 10–99), hundreds (numbers 100–999), etc. Very large numbers and very small numbers can be very tedious to write out in full, so there is a shorthand, which we symbolize as $\times 10^n$, where n can be any number. As examples of using this

shorthand 100 can be written 10^2 (i.e. 10×10), 1000 can be written as 10^3 (i.e. $10 \times 10 \times 10$), 200 this could be written as 2×10^2, 5000 could be written as 5×10^3, etc. The same idea can be used for very small numbers. We introduced the idea of negative powers in E5. Negative powers of 10 can be used as a shorthand when writing small numbers. For example, 0.1 can be written as 10^{-1} as this is telling us that the number is 1/10. So 0.01 is 10^{-2} or 1/100; 0.5 this could be written as 5×10^{-1}, i.e. $5 \times 1/10$, etc. Your calculator (and computer) will report large and small numbers in this way. If you used your calculator to answer Q5(b), it would be reported as 1.28601×10^{-4}.

Apart from the base 10 system of numbering that we are very familiar with, you will encounter other scales each indicated using a particular symbol. The four you need to be able to work with are percentages (%), logarithms to the base 10 ($\log_{10}$), natural logarithms ($\ln = \log_e$), and inverse sine ($\sin^{-1}$). We cannot explain here how these scales are derived, but encourage you to find the relevant buttons on your calculator and feel confident using them.

On a scientific calculator, you will probably not have a % button. This is calculated from a proportion or ratio that is multiplied by 100. For example, 3/4 as a % = $(3/4) \times 100 = 0.75 \times 100 = 75\%$. Values may need to be converted to the $\log_{10}$ scale or converted from the $\log_{10}$ scale, taking an antilog. The first is achieved by pressing the $\log_{10}$ button on your calculator (sometimes labelled log). For example, $\log_{10} 2 = 0.30103$. To take the antilog, for most calculators press shift $\log_{10}$ =. This should return the log value back to the base 10 scale. The same process can be used for $\ln 2 = 0.69315$ and its antiln using shift ln =. When using the sine functions, you again can take the sine (sin) or inverse sine ($\sin^{-1}$) usually by using the same button with or without the shift function first.

Q8
a. What is 5/20 as a percentage?
b. Rewrite 19,080,000 as a number $\times 10^n$.
c. What is (3 − 6)/2 as a percentage?
d. Rewrite 0.000000345 as a number $\times 10^n$.
e. What is log 4?
f. Take the antilog of 0.05.
g. What is ln 4?
h. Take the antiln of 0.05.
i. What is sine (sin) of 90°?
j. What is the inverse sine ($\sin^{-1}$) of 0.6?

E9 Unpacking complex equations

The four basic procedures in maths are addition (+), subtraction (−), multiplication (×), and division (÷). These are straightforward in their own right, but when there are several of these elements in one calculation, it is not always clear in which order to carry out the procedures and this order can be critical in obtaining the correct outcome. To help convey which order should be followed, brackets are often placed around parts of an equation.

To help unpack complex equations, you need to follow this order:

1) Brackets
2) Powers and roots
3) Division (except where considering fractions, see examples 1–3)
4) Multiplication
5) Addition
6) Subtraction

We are going to unpack some of the equations frequently encountered in this book.

Example 1 (e.g. 5.4.3i)

$$\bar{x} = \frac{3+4}{2} = \frac{7}{2} = 3.5$$

If you look at the list above, it seems that, as there are no brackets, powers, or roots, 'division' appears to be the first step. However, this is a simplification. Where you have a fraction and wish to carry out the division step, you first need to simplify the fraction so that the division step may be carried out. In this example the first step then is to add all the values in the numerator (the top of the fraction) together and then divide the total by the denominator (lower part of the fraction).

Example 2 (e.g. 5.7)

$$SE = \frac{10}{\sqrt{4}} = \frac{10}{2} = 5$$

Again, there is a fraction, only this time the denominator has a square root sign ($\sqrt{}$) in front of it. You must first work out the square root of the denominator and then use this to divide the numerator.

Example 3 (e.g. Box 7.3)

$$\chi^2_{calculated} = \frac{(12-6)^2}{4} \quad \frac{(12-8)^2}{2} = \frac{6^2}{4} \quad \frac{4^2}{2} = \frac{36}{4} \quad \frac{16}{2} = 9 \quad 8 = 17$$

In the numerators, there are three different procedures indicated by the symbols, subtraction –, a square ( )², and addition +. The subtraction is within brackets and must be carried out first. The value within the brackets can then be squared. This gives the numerator for each fraction. Each division is then carried out and finally the two numbers can be added together.

Example 4 (e.g. Box 7.6)

$$\chi^2_{calculated} = \frac{(|9-6.5|-0.5)^2}{2} \quad \frac{(|4-6.5|-0.5)^2}{2}$$

This example illustrates the use of another symbol | |. A number within these two vertical lines is called an absolute number. The sign of any number within these lines is ignored. The calculation is carried out following the same rules we have

discussed above: brackets, then powers, then fractions, then addition. Absolute values also occur in t-tests (Box 9.3.).

$$\chi^2_{calculated} = \frac{(|9-6.5|-0.5)^2}{2} + \frac{(|4-6.5|-0.5)^2}{4}$$

$$= \frac{(|2.5|-0.5)^2}{2} + \frac{(|-2.5|-0.5)^2}{4}$$

Ignore the sign within the | |

$$= \frac{(2.5-0.5)^2}{2} + \frac{(2.5-0.5)^2}{4} = \frac{(2.0)^2}{2} + \frac{(2.0)^2}{4} = \frac{4.0}{2} + \frac{4.0}{4}$$

$$= 2 + 1 = 3$$

Example 5 (e.g. 5.5.1)
In some cases we may wish to calculate two values one at the top of the range and one at the bottom of the range. This is indicated by the symbol $\pm$. We use this in both the range for non-parametric data (5.5.1) and confidence limits (5.7).
   For example:

Confidence limits $= \bar{x} \pm (t \times SM)$

This symbol $\pm$ indicates that you need to work out the calculation twice once with an addition and once with a subtraction. If $\bar{x} = 15$, $t = 3$ and SEM = 2, then:

$$\bar{x} \pm (t \times SEM) = 15 \pm (3 \times 2) = 15 \pm 6 = \text{range is } 15 - 6 \text{ to } 15 + 6,$$
$$\text{or 9 to 21}$$

Example 6 (e.g. Boxes 5.1 and 9.6)
A very frequently used equation is the one written here. It is used in parametric ANOVAs (Chapter 9) and is part of the calculation of variance.

$$SS_{sample} = \Sigma x^2 - \frac{(\Sigma x)^2}{n}$$

We have already looked at the terms Σx^2 and $(\Sigma x)^2$ in E5. To complete this calculation, the fraction $(\Sigma x)^2/n$ is calculated first and then the subtraction. We worked out Σx^2, $(\Sigma x)^2$, and n for the x observations in Table 9.2 in E4 and E5, where $n = 13$, $\Sigma x^2 = 475.0$, and $(\Sigma x)^2 = 5867.56$. So for this example:

$$SS = 475.0 - (5867.56/13) = 475.0 - 451.35077 = 23.64923$$

Example 7 (e.g. 5.5.3)
In several places in the book, we have given some complex equations. These can be very off-putting, but when they are unpacked you will see that they are

manageable. The Gaussian equation (5.2.1) has been unpacked for you in 7.1.3 using the data from Example 5.2. Examine the Gaussian equation and decide in what order you would carry out the calculation. How does this compare with the steps in 7.1.3?

Gaussian equation:

$$y = \frac{1}{\sqrt{(2\pi s^2)}}\, e^{-b}$$

where

$$b = \frac{(x - \bar{x})}{2s^2}$$

Answers to questions

A1
a. Two divided by four

b. x multiplied by y

c. a divided by b

d. p is less than 0.001

e. 12.2 divided by 2 approximately equals 6.

A2
a. First work out the values inside the round brackets. So $(10 - 2) = 8$ and $(3 \times 2) = 6$. Then work out the sum within the [] so $[6 - 2] = 4$. Then carry out the division of $8/4 = 2$.

b. First work out the sum within the round brackets where $(x - 1) = 2 - 1 = 1$. Then multiply this by $2 = 2 \times 1 = 2$.

c. First work out the sum within the round brackets $(6 + 3) = 9$. Then carry out the division within the straight brackets $|9/-3|$. When a positive number is divided by a negative number the answer will be negative, so $9/-3 = -3$. The straight brackets tell you this is to be an absolute value so you can ignore the sign. The answer to this sum is +3.

A3
a. $\Sigma y = 3.3 + 6.3 + 6.1 + \ldots + 6.3 + 6.0 + 7.2 = 85.3$mm

b. $\Sigma x = 19.5 + 11.2 + 14.0 + \ldots + 8.2 + 12.4 + 14.2 = 161.2$g

c. $\Sigma y = 7.1 + 3.4 + 4.5 + \ldots + 1.7 + 3.4 + 4.2 = 47.3$ units

A4
a. $(\Sigma y)^2 = 85.3^2 = 85.3 \times 85.3 = 7276.09$

b. $\Sigma y^2 = 3.3^2 + 6.3^2 + 6.1^2 + \ldots + 6.3^2 + 6.0^2 + 7.2^2 = 629.57$

c. $(\Sigma x)^2 = 161.2^2 = 161.2 \times 161.2 = 25985.44$

d. $\Sigma x^2 = 19.5^2 + 11.2^2 + 14.0^2 + \ldots + 8.2^2 + 12.4^2 + 14.2^2 = 2153.88$

e. $(\Sigma xy)^2 = 661.0^2 = 661.0 \times 661.0 = 436921.0$

f. $\Sigma(xy)^2 = 8.4^2 + 13.94^2 + 19.95^2 + \ldots + 77.01^2 + 106.75^2 + 138.46^2 = 51021.652$

A5
a. 7776

b. 0.0001286

c. 0.25

d. 4.0

A6
a. 3628800

b. 720

c. $(13 - 6)! = 7! = 5040$

A7

a. 3.14159

b. 2.71828

c. 7.38906

d. 0.13534

e. $x = 2 \times 2 = 4$, so $e^4 = 54.59815$

A8

a. $0.25 \times 100 = 25\%$

b. 1.908×10^7

c. $-3/2 = -1.5 \times 100 = -150\%$

d. 3.45×10^{-7}

e. 0.60206

f. 1.12202

g. 1.38629

h. 1.05127

i. 1.00 (note: set calculator to degrees not radians)

j. 36.87° (note: set calculator to degrees not radians)

Glossary

a priori   Before the event. You need to be aware when you are designing your experiments or using some statistical tests, such as a chi-squared test, whether you have reason to expect a specific outcome. For example, one *a priori* expectation arises when you are carrying out genetic crosses and you wish to compare your observed data with an expected segregation ratio (Chapter 7). This notion of 'expectation' must not be confused with a personal preference, educated hunch, or prediction.

Absolute   You may find that having calculated the difference between two values you have a negative value. For some parts of some calculations, any negative sign must be removed. In these instances, the number without its sign is called the absolute value and is indicated as |number|. For example |–10| would be used in the calculation as +10.

Aim   A generalized statement about the topic you are investigating.

Association   Where one variable changes in a similar manner to another variable for reasons unknown (see Relationship).

Bar chart   A figure for observations from an investigation with one treatment variable. The data are measured on a nominal or ordinal scale. The *x*-axis is the treatment variable. The *y*-axis is the number of observations (or frequency or percentage). Each category from the nominal or ordinal scale is represented as a discrete bar on the figure. There must be a space between each bar to indicate that the data are not measured on a continuous scale. The tops of these bars should not be joined together by a trend line. If the top of the bar is an average of many observations, then confidence intervals can be used to indicate the variation around the mean.

Bimodal   When a distribution has two high points (peaks).

Binomial distribution   May be obtained if each item examined can have only one or another state. For example, a seed can either germinate or not germinate.

Case   See Item.

Categories   When each observation may fall into only one of two or more mutually exclusive groups (e.g. blood groups A, B, AB, or O). These groups are known as categories. The categories may be either qualitative (e.g. flower colours described as red, pink, blue, green, yellow, purple) or quantitative (e.g. the number of piglets in a litter: 0, 1, 2, 3, 4, 5, 6, etc.). The data in one category may be a sample but not inevitably (see definition of Sample for explanation).

Closed questions   Where you give the participant a limited number of answers and they have to choose one or more or these options. Often used in questionnaires. See Open questions.

Coefficient of variation   The ratio between the mean and the standard deviation expressed as a percentage. Used if you wish to compare variation in one set of data with that of another set of data.

Confidence limits   The values derived from a sample between which the population mean probably falls. These limits are calculated using the standard error of the mean.

Confounding variables   Any factor(s) that may influence your observations but is not the factor(s) being investigated in the experiment. Also called non-treatment variables. See Variables, Sampling error, and Treatment variables.

Continuous   Observations that do not fall into a series of distinct categories and may take any value in the scale of measurement, e.g. height (cm).

Control   An experimental baseline against which any effects of the treatment(s) may be compared. It must be an integral part of an experiment.

Curvilinear   When data are plotted and where part or all of the distribution is found to curve rather than follow a straight line.

Data   A number of observations or measurements on the subject you are investigating.

Degrees of freedom (v)   You may have a sample of three observations and a known total of 25. If two of the three observations are 7 and 15, the third observation has to be 3 so that the total is 25. There is no freedom over what the value of the last observation must be. This limit to choice is

described by the degrees of freedom. The level of constraint results from the experimental design and type of data. In statistical tests, these 'limits' are often allowed for in the calculation process. For many tests, the degrees of freedom is $n - 1$, as in our example. For some tests, there are more constraints, so there are only $n - 2$ degrees of freedom.

Denominator   In an equation where one term is divided by another, the value that is being used to divide is the denominator. For example, in the fraction 2/4, 4 is the denominator.

Dependent and independent variables   These are terms used in investigations with two or more variables. They are both used to imply several different meanings and as such the context of their use has to be considered carefully to appreciate which meaning is intended.

An independent variable is:

1) A factor under investigation where it can reasonably be argued that it causes a change in a second, dependent, variable (i.e. a relationship).
2) A factor under investigation that is taken without sampling error, is usually under the investigator's control, and where there is an association (or relationship) between the independent and dependent variables.
3) A factor under investigation that is taken without sampling error and is usually under the investigator's control.
4) A factor under investigation that for convenience is plotted on the x-axis.

We tend to use the second of these definitions. The independent variable can be ordinal or interval and is plotted on the x-axis.

A dependent variable is:

1) A factor under investigation where it is reasonable to argue that there may be a relationship with an independent variable.
2) A factor under investigation that for convenience is plotted on the y-axis.

The dependent variable is usually measured on an interval scale or is a derived variable such as frequency, proportion, percentage, or rate, and is plotted on the y-axis.

Derived variables   The unit of measure for the observations is the result of a calculation, e.g. ratio, proportion, percentage, or rate.

Discrete   Observations that fall into a series of distinct categories. The number of categories is limited, e.g. the number of eggs in a clutch; colours (red, green, or blue).

Distribution   The shape seen on a graph when data are plotted. See also Normal distribution, Binomial distribution, and Poisson distribution.

Expectation   A mathematical (e.g. Gaussian equation) or biological (e.g. genetic segregation ratio) model of the outcome of an experiment against which the observations can be compared. The model can be consistently applied. Your personal prediction of an outcome is not a statistical expectation.

Expected   The values generated in response to a mathematical or biological model such as the Gaussian equation or a genetic segregation ratio.

Focus group   A discussion-based interview involving more than two people. A theme or focus is provided by the researcher, who also directs the discussion. These discussions are usually recorded as audio- or videotapes and evaluated later.

General hypotheses   The hypotheses are phrased in a general way. For example, there is no difference between the samples. See Hypotheses and Specific hypotheses.

Histogram   A type of figure for observations from an investigation with one treatment variable. The data are measured on a continuous (interval) scale. There are many data points along the x-axis. usually because the treatment variable is not under the investigator's control. Therefore, these observations are organized into classes in a frequency table (e.g. Table 11.4) and these classes and the number of observations in each class are used for the x- and y-axes respectively. Each class is represented as a bar and each bar abuts the next, indicating that the scale of measurement is continuous not discrete. The tops of these bars should not be joined together with a trend line (e.g. Fig. 5.4). If many observations are recorded within each class, the variation in each class may be represented by confidence intervals.

Hypothesis   The formal phrasing of each objective; includes details relating to the experiment and the way in which the data will be tested statistically. The hypotheses can be general (e.g. 'there is no difference...') or specific (e.g. median A is different from median B). In hypothesis testing, you determine which of two hypotheses, the 'null' and 'alternate' hypotheses, is probably correct. The null hypothesis is written in terms of the variation arising probably from sampling error, whilst the alternate hypothesis is written in terms of the variation in the data arising due to the factors under investigation.

Independent observations   All observations are collected in a consistent manner so that no one observation is more similar

than any other observation. For example, in a study comparing the diets of undergraduates during one week from two departments, in one department the investigators recorded the diets for 100 students but in the other smaller department they were only able to sample five students. To increase the data for this smaller department, they therefore recorded the diet for each student on three separate occasions. The diet information from the small department is not independent. Any one student is contributing three results to the study and these results are not independent of each other.

Independent variable   See Dependent variable.

Information-theoretic models   When data are collected and analysed to see whether there are significant groupings. Only then may hypotheses about the factors causing these groupings or the nature of these groupings be tested, i.e. models are constructed from the observations.

Interaction   The response to a treatment shown by one group of items is reversed in a second sample in such a way that it is clear that, for example, two treatment variables are not having the same effect in two groups. For example, the number of rabbits was recorded in two locations A and B at three points in the year. In location 1, the number of rabbits increased during the year, but in location 2, the number of rabbits decreased during the year. The effect of 'time' on the number of rabbits is different for the two samples, so the effect of time is dependent on location. Time and location would be said to be interacting.

Interpolation   The estimation of a value when only nearby values are known, one greater than the required value and the other less than the required value. For a worked example, see 6.3.6.

Interval data   These data are measured on a continuous scale; the data are rankable, and it is possible to measure the difference between each observation. Interval scales include temperature (°C), distance (cm), and time (mins).

Interquartile range   See Range.

Interview   A meeting between the researcher and a participant. Interviews can be structured, in that the interviewer asks the same series of questions to all the interviewees, or unstructured, where the interviewee is left to make their own comments about a topic.

Item   One representative of the statistical population you wish to measure. In other texts, an item may be referred to as an 'experimental unit', subject, or case.

Kurtosis   A measurement that indicates how sharp the peak of the central point is.

Line graph   A figure for observations from an investigation with one treatment variable. The data are measured on a continuous (interval) scale. There are few data points along the x-axis (treatment variable), usually because the treatment variable is under the investigator's control. Each observation or mean from a group of observations for a given x value is plotted as a point on the figure. Variation around a mean can be indicated using confidence intervals (e.g. Fig. 11.4). These points may be joined together with a trend line.

Matched   When two or more observations are recorded for one variable for each item; for example, in a study of heart rate, a resting rate was taken for ten people first, and then following exercise another heart-rate reading was taken for the same ten people. For each person, there is a before and after reading, so these data are matched. An alternative approach would be to assign ten people at random and record a resting heart rate and then assign a different group of ten people to the exercise treatment and take an 'after exercise' heart rate from this second, independent group. These data would not be matched. Matched data can extend to more than two observations. For example, if a mouse was run through a maze once a day for a week and the time taken was recorded, there would be five 'matched' observations. These are more usually called repeated measures.

Mean   One measure of central tendency for interval data where the distribution is known. For a normal distribution, the sample mean ($\bar{x}$) and statistical population mean (μ) are calculated in the same way. The sample mean is used as an estimate of the population mean and a value called the confidence interval (5.7) can be calculated to demonstrate the area around a sample mean in which the population mean will probably fall. The sample mean for normally distributed data is calculated as the sum (Σ) of all observations (x) divided by the number of observations in the data set (n).

Median   One measure of central tendency, calculated as the middle value of an ordered data set.

Mode   One measure of central tendency. When data are organized into categories, the mode is the category that contains the greatest number of observations. The categories may be inherent in the data (for example, the number of blossoms on a rose plant) or imposed on continuous data (for example, the heights of trees placed in ranges of 0.00–0.99m, 1.00–1.99m, etc.).

Nominal   Observations that fall into discrete categories. The categories have no specified order, e.g. colours (red, green, or blue).

Non-parametric data   Data that, when plotted, have a non-normal distribution or, more usually, where the distribution

is unknown. When testing hypotheses and you have non-parametric data, you may either transform the data (5.9) or use non-parametric statistics (Chapters 7–10). Parametric data may be analysed using non-parametric statistics but not the other way round.

Non-treatment variable   See Confounding variables.

Normal distribution   A symmetrical, 'bell-shaped' distribution with a single central peak (unimodal), which can be described by the Gaussian equation (e.g. Fig. 5.3). The data are measured on an interval scale.

Numerator   In an equation where one term is divided by another term, the value that is being divided into is the numerator. For example, in the fraction 2/4, 2 is the numerator.

Objective   Provides succinct and specific details about an experiment that is being carried out in order to examine all or part of an aim. Objectives can be 'clumped' or 'split' and can be experimental or personal.

Observation   A single measurement taken from one item.

One-tailed test   When the hypothesis testing is such that only one end of a distribution is being considered. One-tailed tests are most often encountered when testing specific hypotheses. See also Two-tailed tests.

Open question   Used in questionnaires where the answers are not prescribed. See also Closed questions.

Ordinal   Observations that fall into discrete categories where the categories have an order (they can be ranked), e.g. the number of eggs in a clutch; the ACFOR scale.

Orthogonal   A particular experimental design where every category for one treatment is found in combination with every category for every other treatment.

Outlier   An observation that is noticeably different from all other observations, one that does not follow the apparent trend.

Parametric data   Data measured on an interval scale and so quantitative, continuous, and rankable. The data have a normal distribution, e.g. height (cm) (e.g. Table 5.1). When testing hypotheses and you have parametric data, you should use parametric statistics. See also Non-parametric data.

Pie chart   A figure for observations from an investigation with one treatment variable. The data are measured on a nominal or ordinal scale. A circle is divided up proportionately to reflect the relative proportions (or percentages) of the observations in each category or sample (e.g. Fig. 11.2). If many observations are recorded within each category, you should consider using a bar chart so that the variation in each category may be represented by confidence intervals.

Poisson distribution   A unimodal distribution, which is often very asymmetrical with a protracted tail either to the right (a positive skew) or to the left (a negative skew). A distribution often found where events are randomly distributed in time or space, such as the distribution of cells within a liquid culture or the dispersal of pollen or seed by wind.

(Statistical) Population   All the individual items that are the subject of your research. A statistical population may be the same as an ecological or genetic population but not necessarily.

Power of a test   Table G1 illustrates the four possible outcomes from hypothesis testing. Two of these are where the actual biology accords with the outcome from the hypothesis testing: either there really is a difference and the statistical test detects this, or there is not a difference and

Table G1. The power of a test, Type I and Type II errors

	The analysis does:	
Experimentally there really:	detect a difference. (The null hypothesis is rejected.)	not detect a difference. (The null hypothesis is not rejected.)
is a difference. (The null hypothesis should be rejected.)	Correct outcome The power of a test	Incorrect outcome Type II error
is not a difference (The null hypothesis should not be rejected.)	Incorrect outcome Type I error	Correct outcome

the statistical test confirms this. The probability of the first of these happening is called the power of a test.

There are two other possible outcomes from hypothesis testing neither of which is desirable. The first is that there is not a biological difference but the statistical test indicates that there is, or that there really is a difference biologically but the statistical test does not pick this up. These potential errors from hypothesis testing are called Type I and Type II errors, respectively.

Non-parametric tests are generally considered to be less powerful than parametric tests because they are less rigorous in their conditions of use. This is only true if the assumptions behind the use of a statistical test are met. If not, then the value of using any test is undermined. Other factors that affect the likelihood of detecting a real difference include the use of a larger sample size, which is more likely to detect a real trend even when the difference is small, and a small p value, which decreases the probability of a Type 1 error occurring.

(The term 'power' used here is not the same as that used in maths (Appendix e).)

Qualitative   Observations are assigned to named, descriptive categories that are mutually exclusive and non-numerical, e.g. colours (red, green, or blue). The data are always discrete. Qualitative answers in questionnaires are descriptive answers.

Quantitative   Observations are numerical and consist of named, ordered categories, e.g. number of eggs in a clutch, or height (cm). Data may be either discrete or continuous.

Questionnaire   A collection of questions given to all participants in an investigation. The participant writes down their answer and returns the questionnaire to the researcher. Questionnaires are usually anonymous and confidential. They require very little contact between the researcher and volunteers and usually there is no interaction between participants.

Random sampling   Each item must have an equal chance of being sampled each time.

Range   The range is the difference between the highest and lowest observations. A reduced range may be used (e.g. from the 25th observation to the 75th observation in a data set of 100 observations). This avoids any undue influence by extreme values on the range. These are known as percentiles and the interquartile range.

Rankable   Observations consist of named categories or values that have an order to them; for example, finishing positions in a race, or height (cm).

Relationship   Where one variable is directly responsible for causing the change in another variable. See also Association.

Repeated measures   See Matched.

Replication   Where more than one item is exposed to a treatment, more than one group of items is exposed to one treatment, or more than one item is included in a sample. All replicates should be an integral part of one experiment and should be independent of each other.

Representative sample   See Sample.

Research   The process by which you try to find out more about a particular topic.

Sample   If it is not possible to measure the whole statistical population, then a representative subset (a sample) is required. A representative sample should be obtained by a clearly defined and consistently applied method. The observations recorded by this method should closely match the result that would be obtained if every item in the population was measured. The related concept of 'numbers of samples' can be misleading, however. For example you may sample 400 items from one population and then place these into four categories, each with 100 items. In this example, you would describe your design as having one sample, with four categories, each category having 100 items. An alternative investigation may require you to go to four different locations and at each location sample 100 items. The end result for the design is the same as before in that you will have four categories each with 100 items. However, by the previous definition, we would say we had four samples each with 100 observations. The term 'sample' is often used to describe both the sample (e.g. 400 items from one location) and the number of items in each category (i.e. 100). This can be confusing. We therefore prefer to use the terms 'category' and 'numbers in a category' when referring to experimental designs, and to keep the term 'sample' for referring to sampling methods.

Sampling error   The variation in a set of data that is due to the effects of confounding variables and/or chance. The term 'sampling error' may be used to describe the variation between samples collected from a single population that has occurred by chance, the variation between the population value and sample value that has arisen by chance, or the variation in the experimental observations that may be due to factors not being investigated by the researcher. Other authors may use the terms 'noise' or 'within-treatment variation' when referring to sampling error.

Scatter plot   A figure that can be drawn when observations for two or more variables are made for each item. Each of

these observations relates to one treatment variable (e.g. Fig. 8.15). The data can be measured on any scale, although you must be very careful if you infer any trends in the data for nominal scales. If you have an independent variable, this is plotted on the *x*-axis. The second variable, which may be a dependent variable, is plotted on the *y*-axis. If more than two variables are recorded, the plot will be three (or more) dimensions. The two observations from each item are marked as a point on the figure. If both measurements are interval, these points may be joined together with a trend line. You may carry out regression analysis, and if significant, you may draw a regression line (e.g. Fig. 8.15). If more than one *y* value has been recorded for each *x* value, you may use confidence intervals to reflect the variation around the mean *y* value.

Skew   When observations are plotted, the distribution may not be symmetrical. Where there is a tail of observations to the right, the distribution is said to have a positive skew (e.g. Fig. 5.8). Where there is a tail of observations to the left, the distribution is said to have a negative skew (e.g. Fig. 5.9).

Specific hypotheses   Statistical tests of hypotheses may ask the question, 'Is there a difference?' This is an example of a general hypothesis. Some statistical tests allow you to ask more discerning (specific) questions. In Tukey's test, where you are comparing many samples, you are able to compare all the means of these samples in all possible pairwise combinations. Specific tests of hypotheses are usually one-tailed tests.

Standard deviation   This is a measure of the variation in data where the distribution is known. For data with a normal distribution, it is calculated as the square root of the variance and is a value with the same units as the original observations.

Standard error of the mean   For a given statistical population, you may take many samples and these can all be described in terms of a mean. If you then used these means as your data, you could calculate a standard deviation of these mean values. This standard deviation of the mean is more commonly known as the standard error of the mean.

Sum of squares of *x* or *y*   A common step in the calculation of a variance for parametric data.

Systematic/periodic sampling   A regularized sampling method where an item is chosen for observation at regular intervals.

Transforming data   A mathematical calculation that has the effect of squeezing and/or stretching the scale you used when making your measurements so that the distribution takes on the approximate shape of a bell-shaped curve. The process of transforming data can be carried out when you wish to normalize non-parametric data.

Treatment (definition 1)   When carrying out an experiment, your items are exposed to a particular environment that is manipulated by the investigator.

Treatment (definition 2)   A statistical term for any samples that are being compared.

Treatment variables   The factor(s) under investigation. See also Variables and Confounding variables.

Two-tailed test   Where the hypothesis testing is against both ends of a distribution. Most general tests of hypotheses are usually two-tailed. See also One-tailed test.

Type I and Type II errors   If, on the basis of using a particular *p* value, we reject the null hypothesis even when it is really true, this is called a Type I error. If we 'accept' the null hypothesis even when it is false, this is called a Type II error. See Power of a test.

Unimodal   The description of a distribution that has a single high point or peak. See Bimodal.

Unmatched   See Matched.

Variable   The factor(s) that bring about variation between observations so that not all observations on items, a sample, or in a population are the same. Variables can be divided into two groups: treatment variables and confounding variables. Treatment variables are the factors in your experiment whose effect you are examining and/or the factors you are measuring. Confounding variables are the factors that influence the experiment but are not factors you are investigating and the chance effects that are contributors to sampling error. Where you examine the effect of one variable by measuring the response in another variable, these are commonly called independent and dependent variables, respectively.

Variance   A measure of the variation in the data, which for normally distributed data is calculated as:

$$s^2 = \frac{\sum (x - \bar{x})^2}{n - 1}$$

Variation   When the observations within your data set do not all have the same value.

References

Barnard, C., Gilbert, F., & McGregor, P. (2001). *Asking Questions in Biology*. Longman Scientific and Technical, Harlow.

Bullock, J. M., Clear Hill, B., & Silvertown, J. (1994). Demography of *Cirsium vulgare* in a grazing experiment. *Journal of Ecology* 82:101–111.

Crothers, J. H. (1981). On the graphical presentation of quantitative data. *Field Studies* 5:487–511.

Fielding, A. H. (2007). *Cluster and Classification Techniques for the Biosciences*. Cambridge University Press, Cambridge.

Fowler, J., Cohen, L., & Jarvis, P. (1998). *Practical Statistics for Field Biologists*. John Wiley & Sons, Chichester.

Fresnillo, B. & Ehlers, B. K. (2008). Variation in dispersability of mainland and island populations of three wind dispersed plant species. *Plant Systematics and Evolution* 270:243–55.

Grafen, A. & Hails, R. (2002). *Modern Statistics for the Life Sciences*. Oxford University Press, Oxford.

Hawkins, D. (2005). *Biomeasurement: Understanding, Analyzing and Communicating Data in the Biosciences*. Oxford University Press, Oxford.

Johnson, S. & Scott, J. (2009). *Study and Communication Skills for the Biosciences*. Oxford University Press, Oxford.

Klinkhamer, P. & de Jong, T. (1993). Biological flora of the British Isles: *Cirsium vulgare*. *Journal of Ecology* 81:177–91.

Kruskal, W. H. & Wallis, W. A. (1952). Use of ranks in one-criterion analysis of variance. *Journal of the American Statistical Association* 47:583–621.

Legendre, P. & Legendre, L. (1998). *Numerical Ecology*, 2nd English edn. Elsevier Science, Amsterdam.

Lukacs, P. M., Thompson, W. L., Kendall, W. L., Gould, W. R., Doherty, P. F., Burnham, K. P., & Anderson, D. R. (2007). Concerns regarding a call for pluralism of information theory and hypothesis testing. *Journal of Applied Ecology* 44:456–60.

Meddis, R. (1984). *Statistics using Ranks: a Unified Approach*. Blackwell Publishers, Oxford.

Quinn, G. P. & Keough, M. J. (2002). Experimental design and data analysis for biologists. Cambridge University Press, Cambridge.

Radford, M. (2001). *Animal Welfare Law in Britain*. Oxford University Press, Oxford.

Robson, C. (2002). *Real World Research: A Resource for Social Scientists and Practitioner Researchers*, 2nd edn. Blackwell Publishers, Oxford.

Rodwell, J. S. (ed.) (1991). *British Plant Communities*, Vol. **1**: *Woodlands and Scrub*. Cambridge University Press, Cambridge.

Ruxton, G. D. (2006). The unequal variance *t* test is an underused alternative to Student's *t* test and the Mann Whitney U test. *Behavioural Ecology* **17**:688–90.

Scheirer, C .J., Ray, W. S., & Hare, N. (1976). The analysis of ranked data derived from completely randomised factorial designs. *Biometrics* **32**: 429–34.

Shearer, P. R. (1973). Missing data in quantitative designs. *Journal of the Royal Statistical Society Ser. C. Applied Statistics* **22**:135–140.

Skarpaas, O., Stabbetorp, O. E., Ronning, I., & Svennungsen, T. O. (2004). How far can a hawk's beard fly? Measuring and modelling the dispersal of *Crepis praemorsa*. *Journal of Ecology* **92**:747–57.

Sokal, R. R. & Rohlf, F. J. (1994). *Biometry*, 3rd edn. W. H. Freeman & Co., New York.

Stephens, P. A., Buskirk, S. W., Hayward, G. D., & Martinez del Rio, C. (2005). Information theory and hypothesis testing: a call for pluralism. *Journal of Applied Ecology* **42**: 4–12.

Stephens, P. A., Buskirk, S. W., Hayward, G. D., & Martinez del Rio, C. (2007). A call for pluralism answered. *Journal of Applied Ecology* **44**:461–3.

Student (1908). The probable error of a mean. *Biometrika* **6**:1–25.

Thompson, K., Gaston, K. J., & Band, S. (1999). Range size, dispersal and niche breadth in herbaceous flora of central England. *Journal of Ecology* **87**:150–5.

Westbury, D. B., Woodcock, B. A., Harris, S. J., Brown, V. K., & Potts, S. G. (2008). The effects of seed mix and management on the abundance of desirable and pernicious unsown species in arable buffer strip communities. *Weed Research* **48**:113–23.

Williams, D. A. (1976). Improved likelihood ratio tests for complete contingency tables. *Biometrica* **63**:33–7.

Wright, N. R. (1997). Breath alcohol concentrations in men 7–8 hours after prolonged, heavy drinking: influence of habitual alcohol intake. *Lancet* **349**:182.

Zar, J. H. (2010) *Biostatistical Analysis*, 5th edn (international edition). Pearson Higher Education.

Index

Note: If a definition of a term is required, readers are directed to the glossary (pages 448–453).